RENEWABLE ENERGY APPLICATIONS FOR FRESHWAT

Sustainable Energy Developments

Series Editor

Jochen Bundschuh
University of Southern Queensland (USQ), Toowoomba, Australia
Royal Institute of Technology (KTH), Stockholm, Sweden

Volume 2

Renewable Energy Applications for Freshwater Production

Editors

Jochen Bundschuh

University of Southern Queensland (USQ), Toowoomba, Australia
Royal Institute of Technology (KTH), Stockholm, Sweden

Jan Hoinkis

Institute of Applied Research, Karlsruhe University of Applied Sciences,
Karlsruhe, Germany

CRC Press
Taylor & Francis Group
Boca Raton London New York Leiden

CRC Press is an imprint of the
Taylor & Francis Group, an **informa** business

A BALKEMA BOOK

IWA Publishing

London · New York

Co-published by IWA Publishing, Alliance House,
12 Caxton Street, London SW1H 0QS, UK
Tel. +44 (0) 20 7654 5500, Fax +44 (0) 20 7654 5555
publications@iwap.co.uk
www.iwapublishing.com
ISBN13 9781780401218

First issued in paperback 2018

CRC Press/Balkema is an imprint of the Taylor & Francis Group, an informa business

Typeset by MPS Limited, Chennai, India

Published by: CRC Press/Balkema
              P.O. Box 447, 2300 AK Leiden, The Netherlands
              e-mail: Pub.NL@taylorandfrancis.com
              www.crcpress.com – www.taylorandfrancis.co.uk – www.balkema.nl

Library of Congress Cataloging-in-Publication Data

Renewable energy applications for freshwater production / editors,
Jochen Bundschuh, Jan Hoinkis.
    p. cm. — (Sustainable energy developments ; v. 2)
  "A Balkema book."
  Includes bibliographical references and index.
  ISBN 978-0-415-62089-5 (hardback : alk. paper)
  1. Water—Purification—Energy consumption.   2. Saline water conversion—Energy
consumption.   3. Renewable energy sources.   I. Bundschuh, Jochen.   II. Hoinkis, Jan.
  TD430.R46   2012
  628.1'62—dc23

                                                                      2012003099

ISBN 13: 978-1-138-07521-4 (pbk)
ISBN 13: 978-0-415-62089-5 (hbk)

About the book series

Renewable energy sources and sustainable policies, including the promotion of energy efficiency and energy conservation, offer substantial long-term benefits to industrialized, developing and transitional countries. They provide access to clean and domestically available energy and lead to a decreased dependence on fossil fuel imports, and a reduction in greenhouse gas emissions.

Replacing fossil fuels with renewable resources affords a solution to the increased scarcity and price of fossil fuels. Additionally it helps to reduce anthropogenic emission of greenhouse gases and their impacts on climate change. In the energy sector, fossil fuels can be replaced by renewable energy sources. In the chemistry sector, petroleum chemistry can be replaced by sustainable or green chemistry. In agriculture, sustainable methods can be used that enable soils to act as carbon dioxide sinks. In the construction sector, sustainable building practice and green construction can be used, replacing for example steel-enforced concrete by textile-reinforced concrete. Research and development and capital investments in all these sectors will not only contribute to climate protection but will also stimulate economic growth and create millions of new jobs.

This book series will serve as a multi-disciplinary resource. It links the use of renewable energy and renewable raw materials, such as sustainably grown plants, with the needs of human society. The series addresses the rapidly growing worldwide interest in sustainable solutions. These solutions foster development and economic growth while providing a secure supply of energy. They make society less dependent on petroleum by substituting alternative compounds for fossil-fuel-based goods. All these contribute to minimize our impacts on climate change. The series covers all fields of renewable energy sources and materials. It addresses possible applications not only from a technical point of view, but also from economic, financial, social and political viewpoints. Legislative and regulatory aspects, key issues for implementing sustainable measures, are of particular interest.

This book series aims to become a state-of-the-art resource for a broad group of readers including a diversity of stakeholders and professionals. Readers will include members of governmental and non-governmental organizations, international funding agencies, universities, public energy institutions, the renewable industry sector, the green chemistry sector, organic farmers and farming industry, public health and other relevant institutions, and the broader public. It is designed to increase awareness and understanding of renewable energy sources and the use of sustainable materials. It aims also to accelerate their development and deployment worldwide, bringing their use into the mainstream over the next few decades while systematically replacing fossil and nuclear fuels.

The objective of this book series is to focus on practical solutions in the implementation of sustainable energy and climate protection projects. Not moving forward with these efforts could have serious social and economic impacts. This book series will help to consolidate international findings on sustainable solutions. It includes books authored and edited by world-renowned scientists and engineers and by leading authorities in in economics and politics. It will provide a valuable reference work to help surmount our existing global challenges.

Jochen Bundschuh
(Series Editor)

Editorial board

Table of contents

Contributors

Diego-César Alarcón-Padilla Plataforma Solar de Almería, Tabernas (Almería), Spain, E-mail: diego.alarcon@psa.es

Habib Ben Bacha Laboratoire des Systèmes Electro-Mécaniques (LASEM), National Engineering School of Sfax, Sfax University, Sfax, Tunisia & College of Engineering, Alkharj University, Alkharj, Kingdom of Saudi Arabia, E-mail: habibbenbacha@yahoo.fr

Julián Blanco Plataforma Solar de Almería, Tabernas (Almería), Spain, E-mail: julian.blanco@psa.es

Michal Bodzek Silesian University of Technology, Gliwice, Poland, E-mail: michal.bodzek@polsl.pl

Wieslaw Bujakowski Mineral and Energy Economy Research Institute, Polish Academy of Sciences, Krakow, Poland, E-mail: buwi@min-pan.krakow.pl

Jochen Bundschuh University of Southern Queensland, Australia & Royal Institute of Technology (KTH), Stockholm, Sweden, E-mail: jochenbundschuh@yahoo.com

Fernando Castellano Departamento de Agua, División de Investigación y Desarrollo Tecnológico, Canary Islands Institute of Technology, Santa Lucía, Las Palmas, Spain, E-mail: fcastellano@itccanarias.org

F. Julián Domínguez Departamento de Agua, División de Investigación y Desarrollo Tecnológico, Canary Islands Institute of Technology, Santa Lucía, Las Palmas, Spain, E-mail: jdomínguez@itccanarias.org

Noreddine Ghaffour Water Desalination & Reuse Centre, King Abdullah University of Science and Technology (KAUST), Thuval, Saudi Arabia, E-mail: Noreddine.Ghaffour@kaust.edu.sa

Mattheus Goosen Office of Research and Graduate Studies, Alfaisal University, Riyadh, Saudi Arabia, E-mail: mgoosen@alfaisal.edu

Jan Hoinkis Karlsruhe University of Applied Sciences, Karlsruhe, Germany, E-mail: jan.hoinkis@hs-karlsruhe.de

Marta Irene Litter Gerencia Química, Comisión Nacional de Energía Atómica, San Martín, Prov. de Buenos Aires, Argentina & Consejo Nacional de Investigaciones Científicas y Técnicas, Buenos Aires, Argentina & Instituto de Investigación e Ingeniería Ambiental, Universidad de General San Martín, Prov. de Buenos Aires, Argentina, E-mail: litter@cnea.gov.ar

Hacene Mahmoudi   Faculty of Sciences, Hassiba Ben Bouali University, Chlef, Algeria,
                         E-mail: usto98@yahoo.fr

Jorge Martín Meichtry   Gerencia Química, Comisión Nacional de Energía Atómica,
                         San Martín, Prov. de Buenos Aires, Argentina & Consejo Nacional
                         de Investigaciones Científicas y Técnicas, Buenos Aires, Argentina,
                         E-mail:meichtry@cnea.gov.ar

Baltasar Peñate   Departamento de Agua, División de Investigación y Desarrollo
                         Tecnológico, Canary Islands Institute of Technology, Santa Lucía,
                         Las Palmas, Spain, E-mail: baltasarp@itccanarias.org

Vicente J. Subiela   Departamento de Agua, División de Investigación y Desarrollo
                         Tecnológico, Canary Islands Institute of Technology, Santa Lucía,
                         Las Palmas, Spain, E-mail: vsubiela@itccanarias.org

Barbara Tomaszewska   Mineral and Energy Economy Research Institute, Polish
                         Academy of Sciences, Krakow, Poland,
                         E-mail: tomaszewska@min-pan.krakow.pl

Eftihia Tzen   Wind Energy Department, Centre of Renewable Energy Sources &
                         Saving (CRES), Pikermi, Greece, E-mail: etzen@cres.gr

Guillermo Zaragoza   Plataforma Solar de Almería, Tabernas (Almería), Spain,
                         E-mail: guillermo.zaragoza@psa.es

Foreword

Water is essential for life. Living in one of the driest continents on Earth, we Australians probably know it better than anyone else.

Provision of freshwater in sufficient quantity and quality as well as climate protection are key tasks of every government. I remember attending a Science Day in the Queensland Parliament about more than 10 years ago. These are events where university academics meet with members of the parliament and discuss issues that are of mutual interest. One thing I remember from that day was the result of a survey. This was very simple and asked each participant to name two important challenges facing the society that they thought should be addressed by science. The results showed that most of the responses from the scientists were related to their current research interests. On the other hand, the members of the parliament almost unanimously named water and energy as the two most important challenges facing the State.

We all know of course that we can generate vast amounts of freshwater if we had access to unlimited amounts of energy. Unfortunately, energy is scarce and is getting scarcer. At the same time, the climate change issue place further constraints on the allowable solution space.

Water and energy are not independent commodities. Renewable energy sources can be used for production of clean freshwater for drinking, agriculture and other purposes as shown in this book.

Most of the technology options presented in this timely collection are relevant to Australia. Renewable energy used for freshwater production by desalination can provide important options for Australia's agricultural sector. In inland areas, wind, solar, geothermal, and biomass can be used as energy source to desalinate brackish or saline groundwater, which is found at many places; further residual communal and industrial wastewater can be made usable again through desalination. Freshwater production powered by renewable energy resources is not only an option for areas where decentralized solutions are demanded. They can also provide economically sound and environmentally friendly options for desalination along the coastline. These areas are where most of our country's population live and where natural freshwater resources are not enough to cover the recent or future demand. The seawater desalination plants in Perth and Sydney powered predominantly by wind energy are early examples.

While the world covered a lot of ground in recent years in using renewable energy sources for desalination, there is still need for more R&D. Wind and solar powered desalination units have proven technical and economic viability for freshwater production in off-grid areas, but technologies including energy storage, which are required if wind and solar are used, need to be improved. Desalination using geothermal energy which has the enormous advantage that it provides a stable energy supply needs significant R&D to further optimize technologies regarding efficiency by developing smart grid systems and reduce costs.

It can be expected that the continuing maturation of the renewable energy industry will further decrease the costs whereas the technologies depending on fossil fuels will be faced with the continuously increasing costs for fossil fuels, the increasing external costs related to their combustion, and the insecurity of their supply.

Partnerships between excellent researchers, industry, and key end-users are required towards significant national and global economic, environmental and social benefits.

This book shows how to meet the challenge of providing affordable freshwater by using free and locally available energy sources such as wind, solar, geothermal, and biomass. The book presents information about the most appropriate technological solutions using technical, economic, financial, and socio-cultural considerations to ensure the appropriate choice of technology. Further, it identifies and discusses the technological and financial gaps, and shows appropriate answers to fill in these gaps.

I am sure that this book will be useful for many people seeking solutions to effectively use renewable energy resources for providing freshwater from brackish saline groundwater in inland areas and seawater in coastal areas and making residual communal or industrial water again suitable for agricultural and other purposes such as artificial groundwater recharge. I expect that this book will be an important resource for water supply and energy decision makers, energy and water sector representatives and administrators, policy makers, in governments as well as business leaders, energy engineers/scientists, academicians and power producers, financial sector, land planners, agronomists, and citizens interested in one of the most significant challenges facing our generation.

Professor Hal Gurgenci
Director, Queensland Geothermal Energy Centre of Excellence
The University of Queensland
Brisbane
Australia

Preface

Water is essential for all life on earth, including human life. Access to uncontaminated water is a critical human need, and one that is not available to a large number of people around the world. According to the United Nations[1], about a billion people currently do not have access to safe, clean, drinking water. It is estimated that lack of access to safe drinking water contributes to the deaths of between 1 and 15 million children every year, a mortality rate that is on a par or greater than that caused by many epidemics and catastrophes. For the sake of comparison it is estimated that about 30 Million people have died from AIDS/HIV[2] since it was originally diagnosed in 1981 ("Global Report Fact Sheet", *UNAID,* 2010).

What does it mean to have access to safe, clean, drinking water? Providing "access" may mean moving water to areas of need. Supplying "clean water" may mean that treatment will be required. Both these actions will require energy. Providing access to freshwater for irrigation often will also require energy for distribution and/or treatment. This book examines the possible application of renewable energy sources for freshwater and drinking water production.

In developed countries, people are often worried about chronic exposure to low-level concentrations of contaminants such as pesticides and "emerging contaminants" (e.g. pharmaceutical and personal-care products) that are suspected to cause cancer, endocrine disruption, and other health problems. People in developing countries often do not have the luxury of worrying about low-level contaminant issues, or chronic health issues in general, although they are affected by those issues. For example, the prevalence of deadly pathogens in surface waters led to the construction of shallow tube wells that provided pathogen-free drinking waters for people in the Bengal Delta and resulted in immediate and major reductions in mortality rates, especially infant mortality rates. However, as a result of the policy, it is estimated that about 31 Million people in the Bengal Delta have been exposed to arsenic drinking water concentrations above 50 μg/L and 50 Million people have been exposed to concentrations above 10 μg/L (Chakraborti et al., 2002, Talanta, vol 58)[3]. Smith et al. (1992, Environmental Health Perspectives, vol. 97) estimated that a US population exposed to drinking 1 L/day of water with an arsenic concentration of 50 μg/L could have a resulting lifetime cancer mortality risk from liver, lung, kidney or bladder cancer of up to 13 in 1000, which translates to about 400,000 cancer deaths if a population of 31 Million were exposed.

The presence of pathogens and/or of chemical contaminants in freshwater supplies is often a reflection of poor management of infrastructure, resources, and wastes on the landscape and of poor controls on a society's "water commons": the common good represented by access to water resources in sufficient quantity and quality for basic human needs, as well as for ecological

---

[1]Improvements are being made. In 2008, United Nations Environmental Programme (http://www.unep.org/dewa/vitalwater) cited a figure of 1.1 Billion people without access to clean drinking water. More recently, a figure of "more than one in six people worldwide, or 894 Million people- has been cited as lacking access to "improved water sources" (http://www.unwater.org/statistics_san.html). Another recent press release puts the figure a little lower, stating that at least 11% of the world's population, or 783 Million, are still without access to safe drinking water(http://www.wssinfo.org/fileadmin/user_upload/resources/Press-Release-English.pdf). Unless stated otherwise, water statistics reported in this foreword are mentioned in other chapters of this book (esp. chapters 1, 5, and 8).

[2]Acquired immune deficiency syndrome/human immunodeficiency virus

[3]A drinking water limit for dissolved arsenic of 10 μg/L is currently specified by the World Health Organization, European Union, and the United States.

needs. There are analogies between the degradation of a "water commons" and the degradation of other types of common goods or public goods; the analogies, and some factors responsible for the degradation of common goods, are analyzed in Garrett Harding's article (Harding, 1968, Science, vol 162) on the "Tragedy of the Commons". Water resource systems, or "water commons", are more complex and more difficult to manage than a village's "common cattle pasture", an example discussed by Garrett Harding (1968) that followed an 1833 analysis by William Forster Lloyd. The transdisciplinary processes, time and spatial scales, and dynamics that need to be considered to manage or regulate a "water commons" transcend the understanding needed to manage the comparatively static "common cattle pasture".

In the case of the "common cattle pasture", Garrett Harding advocated shifting of property rights and the implementation of "coercive" measures such as taxes and regulations. Applying such measures to protect a "water commons" is more challenging, not only because of the inherent complexities and dynamics of the "water commons", but also because of linkages to broader issues, such as the need to protect ecological systems or to mitigate global change impacts. These needs often get secondary attention; not giving them attention may provide some short-term benefits, but may cause significant problems in the long-term. For example, in heavily water-stressed environments, water availability for ecological needs may be an issue of secondary importance relative to the immediate water needs of people for drinking water or agriculture. Ignoring ecological needs, however, could cause devastation of many ecosystem services that are provided to the societal "commons".

Better and more intensive efforts are needed to preserve and control our "water commons", as well as our broader "ecosystem commons" that contribute to the natural regulation of our water resources, their availability, and their quality. However, providing such controls and protection is a difficult task in any country, regardless of the degree of water stress that may be present. Timely success in meeting these protection and control needs by our global society is difficult to assure. Consequently, society needs to make progress, not only by striving for better regulation and control of its commons, including its "water commons", but also by seeking help through the development and application of better technologies. The renewable energy technologies discussed in this book for the production of freshwater are excellent examples of highly innovative technologies that can help better manage and use our "water commons". Additionally, the application of these renewable energy technologies may help mitigate human impacts on climate change.

Water problems are severe in many countries around the world. They are likely to become even more so as the world population increases. One-fifth of the world's population currently resides in areas of water scarcity; it is estimated that two thirds of the world's population will live in such areas by the year 2025. Addressing the problem of human society's access to safe, clean drinking water is urgent. Disinfection and remediation of contaminated waters, and desalination of waters too saline for human consumption, are clear ways to address this global and urgent need. However, disinfection, remediation, and desalination require large expenditures of energy. These energy expenditures may be beyond the capabilities of some societies or their use may cause long-term harm and degradation to those societies and/or to our global ecosystem.

This book examines the technical and economic feasibility of using *renewable* energies to increase the availability and improve the quality of freshwater and drinking water resources. Renewable energies are of particular interest, not only because their use may mitigate degradation of our global ecosystem and "global commons" (e.g. by decreasing human inputs of greenhouse gases into the atmosphere), but also because (1) their supply is unlimited by definition, and (2) their cost and spatial availability may be much more favorable than what could be provided through conventional energy resources (e.g. fossil fuels). An interesting array of renewable energies are discussed in this book: geothermal, wind, and a diversity of solar energy technologies. The technologies discussed vary from basic and inexpensive (e.g. solar disinfection, passive solar) to more complex, more expensive, and/or more difficult to manage or operate. Differences result from (1) the characteristics of the energy sources (e.g. intermittent, or more difficult or expensive to access), (2) the possible need for a combined use of energy sources (e.g. wind and solar), and (3) the characteristics of the technologies employed (e.g. membrane filtration or reverse osmosis systems) to produce freshwater through application of the energy systems. This book is unique because it provides a state-of-the-art discussion of these technical factors, and also because it examines the economic, geographic, and societal feasibility of their implementation. It provides a practical study of the possible implementation of using renewable energies for freshwater production. It gives hope that society can finally address the clear and urgent need to provide widespread access to safe freshwater resources to the world's citizens while preserving their global commons.

Pierre Glynn
Chief, National Research Program/Eastern Branch
U.S. Geological Survey, Reston, VA

Editors' Preface

"Change is the law of life. And those who look only to the past or present are certain to miss the future."

John F. Kennedy

Future energy provision, freshwater supply and climate change mitigation are intrinsically linked issues and key challenges for modern society. The complex interdependency becomes obvious if we consider forecasted demographic and economic growth which could double global energy demand in the next 25 years. This growth would require a marked increase in freshwater availability, in particular for food production which presently consumes about 75% of the global freshwater supply. Arid and semiarid regions will be increasingly exposed to water shortages. Many other regions will also face depletions in freshwater availability. Freshwater resources are naturally limited and are increasingly contaminated by human activities. In consequence, the production of freshwater from seawater, saline groundwaters, naturally or anthopogenically contaminated waters and wastewaters, will become increasingly important. Water treatment will be essential in meeting the increased demand and providing a guarantee for the future supply of freshwater of sufficient quantity and quality.

Freshwater production, especially by desalination, requires large amounts of energy. This illustrates the important interconnections between freshwater supply, energy and global climate change. Conventional hydrocarbon energy sources such as coal, oil and natural gas are becoming increasingly limited. Their combustion produces greenhouse gases that affect climate. In addition, the dependence of many countries on fossil fuel imports negatively affects their foreign trade balance. These factors all suggest the critical importance of providing freshwater resources through economically and environmentally sound energy solutions. The combined sustainable provision of these linked commodities, energy and freshwater, is indispensable. It is primordial for human social and economic development and is a primary challenge for the 21st century. Fortunately, many regions worldwide have great potential to cover their pressing water needs in an economically and environmentally sound manner through the use of water treatment processes powered by renewable energies, for example using thermal or membrane based technologies for desalination or solar techniques for emerging water disinfection. Renewable energies have the additional benefit of being domestically available. This may reduce a country's dependence on fossil fuel imports. A further benefit is that renewable energy resources are also available locally in off-grid areas. Additionally, the introduction of energy efficient technologies together with renewable energy sources contributes to economic and environmental sustainability. The use of 'clean' energy resources for freshwater production requires the development and application of sophisticated state-of-the-art tools and technologies. In many cases, existing conventional technologies for water production, originally developed to use conventional energy resources, can be adapted to use renewable energy sources, and/or can be modified to become more energy-efficient.

This book provides an overview of some possible cost-efficient techniques and applications feasible for various scales and situations. It shows why the implementation of these technologies faces numerous technological, economic and policy barriers and it suggests how hurdles can be overcome. The costs of novel treatment units using renewable energy sources are discussed. They are compared with those of other technologies for clean water production. External costs, such as environmental and social costs caused by using fossil fuel based technologies, are also considered. Energy efficiency is emphasized because of its critical importance in systems powered by renewable energy. The applications of water supply systems providing water in emergency conditions are also discussed.

This book seeks to provide knowledge regarding freshwater production through renewable energy sources to a broad readership. It is impossible to cover this subject within the space limitations of any normal book. Consequently, a selection of essential topics and case studies has been made. However, almost all the developments described are discussed in depth. The fundamental concepts, the technical details, and the economic, social and environmental issues are described and presented in a simple and logical manner. The book explains in a didactic way the possible applications, depending on local conditions and scales, and it presents new and stimulating ideas for future developments. Additionally, the book discusses R&D needs and will hopefully stimulate readers to pursue these areas. We believe that our approach will provide the reader with a thorough understanding of the topic and will allow him/her to identify the most suitable solutions for specific tasks and needs.

Chapter 1 discusses the global problems relating to energy and water. It introduces the need and potential ways of meeting the increasing demand for freshwater resources (predominantly due to food production) through desalination. Options are discussed for the desalination of seawater or saline groundwater and for the treatment of municipal/industrial wastewater by using sustainable environmentally friendly energy sources and energy efficient technologies. In Chapter 2, the technologies that can be used for water desalination powered by renewable energy sources (solar, wind, geothermal, wave and tidal power) are explained. They are compared given their applicability under specific situations and considerations of scale, economic factors, market potential, environmental sustainability, regulatory, policy and legal considerations and marketing needs; case studies conclude this chapter. Chapters 3–6 discuss in depth desalination technologies that use either passive solar thermal energy (Chapter 3), solar based humidification/dehumidification processes (Chapter 4), solar photovoltaic-powered reverse osmosis systems (Chapter 5) or wind energy powered systems (Chapter 6) for freshwater production. In Chapter 7, Poland's preliminary experience with geothermal water treatment is presented along with an overview on membrane and hybrid desalination treatment options. The chapter highlights the possibility of using geothermal energy to power these technologies. Chapter 8 presents solar disinfection as a low-cost technology for clean water production. The solar disinfection method (SODIS), heterogeneous photocatalysis (HP) and Fenton and photo-Fenton processes in water disinfection are discussed in detail and illustrated in case studies.

We hope that this book will help all readers, in the professional and academic sectors, as well as key institutions that deal with water resources and freshwater supply planning. We hope it will be used as an introduction to the methods that allow domestic renewable energy resources and energy efficient technologies to be used for freshwater supply to meet a continually increasing freshwater demand in an economically sustainable way, while helping reduce our societal impacts on global climate change.

Hopefully, this book will become a reference that is widely used by educational institutions, research institutions, and research and development establishments. The book should prove to be a useful textbook to senior undergraduate, graduate and postgraduate students, to engineers involved in freshwater production, water treatment, water supply, and water resources management, and also to professional hydrologists, hydrogeologists, hydrochemists, environmental scientists, and others trying to address water resources and freshwater supply.

Jochen Bundschuh
Jan Hoinkis
April 2012

About the editors

Jochen Bundschuh (1960, Germany), finished his PhD on numerical modeling of heat transport in aquifers in Tübingen in 1990. He is working in geothermics, subsurface and surface hydrology and integrated water resources management, and connected disciplines. From 1993 to 1999 he served as an expert for the German Agency of Technical Cooperation (GTZ) and as a long-term professor for the DAAD (German Academic Exchange Service) in Argentine. From 2001 to 2008 he worked within the framework of the German governmental cooperation (Integrated Expert Program of CIM; GTZ/BA) as advisor in mission to Costa Rica at the Instituto Costarricense de Electricidad (ICE). Here, he assisted the country in evaluation and development of its huge low-enthalpy geothermal resources for power generation. Since 2005, he is an affiliate professor of the Royal Institute of Technology, Stockholm, Sweden. In 2006, he was elected Vice-President of the International Society of Groundwater for Sustainable Development ISGSD. From 2009–2011 he was visiting professor at the Department of Earth Sciences at the National Cheng Kung University, Tainan, Taiwan. By the end of 2011 he was appointed as professor in hydrogeology at the University of Southern Queensland, Toowoomba, Australia.

Dr. Bundschuh is author of the books "Low-Enthalpy Geothermal Resources for Power Generation" (2008) (CRC Press/Balkema, Taylor & Francis Group) and "Introduction to the Numerical Modeling of Groundwater and Geothermal Systems: Fundamentals of Mass, Energy and Solute Transport in Poroelastic Rocks". He is editor of the books "Geothermal Energy Resources for Developing Countries" (2002), "Natural Arsenic in Groundwater" (2005), and the two-volume monograph "Central America: Geology, Resources and Hazards" (2007), "Ground-water for Sustainable Development" (2008), "Natural Arsenic in Groundwater of Latin America (2008). Dr. Bundschuh is editor of the book series "Multiphysics Modeling", "Arsenic in the Environment", and "Sustainable Energy Developments" (all CRC Press/Balkema, Taylor & Francis Group).

Professor Jan Hoinkis, born 1957 in Germany, holds a degree in chemistry and a doctorate in the field of thermodynamics from the Technical University Karlsruhe. He has about 7 years work experience in chemical industry, being head of a group for process development focusing on optimizing synthesis of chemical specialties for the textile and paper industry considering technical, environmental and economic constraints. Since 1996 he has been a professor at Karlsruhe University of Applied Sciences where he is teaching and conducting research in the field of applied chemistry and process engineering in combination with sensor/control systems. He is specialized in the areas of water treatment and water recycling by use of membrane technologies. Thereby he is conducting research and development in the fields of decentralized small-scale desalinators powered by renewable energies for drinking water production as well as application of membrane bioreactor technology for waste water treatment and reuse. He has coordinated a variety of national and international R&D projects in co-operation with research institutes and companies among them EU funded projects (AsiaProEco, LIFE, FP7). Since 2008 he has been scientific director of the Institute of Applied Research at the Karlsruhe University of Applied Sciences. The Institute of Applied Research (IAF) is a central facility of the Karlsruhe University of Applied Sciences and works to promote applied scientific research and development projects through an interdisciplinary approach. The IAF main research activities are: applied computer sciences and geoinformatics; intelligent measurement systems and sensor technologies and civil, environmental and process engineering. Jan Hoinkis is author of several peer-reviewed scientific publications and contributions to international conferences.

Acknowledgements

This book would be incomplete without an expression of our sincere and deep sense of gratitude to the reviewers for their careful reading, their valuable comments and suggestions which greatly improved the contents of the chapters and the book.

We express our gratitude to and thank the following advisory board members who reviewed the chapters and made a number of very useful suggestions for its improvement: Guillermo Zaragoza (Chapters 1, 2 and 4), Said Al-Hallaj (Chapters 1 and 4), Hussain Al-Towaie (Chapters 1 and 5), Soteris Kalogirou (Chapter 1, 2, 3 and 5), Santiago Arnaltes (Chapter 6). We further thank Pierre Glynn (US Geological Survey; Reston, VA), Eftihia Tzen (Wind Energy Department, Centre of Renewable Energy Sources & Saving, CRES, Pikermi, Greece), Dina L. López, (Ohio University, USA) and Friederike Eberhard (University of Southern Queensland, Australia) for their comments on Chapter 1 and Alessandra Criscuoli (Research Institute on Membrane Technology, ITM-CNR, Rende, Italy) who was so kind to review Chapter 7.

The editors and authors thank also the technical people of Taylor & Francis Group, for their cooperation and the excellent typesetting of the manuscript.

Jochen Bundschuh
Jan Hoinkis
January, 2012

CHAPTER 1

Addressing freshwater shortage with renewable energies

Jochen Bundschuh & Jan Hoinkis

> *"Global freshwater consumption rose sixfold between 1900 and 1995—more than twice the rate of population growth. About one third of the world's population already lives in countries considered to be 'water stressed'—that is, where consumption exceeds 10% of total supply. If present trends continue, two out of every three people on Earth will live in that condition by 2025."*
>
> Kofi Annan (2000)*

1.1 INTRODUCTION

Accessibility to water in sufficient quantity and quality is critical for human development. However, freshwater resources face increasing stress due to contamination and over-exploitation. The continuously increasing demand requires more and more water resources which are not suitable for drinking, irrigation or other purposes without prior treatment. The desalination of seawater or highly mineralized groundwater, and the treatment of water which contains undesired elements such as arsenic or other toxins are examples. Municipal and industrial residual water, after treatment, can also be used as freshwater resource or for artificial groundwater recharge. Such treatments often require energy, another commodity that is increasingly in short supply, especially for conventional energy sources such as coal, oil and natural gas. Sustainable freshwater provision requires economically and environmentally sound provision of energy solutions. The sustainable provision of both these commodities is essential in the 21st century because they are intrinsically linked to social and economic development.

Sustainability has many aspects. It includes not only protection of water resources against over-exploitation and contamination but also freshwater production methods that are sustainable in a broader sense. This includes methods that are economically sound and ecologically-friendly, e.g. those based on sustainable energy solutions, such as the use of renewable energy sources or energy-efficient, low-carbon methods, that help mitigate global climate change.

Increasing stress on water resources, and the need to better assess "clean" energy resources, require the development and application of sophisticated state-of-the-art tools and technologies. In many cases, existing conventional technologies for water production, which were developed for conventional energy resources, can be adapted to be used in units powered by renewable energy sources, or/and they can be modified so that they become more energy-efficient.

In the first part of this introductory chapter we will examine present-day world water and energy issues. Later, we will provide basic information on global water problems (section 1.2) and the energy-related global warming issue (section 1.3) while highlighting the importance to better address these issues for the development and survival of humanity. In the next part (section 1.4), we will discuss recent developments within the scope of freshwater production from seawater, saline groundwater and municipal/industrial waste water using renewable energy resources and highlighting examples that may help top solve some of our water problems.

---

*Source: Kofi Annan in: *We the Peoples*, 2000.

1

1.2   THE WATER PROBLEM

According to an estimate by Shiklomanov (1993), the earth contains about $1400 \times 10^6 \text{ km}^3$ of water, 97.5% of which is contained in the oceans ($\sim 1365 \times 10^6 \text{ km}^3$). Freshwater resources represent about 2.5% of the global water volume (~ 35 million km^3). These freshwater resources are non-uniformly distributed in space (groundwater, surface water) and time (surface water). Many of the world's largest river basins run through sparsely populated regions. However, 68.9% of freshwater (~ 24 million km^3) is found in the form of ice and perennial snow cover in the mountains and polar regions and only about 30.8% (~ 11 million km^3) is stored in the form of groundwater (shallow and deep groundwater basins up to 2000 m depth), soil moisture and permafrost. Lakes and rivers correspond only to 0.3% (105,000 km^3) of the world's freshwater (Shiklomanov, 1993). So it is not surprising that about 1.5 billion people depend on groundwater for their drinking water supply (UNEP, 2008). Figure 1.1 shows the regional volumes of these three freshwater resources components and their irregular regional distribution.

Clean water remains a major problem, with 20% of the world population lacking access to safe drinking water. Estimates from the United Nations show that 1.2 billion human beings lack access to drinking water and another 2 billion use contaminated water (UNEP, 2008). Water-borne diseases from polluted surface waters are a main cause of illness in developing countries; it is estimated that it affects the health of 1.2 billion people, and contributes to the death of 15 million children worldwide annually (UNEP, 2008).

Freshwater resources are limited. Several regions are already experiencing a water shortage. The following approaches have tried to deal with water shortages during the last decades: drilling of groundwater wells creation of dams and reservoirs, and construction of canals, pipelines and other water diversions. However, these approaches are subject to the natural limits of water availability and to problems associated with the overexploitation of aquifers. Reservoir dams also suffer from great environmental concerns and from the absence of new suitable sites.

Supply of freshwater is a crucial issue in many countries, especially in those that have very limited water resources. Figure 1.2 shows water withdrawal and consumption globally and by region. In the coming decades, the largest increase in water withdrawals is expected to occur in Asia followed by Africa and South America, while the increase will be lower in Europe and North America (Harrison and Pearce, 2001; Shiklomanov, 1999; UNESCO, 1999). Globally, since 1900, water demand increased six-fold (UNEP, 2008). Since 1940 the world population has had an average annual growth of about 2% while water extraction has increased about 3% per year (Leon, 2005). Annual global freshwater withdrawal increased from 3790 km^3 in 1995 (consumption: 2070 km^3 or 61%) to 4430 km^3 in the year 2000 (consumption: 2304 km^3 or 52%) (Shiklomanov, 1999). In 2000, most of the freshwater withdrawal (57% corresponding to 70% of the world freshwater consumption) was in Asia where the largest irrigated lands are located (UNESCO, 1999). In the future, annual global water withdrawal is expected to grow about 10–12% per decade, reaching about 5240 km^3 by the year 2025, corresponding to an increase of 38% since 1995. Water consumption is expected to grow in the same period by only 33% (UNESCO, 1999).

Today, agricultural water use accounts for about 75% of global freshwater consumption, mainly through crop irrigation, while industrial use accounts for about 20% and the remaining 5% is used for domestic purposes (UNEP, 2008). At present, people need an average of 27 to 200 L day^{-1} to satisfy their needs (Leon, 2005). Africa and the Middle East are the two regions of the world where there is less freshwater available. The American continents are the richest in water resources (Leon, 2005). The imbalance in the consumption of water becomes evident by comparing figures of the UN (Leon 2005): Brazil, Canada, China, United States, India and Russia which comprise 40% of rivers and lakes of the planet. In Canada each inhabitant disposes of 91,640 m^3 year^{-1}; in Australia, 26,032 m^3; in Mexico, 4547 m^3; in South Africa, 1109 m^3 and in Egypt, 29 m^3 per year.

According to the UN, (Leon, 2005) the quantity of water that a person needs is about 5000 m^3 year^{-1}. Less than 1700 m^3 person^{-1} year^{-1} is called water stress; less than 1000 m^3 person^{-1} year^{-1} is considered scarcity; while less than 500 m^3 person^{-1} year^{-1} is severe

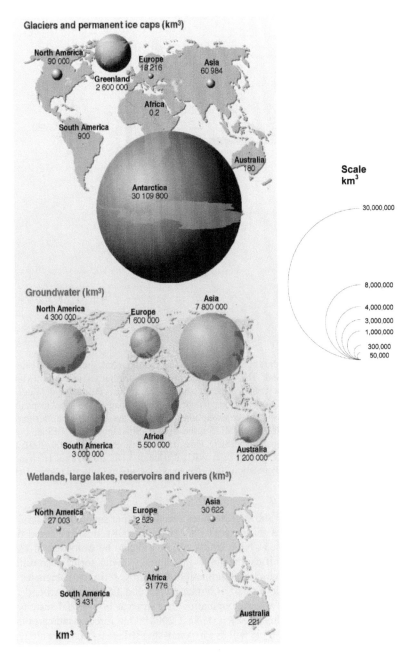

Figure 1.1.   Global freshwater resources: Volumes (km^3) and regional distribution for freshwater from (a) ice and perennial snow cover in the mountains and polar regions, (b) groundwater (shallow and deep groundwater basins up to 2000 m depth), soil moisture and permafrost, and (c) lakes and rivers. Note: Estimates refer to standing volumes of freshwater. From: Vital Water Graphics, http://maps.grida.no/go/graphic/global_freshwater_resources_quantity_and_distribution_by_region. Credit to the cartographer/designer Philippe Rekacewicz, UNEP/GRID-Arendal). Source: Igor A. Shiklomanov, State Hydrological Institute (SHI, St. Petersburg) and UNESCO, 1999; WMO; ICSU; WGMS; USGS.

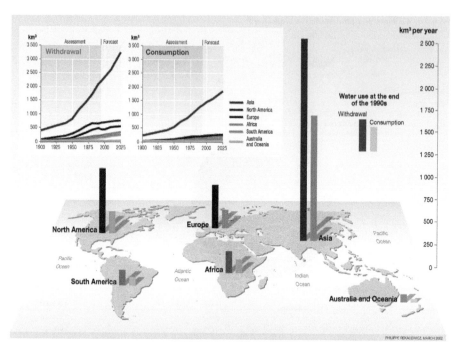

Figure 1.2.   Global freshwater withdrawal and consumption. The diagrams show the respective changes for the 1900–2025 period. The bars in the map show freshwater withdrawal (left column) and consumption (right column) at the end of the 1990s. From: Vital Water Graphics, http://maps.grida.no/go/graphic/water_withdrawal_and_consumption http://maps.grida.no/go/graphic/water_withdrawal_and_consumption; Credit to the cartographer/designer Philippe Rekacewicz (UNEP/GRID-Arendal). Source: Igor A. Shiklomanov, SHI (State Hydrological Institute, St. Petersburg), and UNESCO, Paris, 1999; World Resources 2000–2001: People and Ecosystems: the Fraying Web of Life, WRI, Washington DC, 2000; Paul Harrison and Fred Pearce, AAAS Atlas of Population 2001, AAAS, University of California Press, Berkeley.

scarcity (UNEP, 2008). Figure 1.3 shows freshwater availability per square kilometer for the different world regions and countries for the year 2007 (UNEP, 2008). It is forecasted that by the year 2050 at least 9 billion people (UNEP, 2008) will be affected by water stress. This United Nations estimate assumes no significant technological improvements to increase water supply, such as massive acceleration of seawater desalination. Currently, serious deficiencies in supply and quality of freshwater affect almost half of humanity (around 3.5 billion). Around 1.2 billion people, or almost one-fifth of the world's population, live in areas of physical scarcity, and 500 million people are approaching this situation (UNEP, 2008). One forecast indicates that by the year 2025 two out of every three people will live in water-stressed areas. In Africa alone, it is estimated that 25 countries will be experiencing water stress in the year 2025 (availability $<1700\,\text{m}^3$ per capita per year) (UNEP, 2008).

Freshwater shortage and the related increase in water stress and scarcity is worsening daily because of increasing population, a resulting increase in population density, and the spatial expansion of populated and agricultural areas to supply the increased food demand. In addition, climate change and pollution are also affecting water quality and availability. Many regions with limited or no water resources suitable for drinking or irrigation that were not inhabited in the past (due to the lack of water) are now increasingly populated and also used for food production through increased irrigation. This expansion is possible only if local water is treated or desalinized, or if water is diverted from other regions through pipelines or channels. Finding and implementing sustainable

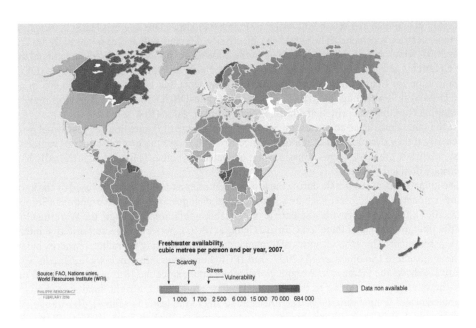

Figure 1.3. Water stress and scarcity by region and country in the year 2007. From Vital Water Graphics, http://www.unep.org/dewa/vitalwater/jpg/0221-waterstress-EN.jpg. Credit to the cartographer/ designer Philippe Rekacewicz (UNEP/GRID-Arendal). Source: FAO, United Nations, World Resources Institute, 2008.

ways to manage water resources, including the production of freshwater from seawater, is a big challenge. Since groundwater is by far the most abundant and readily available source of water in areas far from the sea, economically and environmentally sound treatment of highly saline or contaminated groundwater is of key importance. A number of contaminants can be present, sometimes through natural processes. For example, arsenic occurs in many regions, especially in deeper aquifers. In areas close to the coast, seawater desalination is an option since 70% of the world population lives within 70 km from the sea (El-Dessouky and Ettouney, 2002). Furthermore, the purification of residual domestic or industrial water either for freshwater provision or for artificial recharge of aquifers will be essential.

1.3 THE ENERGY PROBLEM

As "The Forecasts of the International Energy Outlook" show (EIA, 2011) that modern society continues largely to rely on fossil fuel resources to preserve economic growth and a present-day standard of living. However, economically exploitable conventional energy resources have become increasingly limited because of natural limitations. Their use is also questioned by large population groups, especially in industrialized countries, because of their adverse environmental impacts and because of their contribution to global climate change. This assumes that atmospheric greenhouse effect resulting from the use of coal and hydrocarbon resources is not decreased by technologies, such as injection of CO_2 into the deep underground or chemical/biological conversion to other chemical substances (e.g. microbial conversion of CO_2 to methane or methanol; Beecy *et al.*, 2000; Dave, 2008).

The demand for energy is increasing sharply due to the same reasons cited previously for the increase in water demand: population growth and economic growth, which is greatest in the developing world and in countries in transition that have higher population growth rates, fast-expanding emerging economies and related increases in living standards. This can be clearly demonstrated through electricity demand. The world net electricity demand will increase annually by 2.3% from 2008, which means that it will double in the 2008–2035 period, with predicted annual average electricity generation growth rates being only 1.2% in the industrialized countries (OECD: Organization for Economic Cooperation and Development), but as high as 3.3% (4.9% in non-OECD Asia) in developing and transition countries (non-OECD countries). This will make the developing countries the principal emitters of greenhouse gases (EIA, 2011).

At the same time, these countries are especially vulnerable to fluctuations of fossil fuel prices, especially if they depend on fuel imports. Since fossil fuels are the principal energy sources for most countries, greenhouse gas emissions are expected to increase almost proportionally to the electricity demand increase.

The human impact on climate during the last two centuries exceeds impacts caused by the known changes in natural processes, such as solar radiation changes and volcanic eruptions (Solomon *et al.*, 2007). In order to provide a scientific understanding of climate change, the Working Group I of the Intergovernmental Panel on Climate Change (IPCC), performed an authoritative international assessment of how the activities of human industry are affecting the radiative energy balance in the atmosphere (Solomon *et al.*, 2007, and IPCC homepage http://ipcc-wg1.ucar.edu/wg1/). Figure 1.4 shows the increase in average global temperature since the 19th century. The left hand axis shows anomalies relative to the 1961 to 1990 average, and the right hand axis shows the estimated actual temperature (°C). The linear trend fits to the last 25 (yellow), 50 (orange), 100 (purple) and 150 years (red) are shown, and correspond to 1981 to 2005, 1956 to 2005, 1906 to 2005, and 1856 to 2005, respectively. From about 1940 to 1970, increasing industrialization following World War II increased dust pollution (sun light attenuation) in the Northern Hemisphere, and this contributed to some cooling. Increases in carbon dioxide and other greenhouse gas emissions dominate the observed warming after the mid-1970s. (Source: Solomon *et al.*, 2007). Concerning the key question of how temperatures on Earth are changing, the IPCC furnished the following conclusive statements: *"Instrumental observations over the past 157 years show that temperatures at the surface have risen globally, with important regional variations. . . . An increasing rate of warming has taken place over the last 25 years due to increased emission of anthropogenic CO₂, . . . Confirmation of global warming comes from warming of the oceans, rising sea levels, glaciers melting, sea ice retreating in the Arctic and diminished snow cover in the Northern Hemisphere, . . .* [and] *decreases in the length of river and lake ice seasons. . . . the oceans are warming; and sea level is rising due to thermal expansion of the oceans and melting of land ice . . . Expressed as a global average, surface temperatures have increased by about 0.74°C over the past hundred years (between 1906 and 2005; Figs. 1.4 and 1.5)."*

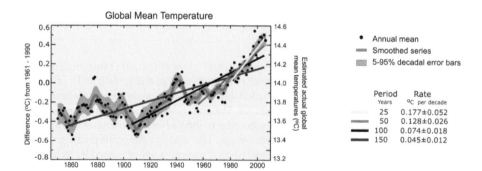

Figure 1.4.   Annual global mean observed temperatures (black dots, from the HadCRUT3 data set) along with simple fits to the data. Reprinted with permission from: Solomon *et al.* (ed.): Figures Climate Change 2007: The Physical Science Basis. Working Group I: Contribution to the Fourth Assessment Report of the Intergovernmental Panel on Climate Change, Figure FAQ 3.1, Figure 1 (top); Cambridge University Press.

The use of domestic renewable energy sources together with the implementation of energy-efficient technologies are sustainable options to overcome the natural limitations of conventional energy sources. These options will contribute to the reduction of greenhouse gas emissions, and will make countries less dependent on energy imports and fossil fuel price fluctuations. Economically and technologically accessible renewable energy resources are much larger than all fossil fuel resources put together. This means that the world's renewable energy resources will remain available for generations long after the last drop of oil is produced. Renewable energy resources can be economically very attractive compared to conventional energy sources especially in off-grid areas. In those areas, wind, solar and geothermal energy can offer economical and environmentally sound and locally available energy supply that can be used for water desalination or treatment at a variety of scales, ranging from small scales for single households and small communities to large scales, for example with desalination plants that have sufficient capacity to irrigate extensive agricultural areas. The provision of energy in off-grid areas may also contribute to rural electrification. Worldwide there are over two billion people living in rural areas, mostly in developing countries, that do not have access to electricity. It is evident that the provision of economically affordable electricity in these rural areas would contribute to their social and economic development. It would increase public health conditions, raise economic productivity and therefore create employment opportunities and increase income and welfare (electricity supply in hospitals and elsewhere, telecommunication, computers, etc.). The use of free and locally available energy sources such as wind, solar, and geothermal allows for the solution of two problems providing affordable electricity and safe water at the same time.

Renewable energy and energy-efficient technologies for water desalination can provide ideal solutions for large-scale desalination and treatment plants, in both off-grid and on-grid areas. The large energy amounts required for desalination can be provided from locally available renewable energy resources independent of whether they are directly produced and applied (e.g. solar thermal, geothermal direct use) or indirectly applied through an interim production of electricity (e.g. solar photovoltaic, electricity generation from wind or geothermal). This electricity can be used, for example, to power a reverse osmosis desalination plant.

The industrialized world should assist developing and transitioning countries in the development of domestic renewable energy resources and in the implementation of energy-efficient

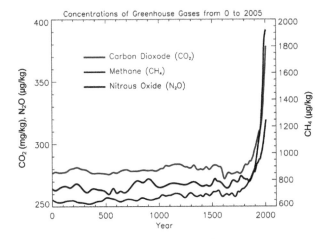

Figure 1.5.   Summary of average concentration *versus* time for the principal components of the radiative forcing of climate change (IPCC, 2008). Reprinted with permission from: Solomon *et al.* (ed.): *Figures Climate Change 2007: The Physical Science Basis.* Working Group I: Contribution to the Fourth Assessment Report of the Intergovernmental Panel on Climate Change, Figure FAQ 2.1, Figure 1; Cambridge University Press.

technologies, e.g. water desalination. There are three reasons for such assistance: (i) to contribute to the provision of safe freshwater, (ii) to improve social and economic development by improving energy security and energy independence, and (iii) to mitigate global climate change. The issues and problems related to freshwater, energy and global warming cannot be solved in a sustainable way unless developing countries also take an active role.

1.4 OVERVIEW ON TECHNOLOGIES BASED ON RENEWABLE ENERGIES FOR FRESHWATER PRODUCTION

The purpose of this section is to give a brief overview of available water treatment technologies based on renewable energies and of their major challenges. More detailed comparisons are given in Chapter 2 of this book and the follow-up chapters describe in detail the individual technologies associated with different energy sources. Renewable energies for use in desalination processes include solar thermal, wind, photovoltaic and geothermal. Our discussion of renewable energy driven desalination falls into two categories. The first category includes distillation processes driven by heat produced by renewable energy systems, while the second includes membrane and distillation processes driven by electrical or mechanical energy produced with renewable resources (Eltawil *et al.*, 2009).

1.4.1 *Sustainable freshwater solutions through wastewater treatment and reuse powered by renewable energies*

Due to increased water scarcity there is a worldwide trend in water-stressed regions to reduce demand of freshwater sources by recycling treated wastewater. In this connection the main applications of reclaimed water are: (i) industrial reuse, (ii) irrigation of agricultural crops, (iii) indirect potable reuse such as through groundwater recharge, (iv) irrigation of parks and lawns, and (v) domestic dual-pipe systems. A study of Bixio *et al.* (2006) identified more than 200 ongoing water reuse projects in Europe as well as many others in their planning phase. Treatment technology encompasses a vast number of options and membrane processes that are key elements of advanced wastewater reclamation and reuse (Wintgens *et al.*, 2005). In this context, membrane bioreactor (MBR) technology has drawn increasing interest in the last decade. MBR technology couples a conventional biological sludge process with a micro- or ultrafiltration membrane system; and hence no settling tank is needed for biomass separation (Judd, 2006). MBR technology offers the advantage of higher product water quality and low footprint. Due to its advantages, membrane bioreactor technology has great potential in a wide range of applications including municipal and industrial wastewater treatment and process water recycling. By 2006, around 100 municipal full scale plants (>500 population equivalent, p.e.; $= 100\,m^3\,day^{-1}$) and around 300 industrial large scale plants ($>20\,m^3\,day^{-1}$) were in operation in Europe (Lesjean and Huislesjow, 2008).

In recent years the wastewater sector has been investigating the feasibility and commercial viability of renewable energy sources for operating wastewater treatment plants. In this regard renewable energy sources can provide power for traditional or conventional water treatment technologies as well as for new emerging technologies. Renewable energy sources, such as solar, wind, biomass, and biofuel sources are becoming more attractive for water supply and wastewater treatment applications.

In 2005 the Atlantic Counties Utilities Authority (ACUA), USA, established a combined solar and wind power farm that provides renewable energy to operate a wastewater treatment plant. This system features a 0.73 MW solar power system, including two roof-top and two ground level arrays with supplementary arrays on solar car ports. Five 1.5 MW wind turbines are also installed, generating up to 7.5 MW of power, sufficient for delivering electricity to 2500 homes in addition to powering the water treatment plant (Coffey, 2008).

The use of a combined biological anaerobic-aerobic treatment process also appears to be a promising technology in particular for highly polluted industrial wastewater due to its high level of chemical oxygen demand (COD). The anaerobic treatment process generates methane gas which can also be utilized as an energy source to drive the aerobic treatment process (i.e. by use of combined heat and power systems, CHP). However, conventional anaerobic-aerobic systems are found to have operational limitations due to long hydraulic retention time (HRT), space requirements and the facilities needed to capture the biogas (Chan *et al.*, 2009). Therefore considerable attention has been directed towards integrated anaerobic-aerobic bioreactors. With simple yet cost effective technology, with their generation of renewable energy and with their outstanding treatment efficiency, it is envisaged that compact integrated bioreactors will be able to treat a wide range of high organic content industrial and municipal wastewaters (Chan, 2009). However, most integrated bioreactors have not yet been implemented at a large scale and further investigations are needed (Chan *et al.*, 2009). In this regard combined membrane based technologies are of special interest due to advantages such as high removal efficiency, high water quality and small footprint (Ahn *et al.*, 2007).

The direct use of solar energy for wastewater treatment is an emerging technology that is widely studied due to its advantages such as low-cost, environmental friendliness and sustainability (e.g. Malato *et al.*, 2009; Chong *et al.*, 2010). In recent years, advanced oxidation processes (AOPs) have been developed to meet the increasing need of an effective wastewater treatment (Ghaly *et al.*, 2011). AOP generates hydroxyl radicals, powerful oxidizing agents, which can eliminate pollutants in waste waters. Heterogeneous photocatalysis through illumination of UV or solar light on a semiconductor surface (e.g. TiO_2) is an attractive advanced oxidation process for the removal of toxic organic and inorganic contaminants from wastewater (Ghaly *et al.*, 2011). Due to cost effectiveness, the development of practical photocatalytic systems needs to focus on the use of solar energy sources. For cost effectiveness, combined biological and solar processes are also being studied and may be of future interest (Oller *et al.*, 2011; Banu *et al.*, 2008). A publication by Rodriguez *et al.* (2004) describes the different collectors used in solar photocatalysis for wastewater treatment and, based on prior experience, the main advantages and disadvantages of each technology are given.

Recently fuel cells (FCs) have been widely studied as power supplies for a variety of applications (Hoogers, 2003). Most of the applications use fuels such as hydrogen or methanol (Hoogers, 2003). However, the use of biomaterials is getting more attractive since FCs directly generate electrical energy without any thermal process and hence have high efficiency factors. FCs coupled with biomass-supplied reactors can convert renewable energy into a useful form in an environmental-friendly and CO_2-neutral manner. It is considered as one of the most promising energy supply systems (Xuan, 2009). Biomass-derived FCs such as ethanol, methanol, biodiesel and biogas can be fed to a fuel processor as a raw fuel for reforming by steam reforming, partial oxidation, or other reforming methods (Xuan, 2009). In particular, solid oxide fuel cells (SOFC) are regarded as very attractive in wastewater treatment plants since they can be fueled by biogas produced through anaerobic digestion processes of high concentration wastewaters or of excess sludge (Farhad, 2010). The electricity and heat required to operate the wastewater treatment plant can be completely self-supplied and the extra electricity generated can be supplied to the electrical grid (Farhad, 2010).

In the last decade, bioelectrochemical systems such as microbial fuel cells (MFCs) have attracted wide attention as an emerging technology that uses microorganisms to directly generate electrical energy (Virdis *et al.*, 2011). A microbial fuel cell is a bioreactor that converts chemical energy stored in chemical bonds in organic compounds to electrical energy through catalytic reactions mediated by microorganisms under anaerobic conditions (Du *et al.*, 2007). However, the development of MFCs is in its infancy and more R&D is needed to develop commercially applicable systems. In addition MFC technology has to compete with mature methanogenic anaerobic digestion technology that already has many commercial applications. However, MFC directly converts biofuel to electrical energy without using any intermediate step (e.g. biogas production) and hence in theory has a higher coefficient of performance (COP).

1.4.2 *Sustainable drinking water solutions through desalination by solar and wind energy*

Apart from treatment and reuse of wastewater, desalination of seawater and brackish water can be regarded as increasingly important technique for safe water production. The seawater desalination industry is growing fast and is constantly evolving through greater cost reductions and more reliable production of water of very high quality. However, desalination processes will always require considerable amount of energy. If conventional energy resources are used, they will contribute to climate change due to high carbon dioxide emissions. To be sustainable, renewable energy sources have to be used at least for part of the power requirement. However, renewable energy driven desalination faces major barriers (see Papapetrou *et al.*, 2010):

(i)   Most systems are not developed with a whole systems approach but are merely a combination of components developed independently and therefore show poor reliability and increased water costs.

(ii) Desalination technology development focuses on large systems; hence it lacks components appropriate for small desalination plants. So there is a need for robust, sustainable stand-alone systems that can be located in difficult environments.

(iii) Current desalination technology has been designed for use with a constant energy supply. However, most renewable energy systems provide variable energy supply (e.g. solar, wind). So there is a need to develop components and control systems that allow desalination technology to better deal with available energy inputs such as hybrid systems and energy storage.

There are a diversity of techniques to drive desalination including distillation and membrane based techniques (Qiblawey and Banat, 2008; Eltawil *et al.*, 2009; Gude *et al.*, 2010; Papapetrou *et al.*, 2010). Their applicability strongly depends on local availability of renewable energy sources and on the quality of water to be desalinated. The selection of the appropriate renewable technology depends on a variety of factors such as plant size, feed water salinity (brackish or seawater), remoteness, availability of an energy grid, technical infrastructure and the type and potential of local renewable energy resources (Tzen and Morris, 2003; Voivontas *et al.*, 2001). Figures 1.6 and 1.7 depict global wind speed and global solar irradiance. A detailed description of treatment technologies is given in Chapter 2 of this book. Solar heat can be used either directly or *via* steam and power generation. In addition, wind energy can be also used for electrical power generation and can therefore drive reverse osmosis or electrodialysis units. Figure 1.8 gives an overview on different techniques which are commercially available or currently tested in large-scale pilot projects.

Solar stills (ST) are the simplest techniques for water desalination and they are called solar distillers when they are the only solar distillation process. Some call solar stills "passive solar distillation". The working principle is based on transparent materials which have the property of transmitting solar radiation through the transparent cover and which is absorbed as heat by a black surface in contact with the salty water to be distilled. The vapor is condensed on the glass cover which is naturally cooled by ambient air. The condensate runs down into a groove and can subsequently be collected.

The solar humidification-dehumidification process (H-DH), also called multiple-effect H-DH or multiple-effect humidification (MEH), mimics the natural water cycle (evaporation and condensation) by using the air as a vapor carrier; in other words, heated air is loaded with water vapor which condenses when air is cooled down. The difference with solar stills is that in MEH both processes take place separately. During re-condensation of the humid air, most of the energy used prior for evaporation can be regained to preheat the incoming water and can also be used in subsequent cycles of evaporation and condensation which makes the entire process very energy-efficient. Solar thermal MEH collectors require temperatures of 70–85°C.

Membrane distillation (MD) is a separation technique which combines a thermally-driven distillation process with a membrane process. The membrane is only permeable for water vapor but not for liquid water: it separates the pure distillate from the retained solution at a typical operating temperature of 60–80°C.

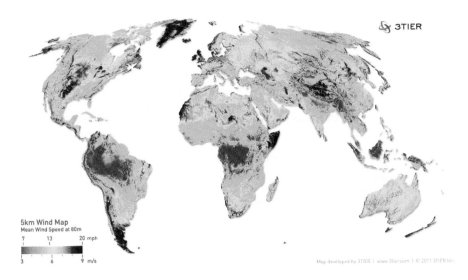

Figure 1.6.   Global mean wind speed at 80 m. Reprinted with permission of 3TIER Inc., Seattle, WA (www.3tier.com).

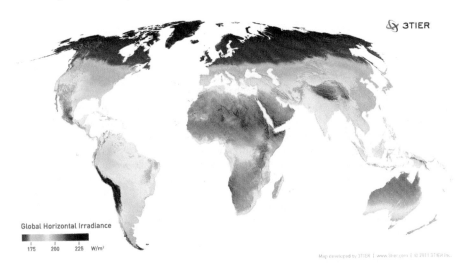

Figure 1.7.   Global mean solar irradiance. Reprinted with permission of 3TIER Inc., Seattle, WA (www.3tier.com).

Multiple-effect distillation (MED) takes place in a series of vessels and uses the principle of evaporation and condensation at reduced ambient pressure. A series of vessels produces freshwater at progressively lower pressure. Since water boils at lower temperatures as pressure decreases, the water vapor of the first vessel serves as the heating medium for the second and so on. The more vessels there are the higher the energy efficiency will be (typically 15–20 vessels).

The multi-stage flash (MSF) distillation process is based on several chambers that generate vapor free of salts. First, the MSF heats the feed water by steam extracted under high pressure. The vapor could be provided by CSP technology (see below). Then the steam is fed into the first so-called flash chamber, where the water boils rapidly through a sudden evaporation process ("flashing"). In each successive stage the "flashing" of part of the input water continues since

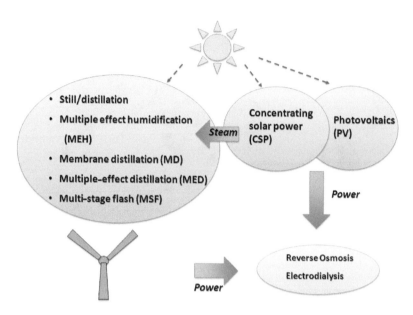

Figure 1.8.   Different desalination technologies for sustainable freshwater production.

the pressure at each stage is lower than in the previous stage. The vapor produced by flashing is cooled by the incoming feed water and hence converted into freshwater by condensation. The MSF system can also be combined with a mechanical vapor compression system (MVC) which has the advantage of smaller equipment and lower operating costs (Mabrouk *et al.*, 2007).

A more efficient way to use solar power is by applying concentrating solar power (CSP) technology. The technology is based on glass mirrors that continuously track the position of the sun to optimize energy production. Currently four configurations are commercially available: central receivers, parabolic troughs, parabolic dishes and linear Fresnel systems. The concentrated heat is transferred to a power cycle that generates high-pressure and high-temperature steam that can drive turbines and that also can subsequently be used for MSF distillation or MED. The electrical power generated can be used to operate reverse osmosis (RO) desalting units. Usually CSP technology needs large amounts of freshwater for operation (cooling system, mirror cleaning) but the integration of CSP and desalination makes the solar power concept fully sustainable since it can provide freshwater for its own cooling system and for the mirror cleaning.

Since reverse osmosis (RO) and electrodialysis/electrodialysis reversal (ED/EDR) are powered by electrical energy, these desalination technologies can be combined with photovoltaics (PV) which is already a mature and well-established technology with a large list of suppliers in many countries (also in newly industrializing countries such as China and India). RO is a physical process, which relies on an imposed pressure gradient generated by high-pressure pumps to force water through a membrane, while retaining most of its dissolved salts behind. ED and EDR use electrical energy directly for desalination. Their cells consist of electrodes in a feed compartment and in a concentrate (brine) compartment formed by an anion exchange membrane and a cation exchange membrane placed between two electrodes that generate the desalinated water and the remaining brine. EDR operates similarly to ED except that the polarity of the poles is reversed several times an hour.

Currently there is more technological experience with the combination of PV-RO than with PV-ED/EDR. Since PV and RO are both mature technologies, their combination is being increasingly used in many countries worldwide. However, there are intensive R&D efforts to improve PV and RO process performance that might get us closer to the whole systems approach that is needed.

Table 1.1. Solar and wind powered desalination technologies and typical properties (modified from Papapetrou *et al.*, 2010).

	Capacity m^3 day^{-1}	Energy consumption kWh m^{-3}	Cost estimate for fresh water Euro m^{-3}	Development stage
Solar still	<0.1	solar passive	1–5	applications
Solar multiple-effect humidification (MEH)	1–100	thermal: 100 electrical: 1.5	2–5	application/ advanced R&D
Solar membrane distillation (MD)	0.15–10	thermal: 150–200	8–15	advanced R&D
Solar/concentrating solar power (CSP) & multiple-effect distillation (MED)	>5000	thermal: 60–70 electrical: 1.5–2	1.8–2–2 (prospective cost)	advanced R&D
Photovoltaics-reverse osmosis (PV-RO)	<100	electrical: BW: 0.5–1.5 SW: 4–5	BW: 5–7 SW: 9–12	application/ advanced R&D
Photovoltaics-electrodialysis reversed (PV-EDR)	<100	electrical, only BW: 3–4	BW: 8–9	advanced R&D
Wind-RO	50–2000	electrical: BW: 0.5–1–5 SW: 4–5	units <100 m^3 day^{-1} BW: 3–5 SW: 5–7 units >1000 m^3 day^{-1} 1.5–4	application/ advanced R&D

BW: Brackish water, SW: seawater.

Wind energy has been used as a power supply for desalination systems especially for RO units. In the majority of cases, wind energy is not directly used but is instead converted to electrical energy to drive the RO pumps. For operational and economic reasons, most of the existing wind-powered desalination systems are connected to the grid and there is not much experience with off-grid wind systems. However, autonomous wind-powered desalination systems are very interesting in remote windy areas particularly on small islands without grid access. Especially in areas with good wind resources and high energy costs, wind-powered RO has proved to be cost-competitive compared to conventional desalination systems (Forstmeier *et al.*, 2007).

Table 1.1 presents basic information about the most common desalination technologies, including current capacities, energy consumption, cost estimate and development stage (Papapetrou *et al.*, 2010).

Figure 1.9 shows the distribution of renewable energy powered desalination systems (Tzen, 2005) with PV-RO having the largest share: about 1/3 of total capacity. Wind-RO systems are second with a share of around 20%, and all other systems are below 20%, including hybrid systems that use complementary renewable energies.

Basically any desalination plant can be operated with renewable energy resources. A future target should be the development of standardized, reliable and robust systems that integrate renewable energy use with a desalination unit and that offer the end user comprehensive performance guarantees (Papapetrou *et al.*, 2010). The major future challenges in R&D include:

- Dealing with the variability of energy supply.
- Combining existing renewable energy into a whole systems approach (e.g. wind turbine with PV technology).
- Hybridization with an electrical grid through intelligent sensor and control systems.
- Improving thermal and electrical storage techniques.

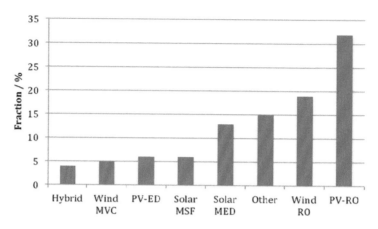

Figure 1.9.   Distribution of renewable energy powered desalination technologies (Tzen, 2005).

- Improving energy minimization strategies.
- Developing efficient and robust small-scale desalination systems.

Until now, conventional desalination plants have been built assuming a constant supply of energy. When applying renewable energies, there is typically a mismatch between energy supply and demand. There are two approaches to tackle this problem. Energy can be stored or alternatively, the impact of a variable energy supply on the desalination plant operation can be minimized. For storage of electrical energy, the only commercially available option is the use of conventional batteries (mostly lead based). These are expensive, have limited storage capacity, pose an environmental threat, and have a relatively short operational life span. In the future, novel electrochemical storage systems, such as redox flow battery systems, may provide an interesting alternative. They appear to offer great promise as a low-cost, high efficiency large-scale energy storage system. In particular, the vanadium redox flow system is currently being investigated (e.g. see Skyllas-Kazacos, 2009; Zhao *et al.*, 2006). For heat storage applications, latent heat storage materials like molten salts and paraffines are increasingly being studied (Michels and Pitz-Paal, 2007; Nallusamy *et al.*, 2007).

To minimize impacts of a variable energy supply, the desalination plant design can be adapted, for example through the development of RO membranes that are less sensitive to variable pressure and flow, and through the design of new control systems that ensure that the available energy is optimally used. For example, wind-powered RO systems without energy storage have been studied. Heijman *et al.* (2009) carried out pilot testing with a direct mechanical drive of an RO pump with an energy recovering system that used a windmill and the storage of clear water during periods of low wind speed. Unlike with most wind-powered RO systems, no electricity is required and the system can therefore be used in remote areas. Based on pilot trials Park *et al.* (2011) concluded that an RO wind-membrane system can be operated in a safe operating window with large power fluctuations, but further control strategies are required to deal with intermittent operations, especially with higher salinity feed waters. Riffel and Carvalho (2009) designed and operated a small-scale PV powered RO plant without batteries at variable flow/pressure conditions for stand-alone applications in equatorial areas to desalinate brackish water.

Several systems have been designed and successfully implemented that combine RO and hybrid solar PV/wind (generating electrical energy) power systems. Reviews of such systems have been published by Charcosset (2009) and Käufler *et al.* (2011). These systems deal with wind-powered RO in an optimized hybrid grid-parallel configuration and with balanced load and/or energy management.

Particularly in the field of RO treatments, minimizing energy usage is critical and has been the focus of a variety of R&D projects. Up to 30% of desalination costs is due to the energy requirements for the production of freshwater (Busch and Mickols, 2004). The factors affecting these energy requirements can be classified as follows: (i) enhanced system design, (ii) high efficiency pumping energy recovery, (iii) advanced materials and (iv) innovative technologies (Subramani *et al.*, 2011). A novel technique, namely adsorption desalination technology, shows great promise due to its low energy consumption and its ability to use low-grade thermal energy (Thu *et al.*, 2009; Wang *et al.*, 2011). In the adsorption desalination process, seawater is fed to an evaporator where it is evaporated by water with low thermal load. The process of vapor uptake is maintained by the adsorbent (e.g. silica gel) in the adsorption process and the heat generated in the process is removed by circulating cool water. Hot water supplied to the desorption beds drives the vapor out and is condensed in a condenser to produce the pure water. Besides its low energy consumption of less than $1.5 \, kWh$ per m^3 of freshwater treated, the process has the advantages of no major moving parts (low maintenance) and low corrosion and fouling due to the evaporation of saline water at low temperature. The process can be operated at a temperature of 65–85°C. The process has been studied at the pilot scale (Wang *et al.*, 2011).

The design and configuration of membrane units have significant effect on the performance and economics of a RO plant (Wilf and Bartels, 2005). Recently, energy demand reduction in RO desalination has occurred through optimal process configuration and control schemes (Subramani, 2011). There are two classes of energy recovery systems (Wang *et al.*, 2004). Class I units use hydraulic power to cause a positive displacement in the recovery device and the hydraulic energy is directly transferred in a single step. Class II devices use the hydraulic energy of the RO concentrate in a two-step process that first converts the energy to centrifugal mechanical energy and then back to hydraulic energy (Subramani, 2011). Great progress has been achieved in the last few decades. In 1980, seawater RO systems consumed more than $26 \, kWh \, m^{-3}$. Today, such systems consume on average only $3.4 \, kWh \, m^{-3}$ (Chang *et al.*, 2008). However, this depends largely on the quality of the raw seawater. The minimum theoretical use at 50% recovery is about $1.08 \, kWh \, m^{-3}$ (Voutchkov, 2010). There is further potential for minimizing energy consumption through application of novel materials such as nanocomposite RO membranes made by combining zeolite nanoparticles dispersed within a traditional polyamide thin film, or by incorporating carbon nanotubes into polymer membrane materials (Subramani, 2011). In both cases, water flux through the membrane can be significantly increased resulting in 20–50% energy reduction. In addition, novel innovative technologies such as forward osmosis which uses a concentrated draw solution for desalination instead of hydraulic pressure, and ion concentration polarization which uses ion-selective membranes, offer great promise for future reduction in energy demand (Subramani, 2011).

Until now, the focus of desalination technologies has been on large-scale units ($> 100 \, m^3 \, day^{-1}$). Consequently there is a lack of technologies appropriate for small-scale applications (capacity $< 10 \, m^3 \, day^{-1}$). These offer a growing promising market for remote areas such as isolated houses, remote holiday homes and hotels, small health centers etc. where electricity and water supply are non-existent, not reliable or too expensive. A variety of desalination technologies are addressing this market segment such as solar stills, solar membrane distillation (MD), and solar multiple-effect humidification (MEH). Koschikowski *et al.* (2009) for example, studied MD for autonomous small-scale freshwater production. Narayan *et al.* (2010) investigated solar humidification process for small-scale decentralized application. However, RO is the primary technology for small units due to its modularity and flexibility.

A RO-based system that can be used regardless of the type of energy source was developed by Schäfer and Richards (2005). It is designed for brackish water treatment in remote areas for a production of about $1 \, m^3 \, h^{-1}$. It was tested in field trials in Australia in a remote area. The specific energy consumption was below $5 \, kWh \, m^{-3}$ when operated at a pressure above 7 bar (0.7 MPa). An ultrafiltration module was applied as pre-treatment.

Qiblawey *et al.* (2011) discuss the ADIRA (Autonomous Desalination in Rural Areas) project on water desalination by RO with electricity generation using PV technology with additional battery storage. The project was funded by the European Community (EC). The project focused on small units powered by autonomous systems with freshwater output of $<10\,m^3\,day^{-1}$. The RO-PV system has been successfully designed, installed and tested in the northern part of Jordan. The specific energy consumption ranged between 2 and $14\,kWh\,m^{-3}$.

Small-scale RO desalinators (known as "watermakers", capacity $<5\,m^3\,day^{-1}$) meanwhile are widely used to produce drinking water from seawater on boats (for overview, see Nauticexpo). This is a proven technology, which works reliably at remote locations under mechanically, climatically and chemically rough conditions. Most of these systems have been optimized in terms of energy efficiency and productivity. Small-scale marine-RO units can also be used for arsenic or heavy metal removal. Some of these units can be powered by sustainable energy sources such as photovoltaic, wind turbines, or can be operated manually by generating pressure. In addition the direct use of solar thermal energy by means of thermodynamic power cycles, such as the Stirling or Rankine process, may be interesting to pursue due to their high efficiency and simplicity since no electrical system is required (Garcia-Rodriguez et al., 2007). Small-scale RO units are currently being tested at Karlsruhe University of Applied Sciences in Germany to test their efficiency for arsenic removal (Geucke et al., 2009). Work is also ongoing to test their suitability for rural conditions in Bihar, India in cooperation with the Anugrah Narayan college. The most salient feature of these desalinators is their very low energy consumption; 80% less power is needed than for conventional RO watermakers. Considering information about chemical composition and grades of mineralization of the water treated, together with information about the available infrastructure, the cost of the treated water was estimated to be around 10–20 Euro m^{-3}. It is disproportionally higher than the cost of large scale systems due to low capacity ($0.25–2\,m^3\,day^{-1}$). However, in many cases, the drinking water alternatives are bottled water, or water delivered by trucks and boats. These alternatives are very costly and consume a large fraction of people's income in developing countries. If water is only used for drinking then 3–5 liters per person per day is enough. If the water is also used for other purposes, such as irrigation, much more water is needed and this needs to be carefully considered due to the high cost of small-scale desalination. Freshwater produced by small-scale decentralized units should probably only be used for drinking and not for agriculture.

Decentralized desalination technologies require more R&D because there is still a lack of adequate whole systems investigations including regarding the use of small efficient pumps, system control algorithms for decentralized application, suitable pre and post treatment, and energy or water storage to reduce or avoid fluctuations in water production. For electrical energy storage, the only commercially available option is the use of batteries that have limited storage capacity, relative short operational life and relatively high costs. Peter-Varbanets et al. (2009) reviewed a variety of membrane-based decentralized systems for producing freshwater. They came to the conclusion that there are good prospects for the application of such systems. However, more R&D is needed to develop systems with low-cost and low maintenance especially for use in developing and transitioning countries.

1.4.3   *Geothermal resources options for desalination*

The geothermal option for electricity generation and direct use of geothermal heat for desalination has recently been recognized as an economically and environmentally sound choice. This coincides with a sharply increased recognition of the potential of geothermal energy to meet much of future electricity demand and to guarantee energy security and energy independence in both developing and developed countries (Chandrasekharam and Bundschuh, 2002, 2008; Aaheim and Bundschuh, 2002; Bundschuh et al., 2002, 2007; Bundschuh and Coviello, 2002; Clauser, 2006; Bundschuh and Suarez Arriaga, 2010, Huenges and Ledru, 2010; further information can be found in the homepages of the International Geothermal Association (IGA 2011a,b) the Geothermal Resources Council (GRC, 2011) and the Geothermal Education Office (GEO, 2011)). The major advantage of geothermal powered desalination is the stability of the energy source, and the fact that no energy storage is required unlike for wind- and solar-powered desalination technologies. This reduces costs. However, the geothermal option has other costs, such as the need for

subsurface drilling. This cost depends strongly on local geological conditions and the local heat flow regime.

There are three principal types of geothermal reservoirs; for more details see e.g. Clauser (2006), Chandrasekharam and Bundschuh (2008), Bundschuh and Suárez Arriaga (2010), Huenges and Ledru (2010) and the previously mentioned webpages of IGA (2011a,b), GRC (2011) and GEO (2011):

(i) Convection-dominated high-enthalpy hydrothermal geothermal systems (vapor- or liquid-dominated) that are located in volcanic and/or tectonically active areas and are therefore limited to active plate boundaries on both continents and oceans. Here, uprising magma or deep-seated intrusives are the main sources of heat for low- ($<150°$) and high- ($>150°$C) enthalpy convective geothermal systems that are recharged by meteoric water.

(ii) Convective low-enthalpy hydrothermal resources ($<150°$C) that are not limited to active tectonic plate boundaries and consequently have a much broader spatial distribution and a considerably larger geothermal potential relative to high-enthalpy systems. However, despite this benefit, low-enthalpy systems are practically unutilized for electricity generation in both developed and developing countries (Chandrasekharam and Bundschuh, 2002, 2008; Bundschuh and Chandrasekharam, 2002; Bundschuh & Coviello, 2002; Bundschuh et al., 2007).

(iii) Conduction-dominated Enhanced Geothermal Systems (EGS, or Engineered Geothermal Systems) that are not limited to volcanic and/or tectonic active areas, or to the presence of natural fluids and that have even greater energy potential relative to the two previously mentioned types of geothermal reservoirs. In these systems, the heat is provided by natural radioactivity of elements such as U, Th, and K, and by conductive heat transport from the earth's mantle to shallower depths through the deep continental crust (Chandrasekharam and Bundschuh, 2008; Huenges and Ledru, 2010). EGS resources, which are available practically in every country, have received great attention in recent years. According to an MIT report (MIT, 2006), the USA alone has an EGS potential of about 13,000,000 Exa Joules (EJ $= 10^{18}$ Joules) (depth 3–10 km), of which 200,000 EJ can be extracted for utilization, which corresponds to about 2000 times the annual primary energy consumption of the country in 2005. By 2050, the USA could economically generate about 100,000 MW$_e$ with only a modest R&D investment (MIT, 2006).

Despite receiving little consideration in the past, desalination is one of the most promising areas for the application of geothermal energy due to the need for constant energy supply, 24 hours a day and 365 days a year. It can be tapped in many places around the world where water scarcity or water quality is a problem (e.g. Mediterranean area, Eastern Africa, Arabian peninsula, northern Chile, Andean range and forelands, etc.), or where domestic or industrial effluents can be treated for freshwater production. Therefore, geothermal desalination should be investigated in more detail, especially widely available low-enthalpy geothermal resources ($<150°$C). Minimum required temperature is 60°C for membrane distillation (MD) and other systems that can be powered by heat (Shih, 2005). Electrodialysis (ED) and reverse osmosis (RO) are not suitable for direct use of geothermal heat but they can be run by electricity produced from geothermal resources that require higher temperature geothermal fluids. Using present-day technology, i.e. applying recent innovations in the form of binary fluids and heat exchangers, fluid temperatures of more than 85°C are suitable for economical geothermal electricity generation (Chandrasekharam and Bundschuh, 2008). Access to geothermal energy generally requires drilling, and applications are therefore restricted to large and middle-sized desalination plants. Geothermal desalination is applicable to small-scale desalination units suitable for water supply to small communities only in the very few cases where no drilling is needed, i.e. where geothermal fluids of sufficiently high temperature discharge naturally or very close to the earth's surface.

Geothermal desalination of offshore seawater using submarine geothermal reservoirs as an energy source is an interesting future technology for predominantly large-scale applications. It has the benefit that the brines produced (e.g. by RO) can be directly discharged into the sea. We can distinguish two types of submarine geothermal reservoirs: deep and shallow ones. Both constitute ideal energy resources for desalination of seawater. Deep submarine geothermal reservoirs (generally >2000 m) are related to the existence of hydrothermal vents emerging in many places along the oceanic spreading centers between tectonic plates that extend over a length of about 65,000 km in the earth's oceanic crust (Marshall, 1979). These have a practically infinite energy potential. Shallow submarine geothermal fields are related to faults and fractures near continental platforms at active plate boundaries and are found at depths between 1 and 50 m. Examples of shallow direct discharge of geothermal fluids close to coastlines or offshore (i.e. suitable for powering small-scale desalination units) are found for example in California (Suárez Bosché et al., 2000; Suárez and Samaniego, 2005, 2006; Bundschuh and Suárez Arriaga, 2010), Dominica (Lesser Antilles: McCarthy et al., 2005), the Aegean Sea (Suárez Arriaga et al., 2008), Papua New Guinea (Pichler, 2005; Price and Pichler, 2005). In the case of the Gulf of California in Mexico, submarine thermal areas are located offshore from the port of Ensenada in Baja California, and temperatures of up to 102°C have been measured at 20 m depth, less than 100 m from the coast. In addition, a submarine geothermal reservoir is located in the Wagner Depression at less than 30 m depth. Rough estimates indicate that the geothermal potential of this submarine zone could be 100 times larger than that of the Cerro Prieto geothermal reservoir which, with an installed capacity of 720 MWe in the year 2009, is one of the largest geothermal reservoirs in terrestrial Mexico (Grijalva, 1986; Mercado, 1990, 1993; Gutiérrez-Negrín et al., 2010).

Existing oil and gas wells may provide another economical option to access geothermal energy resources because the costly drilling has already been done. However, because oil/gas wells have a different design than that of geothermal wells, some modifications may be needed. Existing oil and gas wells can also give good information on underground temperature, geological structure, hydraulic permeability and porosity of the potential reservoir. With state-of-the-art technology, these wells may generate geothermal electricity if temperature is higher than 85°C (or can be used for direct heat desalination if temperature is higher than 60°C). The electricity can be generated from (i) oil and gas waste fluids changing the warm water that is co-produced with the oil and gas into an asset, or from (ii) abandoned unplugged oil and gas wells converted into geothermal energy wells. This applies for both on-shore and off-shore fields. Since the disposal of oil platforms is expensive, their conversion to geothermal platforms may be an interesting option. The huge energy potential becomes evident if we consider that worldwide there are over a million oil and gas wells, and that many of them are drilled at relatively large depths where temperature and pressure are high and therefore represent a high energy potential. Often they produce, or have produced, hot water together with hydrocarbons. Currently, the energy from hot water in oilfields is wasted; it is either reinjected into the subsurface or released to the environment after treatment.

Locally, the water from abandoned deep underground mines may also have temperatures high enough for desalination applications, and these may become an interesting alternative geothermal resource for desalination. However, there are not many mines with temperatures meeting or exceeding the minimum temperature requirements for desalination by direct heat (>60°C) or geothermal electricity production (>85°C).

1.5   CONCLUSIONS AND OUTLOOK

The application of renewable energies is now a reasonable and technically mature option for freshwater production in the fast growing desalination market. The total global capacity of desalination plants in 2009 was around 59.9 million m^3 day^{-1} (about 22 billion m^3 $year^{-1}$) and the annual desalination growth is estimated at 9.5% (IDA Report, 2009, www.desalination.biz, 2011). The newly installed capacity in 2009 represents 6.6 million m^3 day^{-1} and it was the largest amount of desalination capacity installed in a single year (IDA Report, 2009). Among the possible combinations of desalination and renewable energy technologies, solar and wind have been demonstrated as most promising in terms of economic and technological feasibility. In particular, wind and PV-powered membrane processes and direct and indirect solar distillation can be regarded as the most mature technologies. However, their viability depends strongly on feed water quality and local availability of these resources. Desalination processes based on geothermal energy can be operated at reasonable cost wherever a proper geothermal source is available since there is no energy storage required.

More R&D is required to make desalination powered by renewable energies technically and economically competitive. The major challenge is the development of highly efficient and cost effective combined processes coupled with smart, constant, energy supply. Greater efforts are needed in particular to optimize and design novel small-scale desalination systems that can be applied in developing and newly industrializing countries in Africa and Asia where there is great market potential.

Treated wastewater is another source of freshwater. This source is becoming increasingly important in water scarce regions far from coastal areas. The combination of biological anaerobic and aerobic treatments gives the advantage of generating sustainable energy through biogas production that can then be used to drive the process. The application of membranes in this combined process can contribute to reduce the footprint, enhance efficiency and obtain high water quality. With regards to wastewater treatment, the direct use of solar energy with advanced solar oxidation processes can be regarded as a promising technology in the sunny regions of our planet. However, more R&D is needed to develop feasible processes that can be transferred at competitive costs into the market.

REFERENCES

Aaheim, A. & Bundschuh, J.: The value of geothermal energy for developing countries. In: D. Chandrasekharam & J. Bundschuh (eds): *Geothermal energy for developing countries*. Balkema Publisher, Lisse, The Netherlands, 2002, pp. 37–52.

Ahn, Y.T., Kang, S.T., Chae, S.R., Lee, C.Y., Bae, B.U. & Shin, H.S.: Simultaneous high-strength organic and nitrogen removal with combined anaerobic upflow bed filter and aerobic membrane bioreactor. *Desalination* 202 (2007), pp. 114–121.

Banu, J.R., Anandan, S., Kaliappan, S. & Yeon, I.T.: Treatment of dairy wastewater using anaerobic and solar photocatalytic methods. *Solar Energy* 82 (2008), pp. 812–819.

Beecy, D.J., Ferrell, F.M. & Carey, J.K.: Biological methane: a long-term CO_2 recycle concept. 2000, www.netl.doe.gov/publications/proceedings/01/carbon_seq/5a1.pdf (accessed October 2011).

Bixio, D., Thoeye, C., De Koning, J., Joksimovic, D., Savic, D., Wintgens, T. & Melin, T.: Wastewater reuse in Europe. *Desalination* 187 (2006), pp. 89–101.

Bundschuh, J. & Chandrasekharam, D.: The geothermal potential of the developing world. In: D. Chandrasekharam and J. Bundschuh (eds): *Geothermal energy for developing countries*. Balkema Publisher, Lisse, The Netherlands, 2002, pp. 53–62.

Bundschuh, J. & Coviello, M.: Geothermal energy: capacity building and technology dissemination. In: D. Chandrasekharam & J. Bundschuh (eds): *Geothermal energy for developing countries*. Balkema Publisher, Lisse, The Netherlands, 2002, pp. 77–90.

Bundschuh, J. & Suárez Arriaga, M.C.: *Introduction to the numerical modeling of groundwater and geothermal systems: fundamentals of mass, energy and solute transport in poroelastic rocks*. Balkema/CRC Press/Francis & Taylor, Leiden, The Netherlands, 2010.

Bundschuh, J., Birkle, P., Aaheim, A. & Alvarado, G.E.: Geothermal resources for development—valuation, present use and future opportunities. In: J. Bundschuh & G.E. Alvarado (eds): *Central America: geology, resources and hazards* (2 Volumes). Balkema Publisher, Lisse, The Netherlands, 2007, pp. 869–894.

Busch, M. & Mickols, W.E.: Reducing energy consumption in seawater desalination. *Desalination* 165 (2004), pp. 299–312.

Chan, Y.J., Chong, M.F., Law, C.L. & Hassell, D.G.: A review on anaerobic-aerobic treatment of industrial and municipal wastewater. *Chem. Eng. J.* 155 (2009), pp. 1–18.

Chandrasekharam, D. & Bundschuh, J. (eds): *Geothermal energy resources for developing countries*. A.A. Balkema, Lisse, The Netherlands, 2002.

Chandrasekharam, D. & Bundschuh, J.: *Low-middle enthalpy geothermal systems for power generation.* CRC Press/Taylor & Francis/Balkema, Boca Raton, FL, 2008.

Chang, Y., Reardon, D.J., Kwan, P., Boyd, G., Brant, J., Rakness, K. & Furukawa, D.: Evaluation of dynamic energy consumption of advanced water and wastewater treatment technologies. AWWARF Final Report, 2008, http://www.waterrf.org/ProjectsReports/PublicReportLibrary/91231.pdf.

Charcosset, C.: A review of membrane processes and renewable energies for desalination. *Desalination* 245 (2009), pp. 214–231.

Chong, M.N., Jin, B., Chow, C.W.K. & Saint, Ch.: Recent developments in photocatalytic water treatment technology: a review. *Water Research* 44 (2010), pp. 2997–3027.

Clauser, C.: Geothermal energy. In: K. Heinloth (ed.): Landolt-Börnstein, Group VIII: *Advanced materials and technologies*, Vol. 3: *Energy technologies*, Subvol. C: *Renewable energies*. Springer, Heidelberg-Berlin, Germany, 2006, pp. 480–595.

Coffey, M.: Renewable energy: filtration and the green energy revolution. *Filtration and Separation* 45 (2008), pp. 24–27.

Dave, B.C.: Prospects for methanol production. In: J.D. Wall, C.S. Harwood & A. Demain (eds): *Bioenergy.* ASM Press, 2008, pp. 235–248.

Du, Z., Li, H. & Gu, T.: A state of the art review on microbial fuel cells: a promising technology for wastewater treatment and bioenergy. *Biotechnol. Adv.* 25 (2007), pp. 464–482.

EIA (Energy Information Administration): International Energy Outlook 2011. Energy Information Administration, Office of Integrated Analysis and Forecasting, US Department of Energy, Washington, DC, DOE/EIA-0484(2011), 2011, http://www.eia.doe.gov/oiaf/ieo/index.html (accessed October 2011).

El-Dessouky, H.T. & Ettouney, H.M.: *Fundamentals of salt water desalination.* Elsevier Science, 2002.

Eltawil, M.A., Thengming, Z. & Yuan, L.: A review of renewable energy technologies integrated with desalination systems. *Renew. Sustain. Energy Rev.* 13 (2009), pp. 2245–2262.

Farhad, S., Yoo, Y. & Hamdullahpur, F.: Effects of fuel processing methods on industrial scale biogas-fuelled solid oxide fuel cell system for operating in wastewater treatment plants. *J. Power Sources* 195 (2010), pp. 1446–1453.

Forstmeier, M., Mannerheim, F., Amato, F.D., Shah, M., Liu, Y., Baldea, M. & Stella, A.: Feasibility study on wind-powered desalination. *Desalination* 2003 (2007), pp. 463–470.

Garcia-Rodriguez, L. & Delgado-Torres, A.M.: Solar powered Rankine cycles for freshwater production. *Desalination* 212 (2007), pp. 319–327.

GEO: Web page of the Geothermal Education Office, 2011, http://geothermal.marin.org/ (accessed October 2011).

Geucke, T., Deowan, S.A., Hoinkis, J. & Pätzold, Ch.: Performance of a small-scale RO desalinator for arsenic removal. *Desalination* 239 (2009), pp. 198–206.

Ghaly, M.Y., Jamil, T.S., El-Seesy, I.E., Souaya, E.R. & Nasr, R.A.: Treatment of highly polluted paper mill wastewater by solar photocatalytic oxidation with synthesized nano TiO_2. *Chem. Eng. J.* 168 (2011), pp. 446–454.

GRC: Web page of the Geothermal Resources Council, 2011, http://www.geothermal.org/ (accessed October 2011).

Grijalva, N.: Investigación de la energía geotérmica en la Depresión de Wagner en el Golfo de California, latitud 31°00′ al 31°15′ y longitud 113°50′ al 114°10′. Unpublished reports prepared for the Comisión Federal de Electricidad under Contract No. CCP-CLS-002/ 86 (date of 1st report: 10/1/86, date of 2nd report: 11/22/86), Cerro Prieto, B.C., Mexico, 1986.

Gude V.G., Nirmalakhandan, N. & Deng, S.: Renewable and sustainable approaches for desalination. *Renew. Sustain. Energy Rev.* 14 (2010), pp. 2642–2654.

Gutiérrez-Negrín, L.C.A., Maya-González, R. & Quijano-León, J.L.: Current status of geothermics in Mexico. *Proceedings World Geothermal Congress 2010*, 25–29 April 2010, Bali, Indonesia, 2010.

Harrison P. & Pearce F.: AAAS Atlas of Population and Environment, 2001. Victoria Dompka Markham (ed.), American Association for the Advancement of Science and the University of California Press, 2001.

Heijman, S.G.J., Rabinovitch, E., Bos, F., Otthof, N. & van Dijk, J.C.: Sustainable seawater desalination: stand-alone small-scale windmill and reverse osmosis system. *Desalination* 248 (2009), pp. 114–117.

Hoogers, G.: *Fuel cell technology handbook.* CRC Press, 2003.

Huenges, E. & Ledru, P. (eds): *Geothermal energy systems: exploration, development, and utilization.* John Wiley and Sons-VCH, Weinheim, Germany, 2010.

IDA Report, Desalination: Current situation and future perspectives. 2009, www.idadesal.org (accessed Nov. 2011).

IGA: Website of the International Geothermal Association, 2011a, http://iga.igg.cnr.it/index.php (accessed September 2011).

IGA: IGA Geothermal Conference Papers Database. 2011b, http://www.geothermal-energy.org/304, iga_geothermal_conference_database.html (accessed October 2011).

IPCC: Intergovernmental Panel on Climate Change. http://ipcc-wg1.ucar.edu/wg1/FAQ/wg1_faq-2.1.html (accessed October 2011).

Judd, S. (ed.): The MBR Book: *Principles and applications of membrane bioreactors for water and wastewater treatment*. Elsevier, 2006.

Käufler, J., Pohl, R. & Sader, H.: Seawater desalination (RO) as a wind powered industrial process – Technical and economical specifics. *Desal. Water Treat.* 31 (2011), pp. 359–365.

Koschikowski, J, Wieghaus, W., Rommel, M., Ortin, V.S., Suarez, B.P. & Rodriguez, J.R.B.: Experimental investigations on solar driven stand-alone membrane distillation systems for remote areas. *Desalination* 248 (2009), pp. 125–131.

León Diez, F. (ed.): AGUA. Special edition LaJornada, 2005, Mexico.

Lesjean, B. & Huisjeslow, E.H.: Survey of the European MBR market: trends and perspectives. *Desalination* 231 (2008), pp. 71–81.

Mabrouk, A.A., Nafey, A.S. & Fath, H.E.S.: Analysis of a new design of multi stage flash – mechanical vapor compression desalination process. *Desalination* 204 (2007), pp. 482–500.

Malato, S., Fernandez-Ibanez, P., Maldonado, M.I., Blanco, J. & Gernjak, W.: Decontamination and disinfection of water by solar photocatalysis: recent overview and trends. *Catalysis Today* 147 (2009), pp. 1–59.

Marshall, N.B.: Hydrothermal oases. In: *Developments in deep-sea biology*. Blandford Press, London, UK, 1979, pp. 284–307.

McCarthy, K.T., Pichler, T. & Price, R.E.: Geochemistry of Champagne Hot Springs shallow hydrothermal vent field and associated sediments, Dominica, Lesser Antilles. *Chem. Geology* 224:1–3 (2005), pp. 55–68.

Mercado, S.: Geotermoquímica de manifestaciones hidrotermales marinas de alta temperatura. *Geotermia* 9:2 (1993), pp. 155–164.

Mercado, S.: Manifestaciones hidrotermales marinas de alta temperatura (350°C) localizadas a 21°N, a 2600 m de profundidad en la elevación este del Pacífico. *Geotermia* 6:3 (1990), pp. 225–263.

Michels, H. & Pitz-Paal, R.: Cascaded latent heat storage for parabolic trough solar power plants. *Solar Energy* 81 (2007), pp. 829–837.

MIT: The future of geothermal energy—Impact of enhanced geothermal systems (EGS) on the United States in the 21st Century. An assessment by an MIT-led interdisciplinary panel, Massachusetts Institute of Technology, 2006, http://geothermal.inel.gov and http://www1.eere.energy.gov/geothermal/egs_technology.html (accessed July 2011).

Nallusamy, N., Sampath, S., Velraj, R., Experimental investigation on a combined sensible and latent heat storage system integrated with constant/varying (solar) heat sources. *Renewable Energy* 32 (2007), pp. 1206–1227.

Narayan, G.P., Sharqawy, M.H., Summers, E.K., Lienhard, J.H., Zubair, S.M. & Antar, M.A.: The potential of solar-driven humidification-dehumidification desalination for small-scale decentralized water production. *Renew. Sustain. Energy Rev.* 14 (2010), pp. 1187–1201.

Nauticexpo: www.nauticexpo.com/boat-manufacturer/watermaker-1223.html (accessed July 2011).

Oller, I., Malato, S. & Sanchez-Perez, J.A.: Combination of advanced oxidation processes and biological treatment for wastewater decontamination – A review. *Sci. Total Environ.* 409 (2011), pp. 4141–4166.

Papapetrou, M., Wieghaus, M. & Biercamp, C. (eds): Roadmap for the development of desalination powered by renewable energy. ProDes project, 2010, www.prodes-project.org (accessed November 2011).

Park, G.L., Schäfer, A.I. & Richards, B.S.: Renewable energy powered membrane technology: the effect of wind speed fluctuation on the performance of a wind-powered membrane system for brackish water desalination. *J. Membr. Sci.* 370 (2011), pp. 34–44.

Peter-Varbanets, M., Zurbrügg, C., Swartz, C. & Pronk, W.: Decentralized systems for potable water and the potential of membrane technology. *Water Research* 43 (2009), pp. 245–265.

Pichler, T., Veizer, J. & Hall, G.E.M.: The chemical composition of shallow-water hydrothermal fluids in Tutum Bay, Ambitle Island, Papua New Guinea and their effect on ambient seawater. *Marine Chem.* 64:3 (1999), pp. 229–252.

Pichler, T.: Stable and radiogenic isotopes as tracers for the origin, mixing and subsurface history of fluids in submarine shallow-water hydrothermal systems. *J. Volcanol. Geotherm. Res.* 139:3–4 (2005), pp. 211–226.

Price, R.E. & Pichler, T.: Distribution, speciation and bioavailability of arsenic in a shallow-water submarine hydrothermal system, Tutum Bay, Ambitle Island, PNG. *Chem. Geology* 224:1–3 (2005), pp. 122–135.

Qiblawey, H. & Banat, F.: Solar thermal desalination technologies. *Desalination* 220 (2008), pp. 633–644.

Qiblawey, H., Banat, F. & Al-Nasser, Q.: Performance of reverse osmosis pilot plant powered by photovoltaic in Jordan. *Renew. Energy* 36 (2011), pp. 3452–3460.

Riffel, D.B. & Carvalho, P.C.M.: Small-scale photovoltaic-powered reverse osmosis plant without batteries, design and simulation. *Desalination* 247 (2009), pp. 378–389.

Rodriguez, M.S., Galvez, J. B., Rubio, M.I.M., Ibanez, P.F., Padilla, D.A., Pereira, M.C., Mendes, J.F. & Correira de Oliveira, J.: Engineering of solar photocatalytic collectors. *Solar Energy* 77 (2004), pp. 513–524.

Schäfer, A.I. & Richards, B.S.: Testing of a hybrid membrane system for groundwater desalination in an Australian national park. *Desalination* 183 (2005), pp. 55–62.

Shih, M.C.: An overview of arsenic removal by pressure-driven membrane processes. *Desalination* 172 (2005), pp. 85–97.

Shiklomanov I.A.: World water resources: modern assessment and outlook for the 21st century, 1999. Summary of World Water Resources at the Beginning of the 21st Century, prepared in the framework of the IHP UNESCO. Federal Service of Russia for Hydrometeorology & Environment Monitoring, State Hydrological Institute, St. Petersburg, Russia.

Shiklomanov, I.A.: World freshwater resources. In P.H. Gleick (ed.): *Water in crisis: a guide to the world's fresh water resources*. Oxford University Press, New York, 1993, pp. 13–24.

Skyllas-Kazacos, M.: Secondary batteries-flow systems, vanadium redox flow batteries. *Encyclopedia of Electrochemical Power Sources*, 2009, pp. 444–453.

Solomon, S., Quin, D., Manning, M., Marquis, M., Averyt, K., Tignor, M., Miller, H.L. & Chen, Z. (eds): Climate change 2007: the physical science basis. Contribution of Working Group I to the 4th Assessment Report of the Intergovernmental Panel on Climate Change. Cambridge University Press, UK, 2007.

Suárez Arriaga, M.C., Tsompanakis, Y. & Samaniego V., F.: Geothermal manifestations and earthquakes in the caldera of Santorini, Greece: an historical perspective. Stanford Geothermal Program (SGP-TR-185), Vol. 33 No. 1, 1–7 January 2008, Stanford, CA, 2008.

Suárez Bosché, N., Suárez Bosché, K. & Suárez Arriaga, M.C.: A submarine geothermal system in Mexico. *Proceedings of the World Geothermal Congress*, 28 May–10 June 2000, Kyushu-Tohoku, Japan, 2000, pp. 3889–3893.

Suárez, M.C. & Samaniego, F.: A preliminary evaluation of the convective energy escaping from submarine hydrothermal chimneys. *Proceedings, 30th Workshop on Geothermal Reservoir Engineering*, January 31–February 2, 2005, Stanford University, Stanford, CA, SGP-TR-176, 2005, pp. 1–8.

Suárez, M.C. & Samaniego, F.: TOUGH and the boundary element method in the estimation of the natural state of geothermal submarine systems. *TOUGH Symposium Proceedings*, Vol. 1/1, Lawrence Berkeley National Laboratory, Berkeley, CA, 2006, pp. 1–6.

Suárez-Bosche, N.E., Suárez-Arriaga, M.C., Samaniego, F. & Delgado, V.: Fundamental characteristics of hydrothermal submarine systems and preliminary evaluation of its geothermal potential in Mexico. *Proceedings World Geothermal Congress 2005*, Antalya, Turkey, Paper Number: 1604.

Subramani, A., Badruzzaman, M., Oppenheimer, J. & Jacangelo, J.G.: Energy minimization strategies and renewable energy utilization for desalination: a review. *Water Research* 45 (2011), pp. 1907–1920.

Thu, K., Ng, K.C., Saha, B.B., Chakraborty, A. & Koyama, S.: Operational strategy of adsorption desalination systems. *Int. J. Heat Mass Transfer* 52 (2009), pp. 1811–1816.

Tzen, E. & Morris, R.: Renewable energy sources for desalination. *Solar Energy* 75 (2003), pp. 375–379.

Tzen, E.: Successful desalination RES plants worldwide. Centre for Renewable Energy Sources, Hammamet, Tunesia, 2005, www.adu-res.org/pdf/CRES.pdf (accessed November 2011).

UNEP: Vital Water Graphics. United Nations Environmental Programme, UNEP, Nairobi, Kenya, 2008, http://www.unep.org/dewa/vitalwater/ (accessed January 2012).

UNEP/GRID-Arendal. Global freshwater resources: quantity and distribution by region. UNEP/GRID-Arendal Maps and Graphics Library, 2002a. Available at: http://maps.grida.no/go/graphic/global_freshwater_resources_quantity_and_ distribution_by_region (accessed September 2011).

UNEP/GRID-Arendal. Water withdrawal and consumption. UNEP/GRID-Arendal Maps and Graphics Library. 2002b. Available at: http://maps.grida. no/go/graphic/water_withdrawal_and_consumption (accessed September 2011).

UNESCO: Summary of the monograph 'World Water Resources at the beginning of the 21st Century", prepared in the framework of IHP UNESCO, 1999. www.espejo.unesco.org.uy/summary/html (accessed November 2011).

Virdis, B., Freguia, S., Rozendal, R.A., Rabaey, K., Yuan, Z. & Keller, J.: Microbial fuel cells. *Treatise on Water Science* 4, 2011, Elsevier, pp. 641–665.

Voivontas, D., Misirlis, K., Maroli, E., Arampatzis, G. & Assimacopoulos, D.: A tool for the design of desalination plants powered by renwable energies. *Desalination* 133 (2001), pp. 175–198.

Voutchkov, N.: Membrane seawater desalination overview and recent trends, desalination: an energy solution. Presentation at International Desalination Association Conference, 2-3 November 2010, Huntington Beach, CA, 2010.

Wang, L.K., Chen, J.P., Hung, Y.-T. & Shammas, N.K. (eds): Membrane and desalination technologies. In: *Handbook of Environmental Engineering* 13, Chapter 9. Humana Press, 2011, pp. 391–423.

Wilf, M. & Bartels, C.: Optimisation of seawater RO systems design. *Desalination* 173 (2005), pp. 1–12.

Wintgens, T., Melin, T., Schäfer, A., Khan, S., Muston, M., Bixio, D. & Thoeye, C.: The role of membrane processes in municipal wastewater reclamation and reuse. *Desalination* 178 (2005), pp. 1–11.

Xuan, J., Leung, M.K.H., Leung, D.Y.C. & Ni, M.: A review of biomass-derived fuel processors for fuel cell systems. *Renew. Sustain. Energy Rev.* 13 (2009), pp. 1301–1313.

Zhao, P. Zhang, H., Zhou, H., Chen, J., Gao, S. & Yi, B.: Characteristics and performance of 10 kW class all vanadium redox flow battery stack. *J. Power Sources* 162 (2006), pp. 1416–1420.

CHAPTER 2

Overview of renewable energy technologies for freshwater production

Mattheus Goosen, Hacene Mahmoudi & Noreddine Ghaffour

> *"Water security is no less important than the national security"*
> King Abdullah of the Kingdom of Saudi Arabia (2011)*

2.1 INTRODUCTION

With the world's energy demands increasing, much research has been directed at addressing the challenges in using renewable energy to meet the power needs for commercial and domestic use. The economic and industrial potentials of renewable energies, such as geothermal, solar, wave and wind, as well as their environmental advantages have been pointed out in several recent studies (Cataldi *et al.*, 1999; Huang, 2010; Lund, 2007; Mahmoudi *et al.*, 2010; Serpen *et al.*, 2010; Stefansson, 2005). Lund (2007) noted that in ancient times recorded accounts show uses of, for example, geothermal water by Romans, Japanese, Turks, Icelanders, Central Europeans and the Maori of New Zealand for bathing, cooking and space heating. Currently, Iceland is widely considered as the most successful state in the geothermal community (Dorn, 2008; Wright, 1998). The country of just over 300,000 people ranks the highest in the 15 top countries that generate electricity from geothermal resources. Furthermore, Wright (1998) has estimated that given that the worldwide energy utilization is equal to about 100 million barrels of oil per day, the Earth's thermal energy to a depth of 10 kilometers could theoretically supply all of mankind's power needs for the forseable future. As another historical example, ancient Greeks were the earliest to use passive solar design in their homes; as early as 400 BC whole Greek cities were built in this way (Butti and Perlin, 1980; Leipoldt, 2011). The Romans later improved on these designs. They were the first to use glass for windows, in this manner trapping solar heat. The Romans even passed a bylaw that it was an offence to obscure a fellow citizen's access to sunlight. The first glasshouses (i.e. pre-curser to modern day greenhouses) were created by them to produce the correct growing environment for exotic plants and seeds that they brought to Rome from different parts of their vast empire.

Moriarty and Honnery (2009) asked two questions that are of vital importance in discussing the future of energy in today's world: how much energy will we consume annually, and what sources of energy will we use? If, as expected in most forecasts, fossil fuel use continues to grow, the appropriation of vast amounts of CO_2 would be needed if global warming is to be restricted. Furthermore, non-carbon sources face their own uncertainties. In discussing hydrogen's role in an uncertain energy future Moriarty and Honnery (2009) argued that energy use over the next few decades faces deep uncertainties. There are widely conflicting opinions on the size of ultimately recoverable fossil fuel reserves, and the extent to which unconventional resources can be tapped.

While the potential for renewable energy (RE) sources, such as wind and solar is great, they face orders of magnitude scale-up in order to be major energy suppliers. Moriarty and Honnery (2009) predicted that a low energy future is the likely outcome. As shown in Table 2.1, the present global energy system is dominated by fossil fuels. Renewable energy sources have consistently accounted for only 13% of the total energy use over the past 40 years. A key question is for how

---

*Source: King Abdullah of KSA, given on September 25th 2011 at the annual meeting of fifth term of Majlis Al-Shura.

Table 2.1.   Global primary energy use in exajoules (EJ), 1970–2006 (adapted from Moriarty and Honnery, 2009). In describing national or global energy budgets, it is common practice to use large-scale units based upon the joule; $1\,EJ = 10^{18}\,J$.

Energy source	1970	1980	1990	2000	2006
Fossil fuels					
Coal	64.2	75.7	93.7	98.2	129.4
Oil	94.4	124.6	136.2	148.9	162.9
Natural gas	38.1	54.9	75.0	91.8	107.8
Total fossil fuels	*196.7*	*255.1*	*305.0*	*339.0*	*400.1*
Nuclear	0.7	6.7	19.0	24.5	26.6
Renewable	29.4	37.6	48.5	55.6	66.2
All energy	*216.8*	*299.5*	*372.4*	*419.0*	*492.9*
Renewable [%]	*13.6*	*12.6*	*13.0*	*13.3*	*13.4*

long can this continue without running into constraints in the form of limited reserves of fossil fuels, or severe environmental problems from their combustion, including not only global climate change from CO_2 and methane emissions, but also air pollution?

The coupling of renewable energy sources (RES) with desalination systems holds great promise for water scarce regions such as MENA (i.e. Middle East and North Africa) (Goosen *et al.*, 2010, 2011; Goosen and Shayya, 1999; Khamis, 2009; Mahmoudi *et al.*, 2008, 2010; Misra, 2010; Serpen *et al.*, 2010; Stock Trading, 2010; Tester *et. al.*, 2007). Renewable energy-driven desalination systems fall into two categories. The first includes distillation processes (e.g. multi-stage flash (MSF), multiple-effect distillation (MED), thermal vapor compression (TVC)) driven by heat produced directly by the renewable energy system, while the second includes membrane (e.g. reverse osmosis (RO), electrodialisys (ED)) and distillation processes (mechanical vapor compression (MVC)) driven by electricity or mechanical energy produced by RES. Furthermore, hybrid systems also show great promise, for example, geothermal brine being used directly in membrane distillation technology (Houcine *et al.*, 1999). As another example of a hybrid system, brackish geothermal water has been employed to feed a solar still installed in southern Algeria (Bouchekima, 2003). While the focus in this chapter is on seawater desalination, the issues that are discussed also apply (with some adaptions) to desalination of saline groundwater, groundwater containing toxic substances such as arsenic, and residual communal/industrial effluents.

Small decentralized water treatment plants can also be connected to a wind energy convertor system (WECs) (Eltawil *et al.*, 2009). Small and medium water treatment plants, for instance, can be connected to a wind turbine in stand-alone applications. For stand-alone applications it is important to have energy storage which is the main reason to limit the size of the plants. Medium stand-alone wind-RO units are market available, using energy storage systems (e.g. batteries, flywheels) or diesel backup units. The wind turbines as well as the desalination system can be connected to a grid system (Eltawil *et al.*, 2009). The Kwinana desalination plant, for example, located south of Perth in Western Australia, produces nearly $140 \times 10^3\,m^3$ of drinking water per day, supplying the Perth metropolitan area (BlurbWire, 2010). Electricity for the plant is generated by the 80 MW Emu Downs wind farm located in the state's Midwest region. As a final example, solar-driven humidification-dehumidification desalination for small-scale decentralized water production has been reported by Narayan *et al.* (2010) (Fig. 2.1) and Bourouni *et al.* (1999a,b, and 2001). However, many questions remain over scale-up problems and economic viability of renewable energy technology for freshwater production.

The aim of this chapter is to provide a critical review of recent trends in water desalination using RES. After providing an overview of desalination using renewable energies, scale-up and economic factors will be considered, specific case studies will be presented, as well as an assessment of environmental risks and sustainability will be perfomed. The chapter will conclude with

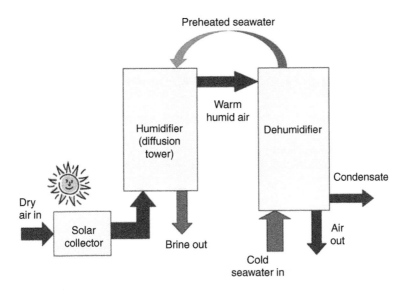

Figure 2.1.   A humidification-dehumidification (H-DH) process (adapted from Narayan *et al.*, 2010).

sections on regulatory and legal considerations, and choosing the most appropriate technology for freshwater production.

2.2   FRESHWATER PRODUCTION USING RENEWABLE ENERGIES

2.2.1   *Applications of solar energy for water desalination*

Desalination by way of solar energy is a suitable alternative to small-scale conventional methods to provide freshwater, especially for remote and rural areas where small quantities of water for human consumption are needed (Al-Hallaj *et al.*, 1998). The blend of renewable energy with desalination systems holds immense promising for improving potable water supplies in arid regions (Mahmoudi *et al.*, 2008, 2009a,b, 2010). Attention has been directed towards improving the efficiencies of the solar energy conversions, desalination technologies and their optimal coupling to make them economically viable for small and medium-scale applications. Solar energy can be used directly as thermal or it can be converted to electrical energy to drive RO units. Electrical energy can be produced from solar energy directly by photovoltaic (PV) conversion or *via* a solar thermal power plant. A variety of possible arrangements can be envisaged between renewable power supplies and desalination technologies (Rodriquez *et al.*, 1996).

Solar stills, which have been in use for several decades, come in a variety of options (Fig. 2.2) (Goosen *et al.*, 2000). The simple solar still (Fig. 2.2A) is a small production system yielding on average 2–5 L day^{-1}. It can be used wherever freshwater demand is low and land is inexpensive. Many modifications to improve the performance of the solar stills have been made. These include linking the desalination process with the solar energy collectors (Fig. 2.2E), incorporating a number of effects to recover the latent heat of condensation (Figs. 2.2D and 2.2F), improving the configurations and flow patterns to increase the heat transfer rates (Figs. 2.2B, 2.2C, 2.2E, and 2.2F), and using low-cost materials in construction to reduce the cost. Nevertheless these systems are not economically viable for large-scale applications. One of the more successful solar desalination devices is the multiple-effect still (Fig. 2.2F) (Al-Hallaj *et al.*, 1998). Latent heat of condensation is recovered, in two or more stages (generally referred to as multi-effects), so as to increase production of distillate water and improve system efficiency. A better understanding of the thermodynamics behind the multiple use of the latent heat of condensation within a

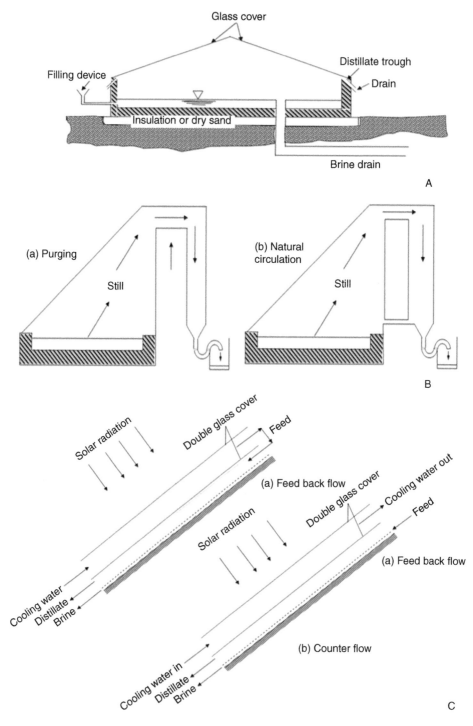

Figure 2.2.   Solar desalination systems (Goosen *et al.*, 2000; adapted from Fath, 1998). A: Single-effect
basin still. B: Single-sloped still with passive condenser. C: Cooling of glass cover by (a)
feedback flow, and (b) counter flow. D: Double-basin solar stills: (a) schematic of single and
double-basin stills and (b) stationary double-basin still with flowing water over upper basin.
E: Directly heated still coupled with flat plate collector: (a) forced circulation and (b) natural
circulation. F: Typical multi-effect multi-wick solar still.

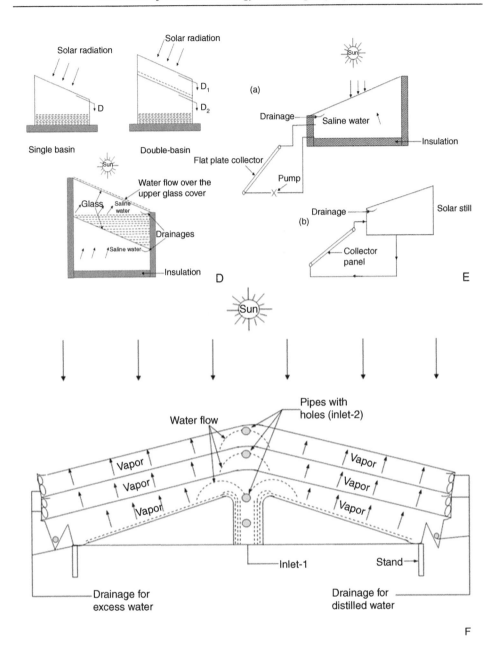

Figure 2.2. Continued

multi-effect humidification-dehumidification solar still is essential in improving overall thermal efficiency (Al-Hallaj *et al.*, 1998). While a system may be technically very efficient it may not be economic (i.e., the cost of water production may be too high) (Fath, 1998). Therefore, both efficiency and economics need to be considered when choosing a desalination system. We can further argue that desalination units powered by renewable energy systems are uniquely suited to provide water and electricity in remote areas where infrastructures are currently lacking.

Solar collectors are usually classified according to the temperature level reached by the thermal fluid in the collectors (Kalogirou, 2005). Low temperature collectors provide low-grade heat, only

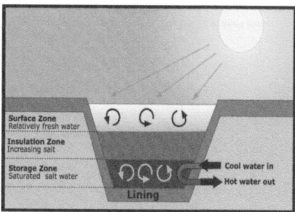

Figure 2.3.   Left: Solar pond for heating purpose demonstration in Australia (http://www.aph.gov.au/
library/pubs/bn/sci/RenewableEnergy_4.jpg). Right: Solar ponds (schematic): The salt content of the pond increases from top to bottom. Water in the storage zone is extremely salty. As
solar radiation is absorbed the water in the gradient zone cannot rise because the surface zone
water above it contains less salt and therefore is less dense. Similarly, cooler water cannot sink,
because the water below it has a higher salt content and higher density. Hot water in the storage
zone is piped to, for example, a boiler where it is heated further to produce steam, which drives
a turbine (Wright, 1982; Energy Education, 2011).

a few degrees above ambient air temperature and use unglazed flat plate collectors. This low-grade
heat is not useful to serve as a heat source for conventional desalination distillation processes
(Fahrenbruch and Bube, 1983; Kalogirou, 2005). Medium-temperature collectors provide heated
liquid at more than 43°C and include glazed flat plate collectors as well as vacuum tube collectors
using air or liquid as the heat transfer medium. They can be used to provide heat for thermal
desalination processes by indirect heating with a heat exchanger. High-temperature collectors
include parabolic troughs or dishes or central receiver systems. They typically concentrate the
incoming solar radiation onto a focal point, from which a receiver collects the energy using a heat
transfer fluid. The high-temperature energy can be used as a thermal energy source in thermal
desalination processes or can be used to generate electricity using a steam turbine. As the position
of the sun varies over the course of the day and the year, sun tracking is required to ensure that
the collector is always kept in the focus of the reflector for improving the efficiency.

Solar ponds can be used to provide energy for many different types of applications. Solar ponds
(Fig. 2.3) combine solar energy collection with long-term storage. The smaller ponds have been
used mainly for space and water heating, while the larger ponds are proposed for industrial process
heat, electric power generation, and desalination. A salt concentration gradient in the pond helps
in storing the energy. Whereas the top temperature is close to ambient, a temperature of 90°C
can be reached at the bottom of the pond where the salt concentration is highest (Fig. 2.3, right).
The temperature difference between the top and bottom layer of the pond is large enough to run
a desalination unit, or to drive the vapor generator of an organic Rankine cycle engine (Wright,
1982). The Rankine cycle converts heat into work. The heat is supplied externally to a closed loop,
which usually uses water. This cycle generates about 80% of all electric power used throughout
the world including virtually all solar thermal, biomass, coal and nuclear power plants (Wright,
1982). An organic Rankine cycle (ORC) uses an organic fluid such as n-pentane or toluene in
place of water and steam. This allows use of lower temperature heat sources, such as solar ponds,
which typically operate at around 70–90°C. The efficiency of the cycle is much lower as a result
of the lower temperature range, but this can be worthwhile because of the lower cost involved in
gathering heat at this lower temperature.

The annual collection efficiency for useful heat for desalination is in the order of 10 to 15% with sizes suitable for villages and small towns. The large storage capacity of solar ponds can be useful for continuous operation of desalination plants. It has been reported that, compared to other solar desalination technologies, solar ponds, even though they are limited to small-scale applications, provide the most convenient and least expensive option for heat storage for daily and seasonal cycles (Kalogirou, 2005). This is very important, both from operational and economic aspects, if steady and constant water production is required. Actually, inland desalination is the only context where solar ponds are competitive, since their efficiency of solar collection is much lower than other solar collectors, so it is only a matter of finding other means for heat storage and brine disposal (as most commercial plants using thermal energy by the seaside do).

The heat storage allows solar ponds to power desalination during cloudy days and night time. Another advantage of desalination by solar ponds is that they can utilize what is often considered a waste product, namely reject brine, as a basis to build the solar pond. This is an important advantage for inland desalination. If high temperature collectors or solar ponds are used for electricity generation, a desalination unit, such as a multi-stage flash system (MSF), can be attached to utilize the waste heat from the electricity production process. Since, the conventional MSF process is not able to operate with a variable heat source, a company ATLANTIS developed an adapted MSF system that is called "Autoflash" which can be connected to a solar pond (Szacsvay *et al.*, 1999). With regard to pilot desalination plants coupled to salinity gradient solar ponds the seawater or brine absorbs the thermal energy delivered by the heat storage zone of the solar pond. Examples of plants coupling a solar pond to an MSF process include: Margarita de Savoya, Italy: plant capacity 50–60 m^3 day^{-1}; Islands of Cape Verde: Atlantis 'Autoflash', plant capacity 300 m^3 day^{-1}; Tunisia: a small prototype at the *Laboratoire de Thermique Industrielle* where a solar pond of 1500 m^2 drives a MSF system with capacity of 0.2 m^3 day^{-1}; and El Paso, Texas: plant capacity 19 m^3 day^{-1} (Lu *et al.*, 2000).

Solar photovoltaic (PV) systems directly convert the sunlight into electricity by solar cells (Kalogirou, 2005). Solar cells are made from semiconductor materials such as silicon. Other semiconductors may also be used. A number of solar cells are usually interconnected and encapsulated together to form a PV module. Any number of PV modules can be combined to form an array, which will supply the power required by the load. In addition to the PV module, power conditioning equipment (e.g. charge controller, inverters) and energy storage equipment (e.g. batteries) may be required to supply energy to a desalination plant. Charge controllers are used for the protection of the battery from overcharging. Inverters are used to convert the direct current from the photovoltaic modules system to alternating current to the loads. PV is a mature technology with life expectancy of 20 to 30 years. The main types of PV systems are the following: Stand-alone systems (not connected to the utility grid) provide either DC power or AC power by using an inverter. Grid-connected systems consist of PV arrays that are connected to the electricity grid *via* an inverter.

In small and medium-sized systems the grid is used as a back-up source of energy; any excess power from the PV system is fed into the grid. In the case of large centralized plants, the entire output is fed directly into the grid. Hybrid arrangements, on the other hand, are autonomous systems consisting of PV arrays in combination with other energy sources, for example in combination with a diesel generator or another renewable energy source (e.g. wind).There are mainly two PV driven membrane processes, reverse osmosis (RO) and electrodialysis (ED). From a technical point of view, PV as well as RO and ED are mature and commercially available technologies at present time. The feasibility of PV-powered RO or ED systems, as valid options for desalination at remote sites, has also been proven (Childs *et al.*, 1999). The main problem of these technologies is the high cost and, for the time being, the availability of PV cells.

Burgess and Lovegrove (2005) compared the application of solar thermal power desalination coupled to membrane *versus* distillation technology. They reported that a number of experimental and prototype solar desalination systems have been constructed, where the desalination technology has been designed specifically for use in conjunction with solar thermal collectors, either static or tracking. To date such systems are either of very low capacity, and intended for applications

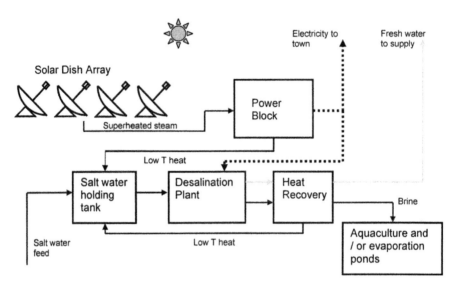

Figure 2.4.   Combined dish-based solar thermal power generation and RO desalination (Burgess and Lovegrove, 2005).

such as small communities in remote regions, or else remain unproven on a larger scale. Several systems which are of some interest were discussed. Schwarzer *et al.* (2001) described a simple system which has flat plate collectors (using oil as a heat transfer fluid) coupled to desalination "towers" in which water evaporates in successive stages at different heights (similar to the multi-effect still shown in Fig. 2.2F). The condensation of vapor in one stage occurs at the underside of the next stage, transferring heat and increasing the gain output ratio. A very similar system (not mentioned by Schwarzer), called a "stacked plate still", is described by Fernández and Chargoy (1990). It is important to remember that a clear distinction should be made between systems where the RO is tailored to be coupled (mechanically) with solar thermal energy and those where electricity is produced by solar thermal energy and standard RO plugged to that electricity.

The Vari-Power Company, based in California, has developed an RO based desalination system which is specifically tailored to solar thermal input (Childs *et al.*, 1999). A patented direct drive engine (DDE) converts heat to the hydraulic power required by RO. Desalinated water production using the DDE is projected to be more than 3 times greater (for an identical dish collector) than that which would be obtained by RO driven by a dish-Stirling electricity generation system or PV power. Burgess and Lovegrove (2005) noted that the project remains at the pilot stage with the DDE not commercially available: it has perhaps become less attractive due to the advances in conventional RO. The choice of the RO desalination plant capacity depends on the daily and seasonal variations in solar radiation levels, on the buying and selling prices for electricity, and on the weight given to fossil fuel displacement. A conceptual layout for a solar dish-based system with power generation and RO desalination is shown in Figure 2.4.

A number of solar desalination pilot plants have been installed and most have been operating successfully with very little maintenance (Reddy and Ghaffour, 2007). Virtually all of them were custom designed for specific locations. Operational data and experience from these demonstration plants can be utilized to achieve higher reliability and cost minimization. Indirect collection systems for these plants comprise solar collectors that produce thermal, or electrical, or shaft energy. These types of energy produced by the collector systems can be used to run conventional desalination processes such as reverse osmosis (RO), electrodialysis (ED), multi-stage flash distillation (MSF), multi-effect distillation (MED), thermal vapor compression (TVC), mechanical vapor compression (MVC), humidification-dehumidification systems (H-HD) and other promising processes which are under development such as membrane distillation (MD) and adsorption

Table 2.2. Energy consumption (using waste heat in thermal processes) in large desalination processes (Mahmoudi *et al.*, 2009b; Ghaffour, 2009).

Process*	Thermal energy [kWh m^{-3}]	Electrical energy [kWh m^{-3}]	Total energy [kWh m^{-3}]	Product water quality [mg L^{-1}]
MSF	7.5–12	2.5–3.5	10–15.5	5–30
MED	4–8	1.5–3	5.5–11	
SWRO	–	3–6	3–6	100–500
BWRO	–	0.5–2.5	0.5–2.5	

*MSF: multi-stage flash; MED: multi-effect distillation; SWRO: seawater reverse osmosis; BWRO: brackish water reverse osmosis.

Table 2.3. Productivity of different desalination processes per square meter of solar collector area (Childs and Dabiri, 2000; Childs *et al.*, 1999).

Desalination process*	Water produced per solar collector surface area [L day^{-1} m^{-2}]
Simple solar still	4–5
H/D process–Medium T solar thermal collector	12
MSF, MED with thermal storage-Med T solar thermal collector	40
SWRO-PV	200
VARI-RO DDE-Dish Sterling solar collector (concept stage)	1200

*H/D: humidification-dehumidification; MSF: multi-stage flash; MED: multi-effect distillation; SWRO-PV: seawater reverse osmosis-photovoltaics.

desalination (AD). However, factors that need to be considered when making a choice of which combination system to pick to go with a specific type of solar energy include product water quality, feedwater quality, size of the unit, power requirements, economics, and operation and maintenance. More than 80% of the relatively small solar desalination capacity, for example, is produced by RO and MED from mostly demonstration plants (Reddy and Ghaffour, 2007).

The energy consumption for different desalination technologies is presented in Table 2.2. Distillation technologies need thermal and electrical energy while membrane processes in large commercial plants need only electrical energy. However, both are energy intensive accounting for up to 50% of the operating cost of each process (Mahmoudi *et al.*, 2009b; Ghaffour, 2009). The total water cost and the energy consumption of these systems depends strongly on the specific parameters of each technology. Details of desalination costing and energy requirement for each type of water desalting system were reported by Reddy and Ghaffour (2007). Furthermore, efforts to reduce energy consumption have been directed not only at reducing the cost of the produced freshwater but also at minimizing the dependence on costly fossil fuel so as to reduce, for example, CO_2 emissions. The amount of desalinated water produced by different desalination processes per square meter of solar collector area required is given in Table 2.3. The characteristics of solar thermal/electric power systems have been described by Malik *et al.* (1982).

Consider for a moment the economics of the solar desalination system. The cost of a poly-crystalline silicon PV cells unit surface is 5 to 6 times higher than that of solar thermal collector surface used for other processes (Mahmoudi *et al.*, 2009b; Ghaffour, 2009). VARI-RO direct drive technologies (Childs and Dabiri, 2000; Childs *et al.*, 1999) and solar powered Rankine cycle (Gracia-Rodriguez and Blanco-Galvez, 2007) have been proposed for producing shaft energy from solar thermal energy to drive RO pumps for improving the efficiency of complete process. It was claimed that this VARI-RO direct drive system will produce about 1200 L day^{-1} m^{-2} solar collector area (Gracia-Rodriguez and Blanco-Galvez, 2007). However, improvements in PV cells required to achieve this include an increase in their efficiency, selection of the most appropriate material and reducing manufacturing costs. Efficiencies of 24% and 30% have been achieved with monocrystalline and polycrystalline silicon and gallium arsenide (GaAs) and its alloys, respectively (Garcia-Rodriguez, 2002).

Finally, it is important to note here that there are still several limitations in the use of solar desalination. Firstly, at present, solar desalination does not appear to be a viable option for very large scale applications, either technically or economically. However, it can be used for small/medium scale applications for supplying water in remote locations where there is no electric grid connection. Secondly solar energy is available only during day time and its intensity changes from morning to evening, with peak intensity in the afternoon. In contrast the energy requirements of any desalination process are constant and continuous and the efficiency of any desalination process is low if operated at variable load; variable operation even affects the plant operating life. An energy storage system or an alternative back-up source of energy is therefore required to run the desalination plant continuously at constant load. Finally, available conventional desalination processes may not be suitable, technically and economically, for operating with solar energy in remote locations since the operation and maintenance of these technologies require skilled operators (Mahmoudi *et al.*, 2009b; Ghaffour, 2009).

2.2.2   *Wind power and desalination*

Wind is generated by atmospheric pressure differences, driven by solar power. Kalogirou (2005) in a rigorous review on renewable energy sources for desalination argued that purely on a theoretical basis, and disregarding the mismatch between supply and demand, the world's wind energy could supply an amount of electrical energy equal to the present world electricity demand. Of the total 173,000 TW of solar power reaching the earth, about 1200 TW (0.7%) is used to drive the atmospheric pressure system (Soerensen, 1979). This power provides a kinetic energy reservoir of 750 EJ with a turnover time of 7.4 days. This conversion process mainly takes place in the upper layers of the atmosphere, at around 12 km height (where the "jet streams" occur). If it is assumed that about 1% of the kinetic power is available in the lowest strata of the atmosphere, the world wind potential is of the order of 10 TW, which is more than sufficient to supply the world's current electricity requirements.

Small decentralized water treatment plants combined with an autonomous wind energy convertor system (WECs) (Fig. 2.5, left) show great potential for transforming seawater or brackish water into pure drinking water (Koschikowski and Heijman, 2008). Also, remote areas with potential wind energy resources such as islands can employ wind energy systems to power seawater desalination for freshwater production. The advantage of such systems is a reduced water production cost compared to the costs of transporting the water to the islands or to using conventional fuels as power source. Different approaches for wind desalination systems are possible. First, both the wind turbines as well as the desalination system are connected to a grid system. In this case, the optimal sizes of the wind turbine system and the desalination system as well as avoided fuel costs are of interest. The second option is based on a more or less direct coupling of the wind turbine(s) and the desalination system. In this case, the desalination system is affected by power variations and interruptions caused by the power source (wind). These power variations, however, have an adverse effect on the performance and component life of certain desalination equipment. Hence, back-up systems, such as batteries, diesel generators, or flywheels might be integrated into the system.

Reverse osmosis is the preferred technology due to the low specific energy consumption. The main electrical energy required is for pumping the water to a relatively high operating pressure. The use of special turbines (i.e. energy recovery devices) may reclaim part of the energy. Operating pressures vary between 10 and 25 bars (i.e. 1–2.5 MPa) for brackish water and 50–80 bars (i.e. 5–8 MPa) for seawater (Eltawil *et al.*, 2009). The Kwinana desalination plant, located south of Perth in Western Australia, is one example where a grid connected wind power and reverse osmosis desalination plant have been successfully combined. The plant produces nearly $140 \times 10^3 \, m^3$

Figure 2.5.   Left: Wind farm (Kalogirous, 2005); Right: Wind turbines and PV cells of Sureste SWRO plant located in Gran Canaria, Canary Islands, of a capacity of 25,000 m³ d⁻¹ (Sadhwani, 2008; IDA Conference, 2008).

of drinking water per day (BlurbWire, 2010). Electricity for the plant is generated by the 80 MW Emu Downs wind farm located in the state's Midwest region. The reverse osmosis plant was the first of its kind in Australia and covers several acres in an industrial park.

Lately, many medium- and large-scale water treatment and desalination plants are partially powered with renewable energy mainly wind turbines, PV cells or both. The energy demand of Sureste seawater reverse osmosis (SWRO) plant located in Gran Canaria, Canary Islands, of a capacity of 25×10^3 m³ day⁻¹ is provided by a combination of PV cells (rooftop) with minor share of RO energy demand and the rest from the grid which consist of an energy mix including wind energy (Fig. 2.5, right) (Sadhwani and Veza, 2008; IDA Conference, 2008).

2.2.3   *Wave and tidal power for desalination*

Most of the work on wave energy conversion has focused on electricity production (Davies, 2005); any such converter could, in principle, be coupled to electrically-driven desalination plants, either with or without connection to the local electricity grid. Worldwide exploitable wave energy resource is estimated to be 2 TW, so it is a promising option for electricity generation. Thus, there is a potential option of coupling wave power with seawater RO. Wave-powered desalination offers an environmentally sensitive solution for areas where there is a shortage of water and sufficiently energetic waves. Energy that can be harvested from oceans includes waves, tides and underwater oceanic currents (Fig. 2.6).

A study by Davies (2005) focused on the potential of linking ocean wave energy to desalination. They found that along arid, sunny coastlines, an efficient wave-powered desalination plant could provide water to irrigate a strip of land 0.8 km wide if the waves are 1 m high, increasing to 5 km with waves 2 m high. Wave energy availabilities were compared to water shortages for a number of arid nations for which statistics were available. The maximum potential to correct these shortages varied; in Morocco, for example, wave-powered desalination could supply 16% of the shortfall towards meeting the total freshwater demand, increasing to 600% for Somalia. In a related study, Magagna and Muller (2009) described the development of a stand-alone, off-grid RO desalination system powered by wave energy. The system consisted of two main parts; a high pressure pump (wave catcher) that allows generation of a high pressure head from low head differences, and a wave-driven pump to supply the necessary head to the wave catcher. The high pressure pump could produce 6 MPa of pressure which is necessary to drive a RO membrane for desalination

Figure 2.6.   Wave energy. The Aquabuoy 2.0 is a large 3 m wide buoy tied to a 70 ft (i.e. 21 m) long shaft. By bobbing up and down, the water is rushed into an acceleration tube, which in turn causes a piston to move. This moving of the piston causes a steel reinforced rubber hose to stretch, making it act as a pump. The water is then pumped into a turbine which in turns powers a generator. The electricity generated is brought to shore *via* a standard submarine cable. The system is modular, which means that it can be expanded as necessary (Chapa, 2007).

of water. We can argue that wave energy technology is still at prototype stage and there is no standard technology available. Wave energy has also an intermittent and variable behavior similar to wind energy.

Foley and Whittaker (2009) developed a techno-economic model of an autonomous wave-powered desalination plant. They indicated that freshwater could be produced for as little as $0.75 \, US\$ \, m^{-3}$. The modeled plant consisted of what was called an Oyster wave energy converter (Cameron *et al.*, 2010), conventional RO membranes and a pressure exchanger-intensifier for energy recovery. In their paper, Foley and Whittaker (2009) indicated that an independent wave-powered desalination system is both possible and may produce freshwater at a relatively low cost; however, a number of technological barriers were reported which must be overcome if wave-powered desalination is to be exploited effectively. The first of these barriers is the development of the energy recovery technology. We can speculate that the energy recovery devices commercially available in RO may be modified for use in wave-powered systems. Furthermore, the operation of the proposed technology is very similar to other technologies such as the Clark pump (i.e. recovers mechanical energy from concentrate flow and returns it directly to the feed flow; see Manolakos *et al.*, 2008). The second barrier is the development of a suitable filtering and pre-treatment technology that can be deployed with the wave energy converter and operated without servicing or significant maintenance for extended periods of time. Finally, the satisfactory operation of the RO membranes under the variable feed conditions provided by the wave energy converter needs to be confirmed.

In a related investigation, Jayashankar *et al.* (2009) reported on experimental results from near shore bottom standing oscillating water column (OWC) based wave energy plants in Japan and India have now been available for about a decade (Fig. 2.7). Historically the weakest link in the conversion efficiency of OWC based wave energy plants built so far has been the bidirectional

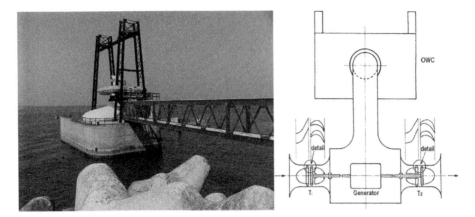

Figure 2.7.   *Left*: The Indian near shore oscillating water column (OWC) plant (circa 1996). *Right*: Twin unidirectional turbine topology (Jayashankar *et al.*, 2009).

turbine. This is possibly because a single turbine has been required to deliver power when the plant is exposed to random incident wave excitation varying by a factor of 10. A new topology that uses a twin unidirectional turbine (which features a high efficiency spanning a broad range) was proposed by Jayashankar *et al.* (2009) (Fig. 2.7 RHS). Using the Indian wave energy plant as a case study, it was shown that the power output from such a module considerably exceeds existing optimal configurations including those based on a fixed guide vane impulse turbine, linked guide vane impulse turbine or a Well's turbine. A wave to wire efficiency of the order of 50% over the incident range was shown to be achievable. We can speculate that the OWC system may be a suitable energy source for a stand-alone or autonomous reverse osmosis desalination plant as described by Foley and Whittaker (2009).

Charlier and Finki (2009) in a recent book on the tidal power of oceans recommended small-scale projects for harnessing the ocean's tidal energy for applications in desalination. Since this is a relatively new area smaller projects would be a lower economical risk. They went on to explain that the environmental benefits such as a reduction in pollution through cleaner air would have to be balanced against the negative aesthetic effects. In a related study, Davis (1997) identified a large number of tidal sites in the world's ocean's which can provide a significant, viable and cost-effective source of reliable energy. A new type of low-head water turbine was also described based on the vertical axis wind turbine proposed by the French hydro engineering Darrieus over 50 years ago.

In a connected study, the potential exploitation of the hydrostatic pressure of seawater at a sufficient operative depth was considered by several investigators in the 1960s in view of increasing the energy efficiency of the then developing RO industrial desalination technology (Drude, 1967; Glueckstern, 1982). More recently, several configurations were proposed for freshwater production from seawater using RO and hydrostatic pressure: submarine, underground and ground-based (Reali *et al.*, 1997). In conventional surface-based industrial desalination plants applying RO technology, the freshwater flow at the membrane outlet is approximately 20–35% of the inlet seawater flow, depending on membrane type and characteristics and feed water salinity. The resulting brine is disposed off in the sea. While RO installations generate the required pressure with high-pressure pumps, the submarine approach uses seawater hydrostatic pressure. The desalinated water, produced at about atmospheric pressure and collected in a submarine tank at the same working depth, is pumped to the sea surface. It was shown that this approach saves about 50% of the electricity consumption with respect to an efficient conventional RO plant (about 2–2.5 kWh m^{-3}) since only the outlet desalinated water is pumped instead of the inlet seawater, thus reducing the pumping flow rate by 55–80% (Pacenti *et al.*, 1999). The advantage of this configuration is also to avoid the pre-treatment of the inlet seawater, therefore saving costs for chemicals and equipment (Charcosset, 2009).

2.2.4   *Geothermal desalination*

Geothermal energy is widely distributed around the world (Dorn, 2008; White and Williams, 1975; Wright, 1998). This energy can be used for heat and electricity generation. Thus, there is a potential use for thermal (MED, MSF, MD, VC) and membrane (RO, EDR) desalination processes. Geothermal reservoirs can produce steam and hot water. Superheated dry steam resources are mostly easily converted into useful energy, generally producing electricity, which can be cheaper than that of conventional sources.

Considering that the energy requirements for desalination continues to be a highly influential factor in system costs, the integration of renewable energy systems with desalination seems to be a natural and strategic coupling of technologies (Tzen *et al.*, 2004). As an example of the potential, the southern part of the country of Algeria consists almost entirely (i.e. 90%) of the great expanse of the Sahara Desert. This district has freshwater shortages but also has plenty of solar energy (Bouchekima, 2003), wind energy (Mahmoudi *et al.*, 2009a,b) and important geothermal reservoirs (Fekraoui and Kedaid, 2005; Mahmoudi *et al.*, 2010). The amalgamation of renewable resources with desalination and water purification is thus very attractive for this region. We will discuss this example in more detail in the case study section (2.4).

When using geothermal energy to power systems such as desalination plants we avoid the need for thermal storage. In addition, the energy output of this supply is generally stable compared to other renewable resources such as solar and wind power (Bourouni and Chaibi, 2005). Kalogirou (2005) has shown that the ground temperature below a certain depth remains relatively constant throughout the year. Popiel *et al.* (2001) reported that one can distinguish three ground zones; surface, shallow and deep, with geothermal energy sources being classified in terms of their measured temperatures as low ($<100°C$), medium ($100–150°C$) and high temperature ($>150°C$), respectively.

Geothermal wells deeper than 100 m can reasonably be used to power desalination plants (Kalogirou, 2005). We can also envisage the utilization of geothermal power directly as a stream power in thermal desalination plants. Furthermore, with the recent progress on membranes distillation technology, the utilization of direct geothermal brine with temperature up to 60°C has become a promising solution (Houcine *et al.*, 1999). Fridleifsson *et al.* (2008) has reported that electricity is produced by geothermal means in 24 countries, five of which obtained up to 22% of their needs from this source. Furthermore, direct application of geothermal energy for heating and bathing has been reported by 72 countries. Fridleifsson *et al.* (2008) goes on to argue that it is considered possible to increase the installed world geothermal electricity capacity from the current 10 GW to 70 GW with present technology, and to 140 GW with enhanced technology.

2.3   SCALE-UP AND ECONOMIC CONSIDERATIONS

2.3.1   *Factors affecting scale-up*

A thorough assessment of the economic, regulatory and technical implications of large-scale solar power deployment was made by Merrick (2010). He reported that despite its current high cost relative to other technology options, a combination of costs reductions and policy support measures could lead to increasing use and application of solar power technologies. For example, the technical and economic characteristics of PV and concentrating solar power (CSP) technology will have implications for the wider electric power system. He noted that concern about climate change is a potential reason why large-scale solar power deployment may occur. Development of the renewable energy sector has the potential to create new industries and associated jobs.

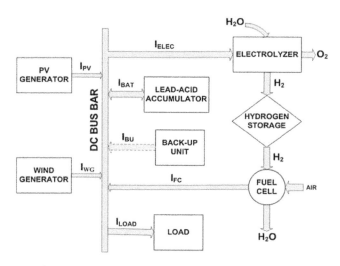

Figure 2.8.   Block diagram of a proposed stand-alone power system (Ipsakis *et al.*, 2009).

If large-scale transformation of the world's electricity systems is to happen, it will likely occur with a combination of renewable energy sources such as wind, wave, solar and geothermal. For large-scale solar power deployment to occur other factors will come into play; transmission infrastructure will be required, electrical energy will be in demand when the sun is not shining, power quality will need to be maintained, other forms of electricity generation will be displaced, and regulatory policies may help or hinder an efficient transformation.

Ipsakis *et al.* (2009) has proposed a stand-alone power system based on a photovoltaic array and wind generators that stores the excessive energy from renewable energy sources in the form of hydrogen *via* water electrolysis. Their aim is to use the power system in a polymer electrolyte membrane (PEM) fuel cell planned for operation at Neo Olvio of Xanthi in Greece (Fig. 2.8). We can argue that this small to medium-scale system has potential applications in providing, for example, electrical energy for running RO units for seawater and brackish water desalination in remote arid, sunny and windy locations.

Another important factor is efficient power management strategy (PMSs) for integrated stand-alone power systems. This was assessed by Ipsakis *et al.* (2009) on the ability of such systems to meet the power load requirements through effective utilization of the electrolyzer and fuel cell under variable energy generation from RES (i.e. solar and wind). The evaluation of the power management strategies was performed through simulated experiments with anticipated conditions over a typical four-month time period for the region where the system would be installed. The key bottle necks or decision factors in the power management strategies were found to be the level of the power provided by the renewable energy sources and the state of charge of the accumulator. Therefore, the operating policies for the hydrogen production *via* water electrolysis and the hydrogen consumption at the fuel cell depended on the excess or shortage of power.

The potential of solar-driven humidification-dehumidification (H-DH) desalination for small-scale decentralized water production has been reviewed in detail by Narayan *et al.* (2010) (Fig. 2.9). It was found that among all H-DH systems, the multi-effect closed air open water (CAOW) water-heated system is the most energy efficient. For this system, the cost of water production was about US\$ 3–7 m^{-3}; even though this is higher than that for reverse osmosis (RO) systems working at similarly small capacities (5–100 m^{3}day^{-1}), the H-DH system has advantages for small-scale decentralized water production. These advantages include much simpler brine pre-treatment and disposal requirements and simplified operation and maintenance. However, it remains to be seen whether or not H-DH systems can be scaled-up successfully.

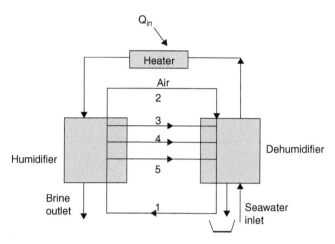

Figure 2.9.   Multi-effect closed air open water (CAOW) water-heated system for freshwater production (Narayan *et al.*, 2010).

2.3.2   *Cost-efficiency compared to conventional energy sources*

Desalination using renewable energy sources such as solar, geothermal, wind and wave consist essentially of a combination of two separate and distinct technologies: energy production and water desalination. The challenge lies in finding an optimal design for combining the two through a system-oriented approach (Mathioulakis *et al.*, 2007). Although solar energy, for instance, is abundant and free, the hardware for using the energy economically, for capturing it in an efficient way, converting it to useful forms, and storing it is not free of charge (Gude and Nirmalakhandan, 2010). We can argue that it is not possible to reduce the cost of desalination using direct heat solar energy to a comparable range with conventional desalination at least not in the near future. Currently, solar energy desalination will probably find only remote location applications where there is no electrical grid connection. Solar stills and the direct use of geothermal energy, for example, are included in this scenario. In solar stills the energy required to produce desalinated water is high since latent heat of condensation is not recovered. This necessitates the requirement of a large surface area which contributes to the increase in capital cost of the unit. It is noteworthy that a geothermal plant has a much higher installation cost compared to a fossil fuel plant (i.e. installation costs per MW installed) but its electricity generation cost is much less than for fossil fuel plants.

For indirect applications, additional equipment is required to convert renewable energy to thermal or shaft work or electrical energy. At the same time the conversion efficiencies from renewable energy to thermal, or shaft work, or electrical energy is low. This contributes to the increase in the capital cost. Also, electrical power is subsidized in many countries where there is lack of freshwater. Furthermore, solar and wind are not continuous sources of energy. However, the operation of desalination processes for technical and economic reasons require continuous operation. This necessitates additional equipment to store thermal and electrical energy generated by, for example, solar and wind, which contributes to an increase in the investment cost of desalination and water treatment plants. Presently, high temperature solar collector systems offer the lowest cost solar electricity for large-scale power generation (>10 megawatt-electric). The installation costs are about US$ 2–3 per watt, which results in a cost of solar power of US$ 0.1–0.20 per kWh. Hybrid systems that combine large concentrating solar power plants with natural gas combined cycle or coal plants can reduce installed costs to US$ 1.5 per watt and reduce the cost of solar power to below US$ 0.08 per kWh (Fig. 2.10).

Ghermandi and Messalem (2009) investigated the current developments in the field of solar-powered RO desalination on the basis of the analysis of 79 experimental and design units

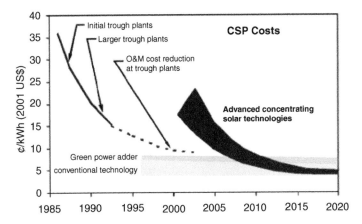

Figure 2.10.   Cost trends of high temperature solar collector systems (Childs and Dabiri, 2000; Childs *et al.*, 1999).

worldwide. They argued that PV-powered RO desalination is mature for commercial implementation. However, DeCanio and Fremstad (2011) noted the high cost of electricity generated by PV technology compared to conventional energy sources. Although no standard design approach has been developed, the technical feasibility of different design concepts was demonstrated in a relatively large number of case studies. Systems that directly couple the PV modules to variable speed DC pump motors seemed to have the highest potential for energy-efficient and cost-effective small-scale PV-RO desalination. Some concern was expressed about their long-term performance.

The combination of solar power with additional power sources, as noted by Ghermandi and Messalem (2009), may be beneficial both in small- and large-scale desalination. In small-scale systems, PV panels may combine favorably with wind turbines, achieving lower overall costs where the complementary aspects of the two renewable sources can be exploited. In large-scale systems, concentrating solar power (CSP) and fuel co-firing of desalination plants can help to bestow stability on desalination unit operation during night-time or periods of low sun shine. In both small- and large-scale systems, connection to the electrical grid for combined power and water generation will promote stable operation of the system. Ghermandi and Messalem (2009) concluded that state-of-the-art solar-RO desalination is cost-competitive with other water supply sources only in context of remote regions.

Delucchi and Jacobson (2011) estimated the cost of electricity transmission in a conventionally configured system, over distances common today (Table 2.4). However, the more that dispersed wind and solar generating sites are interconnected, the less the variability in output of the whole interconnected system. A system of interconnections between widely dispersed generators and load centers has been called a "supergrid". The configuration and length of transmission lines in a supergrid will depend on the balance between the cost of adding more transmission lines and the benefit of reducing system output variability as a result of connecting more dispersed generation sites. The optimal transmission length in a supergrid is unknown. Delucchi and Jacobson (2011) speculated that the average transmission distances from generators to load centers in a supergrid will be longer than the average transmission distance in the current system. They noted that the cost of this additional transmission distance is an added cost of ensuring that wind, wave and solar power generation always matches demand.

In a related case study, Palermo (2009) conducted a net present value analysis of installing a solar power generation system on turkey ranches in California. The research although not directly connected to desalination provides useful information regarding solar systems power production capacity, investment cost, maintenance requirements, amount of energy saved, useful life of

Table 2.4.   Approximate fully annualized generation and conventional transmission costs for wind, wave and solar power (adapted from Delucchi and Jacobson, 2011). Electric power transmission or "high-voltage electric transmission" refers to the bulk transfer of electrical energy from generating power plants to substations located near population centers.

| | Annualized cost (2007 US$ kWh^{-1} delivered) | |
Energy technology	Present (2005–2010)	Future (2020+)
Wind onshore	0.04–0.07	≤0.04
Wind offshore	0.10–0.17	0.08–0.13
Wave	≥0.11	0.04
Geothermal	0.04–0.07	0.04–0.07
Hydroelectric	0.04	0.04
Concentrated solar power (CSP)	0.11–0.15	0.08
Solar photovoltaic (PV)	>0.20	0.10
Tidal	>0.11	0.05–0.07
Conventional (mainly fossil) generation in US	0.07	0.08

equipment, and tax incentives for a company. The investment cost of the system included the price of the equipment and installation service. He noted that many of the system costs may be offset by rebates, tax credits and grants from various government agencies. These must also be included in the financial analysis as they can greatly affect the financial viability of the project. The system was projected to have a useful life of 30 years. Four scenarios were evaluated using two levels of rebates and two electrical rate inflation levels. Palermo (2009) commented that the use of solar energy also gave the company an advantage over the competition when used as a marketing tool due to the use of green technology in company production practices. It was recommended that future research should focus on emerging technologies, enhanced rebate opportunities and evaluating different sized systems. As production of system components becomes more efficient, costs will continue to decrease which will benefit consumers. Government focus on green technology may provide more financing options in the future as well.

DeCanio and Fremstad (2011) argued that solar power is the ultimate technology for energy supply because of its abundance, but currently it is the most expensive of the alternative energy sources. Table 2.5 gives a range of estimates of the cost of electricity from a variety of studies. The highest average value is from PV at 0.491 US$ KWh^{-1} (see bottom of Table 2.5). DeCanio and Fremstad (2010) noted that there is every expectation that solar costs will decrease over time with research and experience, and a number of efforts to estimate the rate of decline have appeared in the literature.

Water desalination cost literature has also been reviewed and assessed by Karagiannis and Soldatos (2008). They reported that desalination cost has decreased over the last years due to technical improvements in a world of increasing fossil fuel prices. For conventional systems the cost for seawater desalination ranges from 0.6 US$ m^{-3} to more than 4 US$ m^{-3}, while for brackish water desalination the cost is almost half, due presumably to lower salt concentrations in the feed water. When renewable energy sources are used the cost is much higher, and in some cases could even reach 20 US$ m^{-3}, due to expensive energy supply systems. However, Karagiannis and Soldatos (2008) argued that this cost is counter-balanced by the environmental benefits.

Finally, the choice of desalination method affects significantly the water desalination cost. Thermal methods are used mostly in medium and large size systems, while membrane methods, mainly RO, are used by medium and low capacity systems. Nevertheless, during the last few years, RO has become the optimal choice in even larger units. Under special conditions hybrid systems such as solar and wind can offer increased and more stable production of freshwater.

Table 2.5. Estimates of levelized cost (US$ kWh⁻¹) of electricity by source (adapted from DeCanio and Fremstad, 2011). Levelised energy cost (LEC) is the price at which electricity must be generated from a specific source to break even. It is an economic assessment of the cost of the energy-generating system including all the costs over its lifetime: initial investment, operations and maintenance, cost of fuel, cost of capital, and is very useful in calculating the costs of generation from different sources.

Study reference	Measure	Coal	Nuclear	Wind	Geothermal	Solar PV	Solar thermal
U.S. Energy Information Administration (2010a,b)	US$2008 per kWh, 2016	0.10	0.12	0.15	0.12	0.40	0.27
RETI Stakeholder Steering Committee (2010)	Levelized cost ranges in US$ kWh⁻¹			0.9–0.13	0.1–0.16	0.25–0.35	0.24–0.29
Lazard Ltd. (2009)	Levelized cost ranges in US$ kWh⁻¹	0.78–0.14	0.1–0.14	0.8–0.14	0.9–0.12	0.21–0.30	0.20–0.33
Borenstein (2008)	US$ kWh⁻¹, annual real interest rate from 1–7%					0.34–0.57	
Renewables 2010 Global Status Report (REN21)	Typical energy cost, US$ kWh⁻¹			0.05–0.09 (onshore) 0.1–0.14 (offshore)			
Benson and Orr (2008)	US$ kWh⁻¹ (2007 $)	0.04		0.08		0.28	0.20
Greenpeace Int., SolarPACES, and ESTELA (2010)	US$ kWh⁻¹ at sites with very good solar radiation						0.15
IEA/NEA (2005)	US$ kWh⁻¹, see IEA/NEA 2010 for details	0.03–0.08 (pulverized)	0.05	0.05–0.16 (onshore) 0.07–0.13 (offshore)		0.23–2.0	0.29
European Commission (2008)	US$ kWh⁻¹, see IEA/NEA 2010 for details	0.05–0.07 (pulverized)	0.07–0.1	0.1–0.14 (onshore) 110–181 (offshore)		0.67–1.14	0.22–0.32
EPRI (2008)	US$ kWh⁻¹, see IEA/NEA 2010 for details	0.06 (pulverized)	0.07	0.09			0.18
IEA/NEA (2010)	US$ kWh⁻¹	0.05–0.12 (r = 5%) 0.07–0.14 (r = 10%)	0.03–0.1 (r = 5%) 0.04–0.1 (r = 10%)	0.05–0.16 (onsh. r = 5%) 0.1–0.19 (offsh. r = 10%)		0.22–0.33 (high load) 0.60 (low load)	0.14–0.24
Average	US$ kWh⁻¹	0.079	0.084	0.112	0.099	0.491	0.225

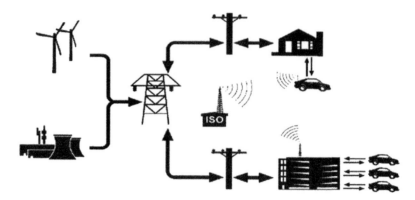

Figure 2.11.   Concept of vehicle-to-grid power (V2G) as a storage resource for large-scale wind power (Kempton and Dhanju, 2006).

2.3.3   *Market potential*

Delucchi and Jacobson (2011) reported on the feasibility of providing all energy for all purposes in all regions of the world using wind, water and solar (WWS) power. They concluded that a large-scale wind, water, and solar energy system can reliably supply all of the world's energy needs, with significant benefit to climate, air quality, water quality, ecological systems, and energy security, at reasonable cost. To accomplish this, about 4 million 5-MW wind turbines, 90,000 300-MW PV plus concentrating solar power (CSP) plants, 1.9 billion 3 kW solar PV rooftop systems, and lesser numbers of geothermal, tidal, wave, and hydroelectric plants are required. In addition, there is a need to significantly expand the transmission infrastructure to accommodate the new power systems and expand production of battery-electric and hydrogen fuel cell vehicles, ships that run on hydrogen fuel-cell and battery combinations, liquefied hydrogen aircraft, air- and ground-source heat pumps, electric resistance heating, and hydrogen production for high-temperature processes.

Delucchi and Jacobson (2011) went on to conclude that the private cost of generating electricity from onshore wind power is less than the private cost of conventional, fossil-fuel generation, and is likely to be even lower in the future. By 2030, the social cost of generating electricity from any WWS power source, including solar photovoltaics, is likely to be less than the social cost of conventional fossil-fuel generation, even when the additional cost of a supergrid and V2G storage (probably on the order of 0.02 $ kWh^{-1}, for both) is included. The world's wind resources are huge. But as wind becomes a larger fraction of electricity generation, grid integration must be resolved, particularly to smooth fluctuations in wind power output. Adding energy storage or back-up has been proposed as a solution by Kempton and Dhanju (2006) but dedicated storage or back-up adds capital cost to wind power. Their article proposes vehicle-to-grid power (V2G) as a storage resource for large-scale wind power (Fig. 2.11).

The social cost of electric transportation, based either on batteries or hydrogen fuel cells, is likely to be comparable to or less than the social cost of transportation based on liquid fossil fuels. Kempton and Dhanju (2006) explained that the complete transformation of the energy sector would not be the first large-scale project undertaken in US or world history. During World War II, the US transformed motor vehicle production facilities to produce over 300,000 aircraft, and the rest of the world was able to produce an additional 486,000 aircrafts. To improve the efficiency and reliability of a wind, water and solar power infrastructure, advanced planning is needed. Ideally, good wind, solar, wave, and geothermal sites would be identified in advance and sites would be developed simultaneously with an updated interconnected transmission system. Interconnecting geographically dispersed variable energy resources is important both for smoothing out supplies and reducing transmission requirements. The obstacles to realizing this transformation of the energy sector are primarily social and political, not technological.

Ghermandi and Messalem (2009) also addressed market concerns by providing an extensive assessment of the experience gathered from solar-powered RO desalination. The prospects for commercial penetration and for further development of the principal technological solutions were identified and discussed. They concluded that concentrating solar power reverse osmosis desalination is the most promising field of development for medium- and large-scale solar desalination. Preliminary design studies suggested that CSP–RO systems may compete in the medium-term with conventional RO desalination and as a result, gain large market shares. They noted that there is a need for testing such promising potential in demonstration and full-scale plants. Ghermandi and Messalem (2009) went on to report that solar-RO desalination is cost-competitive with other water supply sources only in framework of remote regions where grid electricity is not available and freshwater demand is met by water imports or small-scale fuel-driven desalination plants. In this market, the share of PV-RO and hybrid PV-RO desalination plants will likely increase in the near future. They noted that an infiltration of solar-RO desalination in other markets seems unlikely in the short-term. According to the authors, the rapid advancement of both CSP and PV solar technologies offers the best outlook for the wider implementation of these potentially sustainable water supply technologies.

Three potential international desalination markets were studied for export opportunities to European Union (EU) technology developers. The project called *Promotion of Renewable Energy for Water production through Desalination* (ProDes) brought together 14 leading European organisations in order to support the market development of renewable energy desalination in Southern Europe (ProDes, 2011). The analysis will help companies to expand their international activities and to support their case to investors for backing their expansion efforts. The markets analyzed were the Middle East and North Africa (MENA) regions, with a profile on Morocco; the Oceania region, with a profile on Australia; and South Africa, with an in-depth report. Each report provided details on the economic and governmental structures, the power sector, renewable energy legislation, the water sector, recent investment and a summary of the renewable desalination market potential.

Oceania is a very diverse region containing highly developed countries such as Australia as well as the less developed Pacific island nations. Although in some countries there is little water scarcity in general the region suffers from a lack of freshwater. Therefore, there is a high demand for desalination projects in the region which is supported politically due to concerns regarding the impact of climate change. A consequence of this political support is that there are a large number of sources of potential funding for renewable energy desalination projects, which are in many cases supported by the government.

The diversity of the Pacific island's region is matched by the diversity of potential renewable energy desalination projects. Although there is some variation across the region, there is in general a good supply of renewable energy including solar, wind and wave power. Large-scale projects are supported for municipal water supplies, where the presence of a firm electricity grid means that the renewable energy and desalination processes do not need to be intimately tied together. There is also potential for small-scale projects for off-grid water supplies because of the large number of off-grid communities. Furthermore, suitable small-scale projects may range from technologically complex solutions for remote Australian communities to less complex solutions for communities on the smaller Pacific islands.

South Africa is the most developed country in sub-Saharan Africa. However, the political history of the country means that there is a wide range of living standards across the country. Although it is recognized that desalination will have a significant role in providing universal access to freshwater, there is at present less enthusiasm for renewable energy desalination projects due to the perceived high project costs. This is at this time frustrated by the very low cost of electricity against which renewable energy projects must compete. In addition, the region suffers from a significant shortage of skilled personnel, at both the technical and managerial levels, which creates challenges for the successful completion of new projects.

The Middle East and North Africa (MENA) region contains many states where the cultural and political systems are very different from those in Europe. This can generate trouble when trying to initiate renewable energy desalination projects in the region (e.g. difficulty in trying to get agreement between states). Furthermore, many countries in the region have vast reserves of oil and gas and are currently heavily dependent on this energy source. This means that in many countries in the region there is limited support for renewable energy and research in this area is underfunded. On the other hand, this perspective is slowly changing and the region is starting to accept that it must diversify away from oil. An outcome of the huge revenues from oil and gas means that there is potentially a large funding pool for the development of new technologies. The MENA region has a large renewable energy resource, primarily solar and wind energy that could be exploited to power desalination projects. Consequently, the region is ideally suited for the development of renewable energy desalination projects where there is political will. Morocco is a case in point of a state in the area that has recognized the need for augmented utilization of its renewable energy resources.

2.3.4   *Process selection and risk management*

When considering which renewable energy desalination system to select, stand-alone electric generation hybrid systems are generally more suitable than systems that only have one energy source for the supply of electricity to off-grid applications, especially in remote areas with difficult access. However, the design, control, and optimization of the hybrid systems are usually very complex tasks. Bernal-Agustin and Dufo-Lopez (2009) determined in a thorough literature review that the most frequent systems are those consisting of a PV generator and/or wind turbines and/or a diesel generator, with energy storage in lead-acid batteries. Bernal-Agustin and Dufo-Lopez (2009) argued that energy storage in hydrogen, although technically viable, has a drawback in terms of its low efficiency in the electricity-hydrogen-electricity conversion process, besides the fact that, economically, it cannot compete with battery storage at the present time.

Reif (2008) performed a profitability analysis and risk management of geothermal projects being implemented in Bavaria, Germany. Reif's study concluded that the sensitive response of the project's rate of return to changes in the parameters of their computer simulations made it clear that geothermal projects are financially risky. For instance, every project faced the usual business risks, such as budget over-runs, increases in interest rates, and delays. Project management was used to limit these risks. It was recommended that the initiators of a plan must run profitability simulations in order to analyze varying scenarios before implementing the project. The results needed to be updated as the project progressed. Reif (2008) argued that reserves must always be planned for in the financing. In addition, business risks could also be limited by suitable structuring of the contracts with the partners in the project (e.g. drilling companies, power plant supplier, and civil engineering companies).

Lerner *et al.* (2009) in assessing risks discussed how and why accurate wind forecasting is a necessary ingredient for the continued growth and penetration of wind power on a global scale. They noted the critical need for proper collaboration in planning the distribution and size of wind projects, and the electricity transmission grid from the municipal to international scale. A wide range of variables need to be considered when evaluating the suitability of a potential wind energy project location. Site characteristics such as accessibility during construction, the distance to transmission and load can determine if a site is ideal for development. Project location is therefore the single most important, controllable factor in determining the economic viability of a wind energy project. Lerner *et al.* (2009) recommended that computer simulations of the dynamics of the atmosphere (e.g. numerical weather prediction models or NWP) can provide vital information on the wind resources at a site. Proper assessment techniques using NWP modeling can provide valuable information on the expected diurnal and seasonal load for a project as well as a long-term evaluation of the site's potential. Furthermore, Lerner *et al.* (2009) noted that there are three types of error associated with weather forecasting (Fig. 2.12): climatology represents the error associated with using a constant value forecast that is calculated from the average power

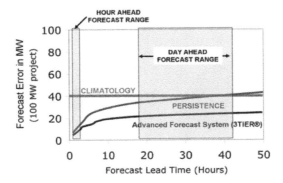

Figure 2.12.   Forecast error (MW) as a function of the forecast lead time (hours) for a wind power plant with 100 MW capacity (adapted from Lerner *et al.*, 2009).

produced for that project based on actual or reconstructed historical data. A persistence forecast utilizes the current power value to produce the constant power forecast for the rest of the forecast period (e.g., 50 hours in Fig. 2.12). A more advanced forecast system is based upon the output of both statistical and NWP models that integrate project power or meteorological observations. Figure 2.12 illustrates that the persistence forecast is not bad on average, particularly for the short-range period, and forecast error, in general, increases with lead time, eventually approaching climatology.

2.3.5   *Promotion of renewable energy policy and reduction in reliance on conventional power generation*

The following was derived from a general investigation of promotion of renewable energy technology. Government sponsorship of renewable energy projects in Germany is often cited as a model to be imitated in other states, being based on energy and environmental laws that go back nearly two decades (Frondel *et al.*, 2010). Frondel *et al.* (2010) critically reviewed the current centerpiece of this effort, the Renewable Energy Sources Act (EEG), focusing on its costs and the associated implications for job creation and climate protection. They argued that German renewable energy policy has failed to harness the market incentives needed to ensure a viable and cost-effective introduction of renewable energies into the country's energy portfolio. They went on to explain that the government's support mechanisms have in many respects underminded and subverted market incentives, resulting in massive expenditures that show little long-term promise for stimulating the economy, protecting the environment, or increasing energy security.

On the positive side, Germany has more than doubled its renewable electricity production since 2000 and has already significantly exceeded its minimum target of 12.5% for renewable energy usage (i.e. wind, biomass and others in Fig. 2.13) as a percentage of total energy usage set for 2010. This increase came at the expense of conventional electricity production, whereby nuclear power experienced the largest relative loss between 2000 and 2008 dropping from 30.2% to 23.3%. Currently, wind power is the most important of the supported renewable energy technologies: in 2008, the estimated share of wind power in Germany's electricity production amounted to 6.3% (Fig. 2.13), followed by biomass-based electricity generation and hydropower, whose shares were around 3.6% and 3.1%, respectively. In contrast, the amount of electricity produced through PV was negligible: its share was as low as 0.6% in 2008.

In their article, Frondel *et al.* (2010) argue that Germany's principal mechanism of supporting renewable technologies through feed-in tariffs (FIT), in fact, imposes high costs without any of the alleged positive impacts on emissions reductions, employment, energy security, or technological innovation. A feed-in tariff is a policy mechanism designed to accelerate investment in renewable energy technologies. It achieves this by offering long-term contracts to renewable energy

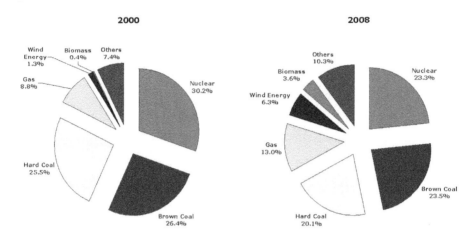

Figure 2.13.   Gross electricity production in Germany in 2000 and 2008 (Frondel *et al.*, 2010).

producers, typically based on the cost of generation of each different technology. Technologies like wind power, for instance, are awarded a lower per-kWh price, while technologies like solar-PV and tidal power are offered a higher price, reflecting their higher costs. Frondel *et al.* (2010) explained that it is most likely that whatever jobs are created by renewable energy promotion would vanish as soon as government support is terminated.

Rather than promoting energy security, the need for backup power from fossil fuels means that renewables increase Germany's dependence on gas imports, most of which come from Russia. The system of feed-in tariffs may also stifle competition among renewable energy producers and may create unwanted incentives to force a company to lock into existing technologies.

As other European governments imitate Germany by enhancing their promotion of renewables, policy makers should scrutinize the logic of supporting energy sources that cannot compete on the market in the absence of government assistance. Nonetheless, government intervention can serve to support renewable energy technologies through other mechanisms that harness market incentives such as funding for research and development (R&D), which may compensate for underinvestment from the private sector.

Consider now the controversy surrounding conventional energy production using for instance nuclear technology. The feasibility of integrated nuclear desalination plants has been proven with over 150 reactor years of experience, chiefly in Kazakhstan, India and Japan (Kadyrzhanov *et al.*, 2007; Khamis, 2009; Misra, 2010; Stock Trading, 2010). However, Jacobson and Delucchi (2011) argued that nuclear energy should not be considered as a long-term global energy source. The growth of nuclear energy has historically increased the ability of nations to obtain or enrich uranium for nuclear weapons and a large-scale worldwide increase in nuclear energy facilities would aggravate this dilemma, putting the world at greater risk of a nuclear incident. The historic link between energy facilities and weapons is evidenced by the development or attempted development of weapons capabilities secretly in nuclear energy facilities in Pakistan, India (Federation of American Scientists, 2010), Iran (Adamantiades and Kessides, 2009), and North Korea. Kessides (2010) asserted that a vigorous global expansion of civilian nuclear power will significantly increase proliferation. Similarly, Miller and Sagan (2009) reported that some new entrants to nuclear power will probably emerge in the coming decades. In Japan, some ten desalination facilities linked to pressurized water reactors operating for electricity production have yielded 1000–3000 m^3 day^{-1} each of potable water, and over 100 reactor years of experience have accrued. However, the recent earthquake and tsunami catastrophe in Japan (11 March 2011) and its effects on their nuclear reactors have shown the susceptibility of this technology and the need for better safeguards.

2.4   CASE STUDIES

2.4.1   *Desalination using renewable energies in Algeria*

Among the major challenges facing the north African countries such as Algeria are limited water and energy resources as well as risk management of the environment (Mahmoudi *et al.*, 2009b; Laboy *et al.*, 2009). Mahmoudi *et al.* (2010) has noted that due to the world economic crisis and the decreasing oil and gas reserves, decision makers in arid countries such as Algeria, need to review their policies regarding the promotion of renewable energies. Algeria is an oil and gas producer; hence decision makers believed that encouraging using renewable energies can affect the country's oil exports (Mahmoudi *et al.*, 2009b). Renewables, for example, can be used within the country allowing for export of more oil, so that the foreign trade balance is improved. This is the economic key which promotes renewables for the national market.

In 1988, an ambitious program was established with the aim to expand the utilization of geothermal heated greenhouses in regions affected by frost; sites in eastern and southern region of the state. Unfortunately, this program has been hampered by national security concerns (Fekraoui and Kedaid, 2005). In the last few decades, much effort has also been expended to exploit the numerous thermal springs of the north and the hot water wells of the Saharian reservoir (Figs. 2.14a, b). More than 900 MWt is expected to be installed in the future (Fekraoui and Kedaid, 2005). Geothermal energy represents one of the most significant sources of renewable energy in the case study area (Figure 2.14a). This can be divided into two major structural units by the South Atlas fault; with Alpine Algeria in the north and the Saharan platform in the south. The northern region is formed by the Tellian Atlas, the High Plains and the Saharian Atlas. This part is characterized by an irregular distribution of its geothermal reservoirs.

The Tellian Nappes, constitute the main geothermal reservoirs. Hot groundwater is generally of about neutral pH, total dissolved solids (TDS) are up to $10\,g\,L^{-1}$ and temperature is in the range of 22 to 98°C (Fekraoui and Kedaid, 2005). The southern region formed by the Algeria northern Sahara is characterized by a geothermal aquifer which is commonly named "Albian reservoir". The basin extends to Libya and Tunisia in the East and covers a total surface of 1 million km^2. This part of Algeria is estimated to cover about 700,000 km^2 and contains approximately 40 thousand billion m^3 of brackish groundwater. The depth of the reservoir varies between 200 m in the west to more than 1000 m in the east. Deeper wells can provide water at 50 to 60°C temperature, 100 to 400 L s^{-1} flow rate and average TDS of $2\,g\,L^{-1}$. Mahmoudi *et al.* (2010) in a recent report proposed the application of direct geothermal sources to power a brackish water greenhouse desalination system for the development of arid and relatively cold regions, using Algeria as a case study (Figs. 2.15a). Geothermal energy is employed for desalination. However it can also be used for heating the greenhouse in winter. This double benefit makes it very economic *versus* fossil fuels use. These authors noted that countries which have abundant sea/brackish water resources and good geothermal conditions are ideal candidates for producing freshwater from sea/brackish water. The establishment of human habitats in these arid areas strongly depends on availability of freshwater. Geothermal resources can both be used to heat the greenhouses and to provide freshwater needed for irrigation and heating in winter of the crops cultivated inside the greenhouses.

University of Queensland's Geothermal Energy Center's director Hal Gurgenci was quoted as saying that geothermal-powered desalination systems could be a boon for small towns facing water shortage (Wash Technology, 2009). He went on to state that this is a clever combination where desalination is coupled with an agricultural function which is both cost-efficient and environmentally-friendly. Gurgenci said that while some of the geothermal resources may not be hot enough for electric power generation, they would be a perfect fit for thermal desalination of underground brackish aquifers. Studies indicate that for plants in the range of one to 100 megaliters (megaliter is one million liters) per day, thermal desalination technologies are more suitable than reverse osmosis especially if there is a cheap and abundant supply of heat. Geothermal heat can be used to heat and to humidify a greenhouse and produce freshwater at the same time.

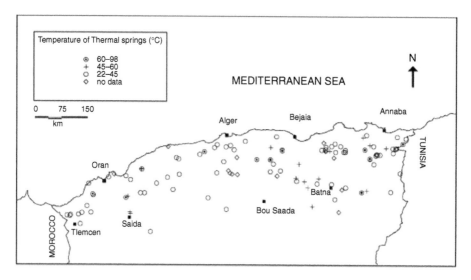

Figure 2.14a.   Main thermal springs in northern Algeria (Mahmoudi *et al.*, 2010; Kedaid, 2007).

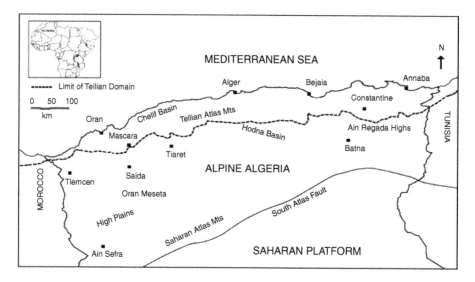

Figure 2.14b.   Geological units in northern Algeria (Mahmoudi *et al.*, 2010; Kedaid, 2007).

The innovative idea of a seawater greenhouse was originally developed by Seawater Greenhouse Ltd in 1991 (Paton and Davies, 1996; Sablani *et al.*, 2003). The first pilot was built and tested in the Canary Island of Tenerife in 1992, once known as the "Garden of the Gods", but now arid and gravely damaged by excessive abstraction of groundwater (Paton and Davies, 1996). The early results were promising and demonstrated the possibility to develop the technology in other arid regions. A modified and improved novel seawater greenhouse was constructed on Al-Aryam Island, Abu Dhabi, United Arab Emirates in 2000 (Davies and Paton, 2005). For both pilot studies the production of crops was excellent, and freshwater was successfully produced for the greenhouse irrigation proposes. In 2004 Seawater Greenhouse Ltd in collaboration with Sultan Qaboos University built a new pilot Seawater Greenhouse near Muscat, Oman (Fig. 2.15b)

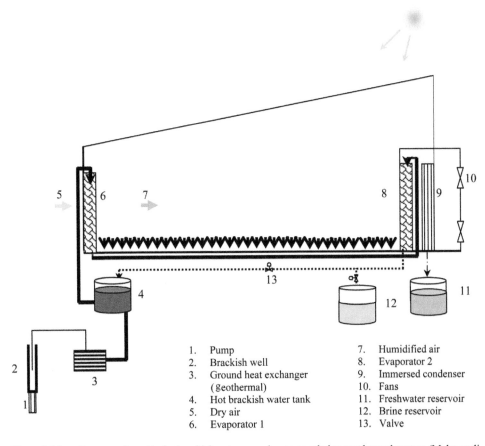

1. Pump	7. Humidified air
2. Brackish well	8. Evaporator 2
3. Ground heat exchanger	9. Immersed condenser
(geothermal)	10. Fans
4. Hot brackish water tank	11. Freshwater reservoir
5. Dry air	12. Brine reservoir
6. Evaporator 1	13. Valve

Figure 2.15a.   Process schematic for brackish water greenhouse coupled to geothermal system (Mahmoudi *et al.*, 2010).

Figure 2.15b.   The seawater greenhouse at Al-Hail, Muscat, Oman (Mahmoudi *et al.*, 2008).

(Mahmoudi *et al.*, 2008). The aim of the project was to demonstrate the technology to local farmers and organizations in the Arabian Gulf.

As mentioned previously, Mahmoudi *et al.* (2010) proposed the application of direct geothermal sources to power a brackish water greenhouse desalination system (Figs. 2.15a). The brackish water is pumped and filtered from a well and sent into a ground heat exchanger where it absorbs heat from a geothermal fluid. This heat exchanger can be built of polyethylene to conserve costs. The heated brackish water is then fed in a cascade to the first evaporator then to the second evaporator. The brine can be circulated in the circuit several times until its concentration increases over an acceptable dissolved salt concentration. The concentrated brine is finally collected in a tank, where it is stored for later treatment or processing or reinjection. The evaporator is the entire front wall of the greenhouse structure. It consists of a cardboard honeycomb lattice and faces the prevailing wind. Hot brackish water trickles down over this lattice, heating and humidifying the ambient cooler air passing through into the planting area and contributing to the heating of the greenhouse. Fans draw the air through the greenhouse. Air passes through a second evaporator and is further humidified to saturation point. Air leaving the evaporator is nearly saturated and passes over the passive cooling system with a condenser (IC) immersed in a water basin. The freshwater condensing from the humid air is piped for irrigation or other purposes. This design can be scaled up to provide 10–20 m^3 day^{-1} of freshwater while also helping greenhouse plant growing.

2.4.2   *Seawater greenhouse development for Oman in the Arabian Gulf*

A key feature in improving overall efficiency is the need to gain a better understanding of the thermodynamics of the processes and how the designs can be made more efficient. Goosen *et al.* (2003) determined the influence of greenhouse-related parameters on a desalination process that combines freshwater production with the growth of crops in a greenhouse (Figs. 2.15a,b). A thermodynamic model was used based on heat and mass balances. A software program developed by Light Works Limited, England was applied to model thermodynamic processes of the humidification/ dehumidification seawater greenhouse system. The computer program consisted of several modules: seapipe, airflow, evaporator 1, roof, planting area, evaporator 2, and condenser (air/water heat exchanger).

Weather data for the year 1995 obtained from the Meteorological Office situated at Muscat, Sultanate of Oman, were used. The software requires a weather data file and a bathymetric (seawater temperature) file. These are specific to a location. The file contained transient data on solar radiation on a horizontal surface, dry bulb temperature and relative air humidity, wind speed and wind direction. The bathymetric file contained temperature of the seawater at distance along the seafloor from the coast. The program predicts the inside air conditions and water production for a given configuration/dimension of the greenhouse, and weather and bathymetric data. The program allows many parameters to be varied. These variables can be grouped into following categories: (i) greenhouse (i.e. dimension of the greenhouse, and its orientation, roof transparency of each layer, height of front and rear evaporative pads, height of the planting area, condenser); (ii) seawater pipe; and (iii) air exchange. Three parameters i.e. dimension of the greenhouse, roof transparency and height of the front evaporator were taken as variables. These parameters were varied as follows: (i) dimensions of greenhouse (width × length): area was kept constant at 10^4 m^2: 50 × 200, 80 × 125, 100 × 100, 125 × 80 and 200 × 50 m; (ii) roof transparency 0.63 × 0.63 and 0.77 × 0.77; and (iii) height of the front evaporator 3 and 4 m. The parameters kept constant were: height of planting area = 4 m; height of the rear evaporator = 2 m; height of the condenser = 2 m; orientation of greenhouse = 40°N; seawater pipe diameter = 0.9 m, length = 5000 m; volumetric flow = 0.1 m^3 s^{-1}; pit depth = −3 m, height = 7.5 m, wall thickness = 0.1 m; air change = 0.15 (fraction)/min; fin spacing = 0.0025 m and depth = 0.1506 m.

Three climatic scenarios were considered. In the *temperate version* the temperature in the growing area is cool and the humidity high. This version is suited to lettuces, French beans, carrots, spinach, tomatoes, strawberries and tree saplings, for example. In the *tropical version* the temperature in the growing area is warm and the humidity very high. Examples of suitable crops include aubergines, cucumbers, melons, pineapples, avocados, peppers and pineapple. The design is similar to that of the temperate version but the airflow is lower. The *oasis version* allows for a diversity of crops. This version is separated into temperate and tropical sections of equal areas. The areas covered by these greenhouses were 1080 m² (*temperate* and *tropical*) and 1530 m² (*oasis*).

The overall water production rate increased from 65 to 100 m³ d⁻¹ when the width to length ratio increased from 0.25 to 4.00. Similarly the overall energy consumption rate decreased from 4.0 to 1.4 kWh m⁻³ when the width to length ratio increased from 0.25 to 4.00. Analyses showed that the dimensions of the greenhouse (i.e. width to length ratio) had the greatest overall effect on water production and energy consumption. The overall effects of roof transparency and evaporator height on water production were not significant. It was possible for a wide shallow greenhouse, 200 m wide by 50 m deep with an evaporator height of 2 m, to give 125 m³ day⁻¹ of freshwater. This was greater than a factor of two compared to the worst-case scenario with the same overall planting area (50 m wide by 200 m deep) and same evaporator height, which gave 58 m³ day⁻¹. For the same specific cases, low power consumption went hand-in-hand with high efficiency. The wide shallow greenhouse consumed 1.16 kWh m⁻³, while the narrow deep structure consumed 5.02 kWh m⁻³.

While the *tropical version* produced water most efficiently (i.e. lowest power consumption), it also produced the smallest surplus of water over that transpired by the crop inside. In this case the fan and pumps are run at their lowest rate to maintain high temperatures and humidity while still producing water in all conditions. Total freshwater production for the three climate scenarios was also calculated. One year's detailed meteorological data from Seeb Airport, Muscat were entered into the model to test the performance sensitivity for the various designs. The model results predicted that the seawater greenhouse would perform efficiently throughout the year, but with measurable variations in performance between the alternative versions. For example, the water production rate and energy efficiency results from the simulations using optimized and constant values for fan and pump speeds showed that the *temperate scenario* had almost double the water production rate per hectare compared to the *tropical scenario* (i.e. 20,370 m³ /hectare compared to 11,574 m³/hectare) while the power consumption for the former was only slightly higher (i.e. 1.9 and 1.6 kWh m⁻³, respectively).

2.4.3 *Water desalination with renewable energies in Baja California Peninsula in Mexico*

In the arid region of Baja California, in Mexico, there is not only an abundance of traditional renewable resources like sun and wind but also hot springs, tidal currents and tidal amplitudes of over six meters in the upper part of the Gulf of California (Alcocer and Hiriart, 2008). The National Autonomous University of Mexico (UNAM) assessed the extent of these renewable resources and looked at ways to use them for desalinating seawater. The project had three specific goals:

- To develop solutions to the scarcity of water in northwestern Mexico, considering the environment, costs and social impact of desalination.
- To form a group of engineers and researchers who would master the topics related to this project, and who would be able to transform science into applied solutions with a high degree of knowledge about renewable energies.
- At the end of this process, knowledge and expertise must be disseminated to society *via* courses, books, seminars and on-site training.

It was established that at only 50 m depth in a well (see Fig. 2.16), very high temperatures could be obtained, sufficient for use in binary geothermal power plants to generate electricity for desalination. It was also found that the amount of electrical power that could be generated with tidal storage and from deep sea hydrothermal vents was of the order of several thousands of MW. Many locations with hot seawater were discovered (Fig. 2.16), the best being at Los Cabos. As soon as the water table was reached, a temperature of 85°C was found at 50 m from the seashore. Alcocer and Hiriart (2008) went on to claim that to have seawater at 90°C is a real advantage for thermal desalination; because this suggests temperatures around 150°C as one drills deeper.

Using satellite imaging, hundreds of anomalous "hot spots", where large amounts of hot water reach the surface through geological fractures, were identified. The information was corroborated

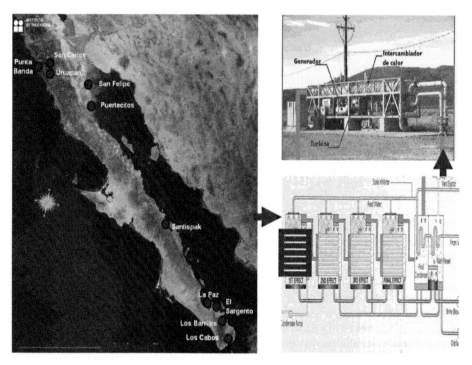

Figure 2.16.   *Left:* Locations with hot seawater at 50 m depths in the Baja Peninsula in Mexico (dots on map) with the Pacific Ocean on western side and the Gulf of California on the eastern side (Alcocer and Hiriart, 2008). *Lower right:* Schematic of multiple effects boiling/distillation (MEB) (or flashing) thermal desalination system with first effect highlighted in black. *Upper right:* Actual boiling or flashing effect (highlighted in black in previous schematic).

by measuring the water temperature directly in the field, and obtaining samples of the water to determine its isotopic composition and to better understand its origin. Although some of these hot springs and wells were on-shore, many were underwater, close to the coast and at very shallow depths. The most important of these shallow, hot seawater vents were near Puertecitos, Bahía Concepción and Ensenada. In those cases field measurements and sampling required some diving. Each hot spring was different; some had high amounts of dissolved gases; others had lower salinity than the surrounding sea water; others had high sulfur content. This information was very valuable for the group in charge of designing the thermal desalinating equipment where the availability and the quality of the hot seawater were very important.

Hot water was used in a heat exchanger to heat clean seawater and then to decrease the pressure to produce instantaneous evaporation in a multi-stage set of chambers (Fig. 2.16, *lower right*). The innovation introduced in the design was the use of hot seawater to heat all the chambers, not just the first one as in a conventional multi-stage flash (MSF) plant. The innovation can be considered as a combination of multi-effect distillation/boiling (MED or MEB) and multi-stage flash (MSF), called "multi-flash with heaters" (MFWH). Preliminary results indicated that for an initial temperature of 150°C, 4 m^3 of seawater were required to produce 1 m^3 of desalinated water. At an initial temperature of 80°C, 14 m^3 were required. For additional information on this topic see Rodriguez *et al.* (1996).

2.4.4   *Geothermal energy in seawater desalination in Milos Island, Greece*

Milos Island is located in the Aegean volcanic arc and is characterized by abundant geothermal resources (Karytsas *et al.*, 2004). Early geothermal exploration undertaken by the Institute of

Geological and Mining Research of Greece indicated that the eastern part of the island and especially the plain of Zefyria was the region with the highest temperature gradients, hence most promising for high-enthalpy geothermal potential. Later drilling exploration identified geothermal fluids with temperature of 300–323°C at depths of 800–1400 m below sea level. In addition, Karytsas *et al.* (2004) found that the region of the island most promising for exploitation of shallow, low-enthalpy (<100°C) geothermal resources was the eastern half of the island. The deep geothermal fluids corresponded to boiled seawater of 80,000 mg kg^{-1} salinity. It was determined that the upper 2 km of the hot rocks below Zefyria could support a 260 MWe geothermal power plant.

The main objective of the Milos Island project reported by Karytsas *et al.* (2004) was to construct and operate a low-enthalpy geothermal energy-driven water desalination unit producing 80 m^3 h^{-1} of drinking water and a 470 kWe of electrical power. The plant consisted of a dual system with the hot water from the deep geothermal wells being employed to run organic rankine cycle (ORC) turbines for electricity generation and at the same time being used directly in a multiple-effect boiler or distillation unit (MED). The working principle of the ORC is the same as that of the Rankine cycle; the heat from the geothermal fluid is exchanged to the working fluid which is then pumped to a boiler where it is evaporated, passes through a turbine which generates the electricity and is finally re-condensed. Multiple-effect distillation (MED) is basically a boiling process often used commercially for seawater desalination. It consists of multiple stages or "effects". In each effect the feed water is heated by steam in tubes. Some of the water evaporates (this is pure water), and this steam flows into the tubes of the next effect, heating and evaporating more water. Each effect essentially reuses the energy from the previous one. The tubes can be submerged in the feed water, but more typically the colder salty feed water is sprayed on the top of a bank of horizontal hot tubes (containing the pure water vapor). The vapor inside the tubes then condenses and drips from tube to tube until it is collected at the bottom of the unit.

The only source of energy for the Milos Island project reported by Karytsas *et al.* (2004) was geothermal heat. The unit is anticipated to be entirely self-sufficient in thermal energy and to have the potential to become self-sufficient in electricity as well. The local community would benefit from the production of desalinated water, which will be produced at a very low cost (i.e. 1.5 EURO per m^3) and from the utilization of a sustainable and environmentally friendly low-enthalpy geothermal energy source. The geothermal power and desalination plant of the project consisted of the following components:

- Geothermal production wells: production will be derived from the wells located closer to the sea, due to their high energy yield and the corresponding hot water transmission costs.
- Geothermal submersible pumps and inverters installed at the production wells.
- Piping network conveying the geothermal water to the main plant.
- Organic rankine cycle (ORC) unit, transforming approximately 7% of geothermal energy to electricity designed to generate approximately 470 kWe.
- Multi-effect-distillation – thermal-vapor-compression (MED-TVC) seawater desalination unit providing 75–80 m^3 h^{-1} desalinated water.
- Main heat exchanger, transferring the energy from the hot geothermal water exiting the ORC unit to the MED-TVC desalination unit.
- Reinjection wells (RE I and II) located at the margin of the geothermal field, close to the coast, downstream and at lower elevation of the main plant.
- Geothermal water transmission lines from main heat exchanger to reinjection wells.
- Seawater transmission lines conveying 1000 m^3 h^{-1} cooling seawater to the MED-TVC unit plus 200–575 m^3 h^{-1} cooling water for the ORC unit.
- Desalinated water transmission line from plant to water tanks near adjacent town.
- Power substation for power provision or delivery to the local power net with a capacity of 500 kWe.

Andritsos *et al.* 2010 reported that the fate of this novel desalination project in Milos Island is still unclear despite the completion of eight production and reinjection wells. The degree

Figure 2.17.　Upper left: Kwinana seawater reverse osmosis desalination (SWRO) plant & wind farm in Perth, Western Australia; Upper right: SWRO plant seaside location in Perth; Lower: Emu Downs wind farm consisting of 48 Vestas wind turbines each with 1.65 MW generating capacity; desalination plant, with 12 SWRO trains with a capacity of 160×10^3 m^3 per day and six BWRO trains delivering a final product of 144×10^3 m^3 per day. (Pankratz, 2008).

of exploitation appeared to be limited due to various non technical obstacles such as licensing problems, lack of proper advising to prospective users, lack of incentive measures by the state, and the small scale of Greek farming.

2.4.5　*The Kwinana desalination plant and wind farm in Perth, Western Australia*

A larger-scale seawater reverse osmosis (SWRO) plant located in Perth at Kwinana, with a capacity of 140,000 m^3 day^{-1}, is one of the largest plant worldwide using renewable energy (BlurbWire, 2010; Pankratz, 2008). The energy demand of the plant is 3.5 KWh m^{-3} (185 GWh year^{-1}) which is met by wind energy. The plant's total energy consumption is offset by energy production from an 82 MW wind farm (Fig. 2.17) with a power output of 272 GWh year^{-1}. The plant runs continuously generating a constant base load electricity demand (Water Corporation, 2002). The reverse osmosis plant was the first of its kind in Australia and covers several acres in an industrial park near the suburb of Kwinana. The wind farm is a joint development between Stanwell Corporation and Griffin Energy. Construction of the project commenced in November 2005, and the project was commissioned in October 2006. Emu Downs wind farm consists of 48 Vestas wind turbines each with 1.65 MW generating capacity, a substation, interconnection to the main 132 kV electricity grid, administration and stores buildings, and a network of access roads.

　　The wind farm is close to the coast, with a good quality wind resource that has increased wind speeds and reliability aligning with periods for peak power demand. Emu Downs wind farm is accredited under the Australian Government's Renewable Energy Electricity Act 2000

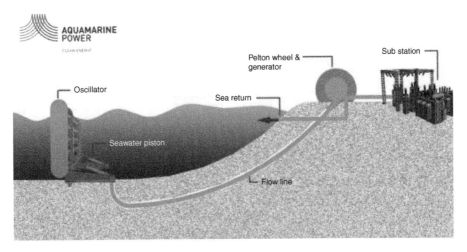

Figure 2.18.   General arrangements of Oyster™, Whittaker *et al.* (2007).

and as a Green Power Generator by the Sustainable Energy Development Authority. The wind farm contributes 270 GWh year^{-1} into the general power grid, offsetting the 180 GWh year^{-1} requirements from the desalination plant. The desalination plant, with 12 SWRO trains with a capacity of 160 ML (i.e. 160×10^3 m^3) per day and six BWRO trains delivering a final product of 144 ML (i.e. 144×10^3 m^3) per day, has one of the world's lowest specific energy consumptions, due in part to the use of pressure exchanger energy recovery devices. The devices are isobaric chamber types which recover energy in the brine stream and deliver it to water going to the membrane feed at net transfer efficiency at up to 98%. The Perth plant has a comprehensive environmental monitoring program, measuring the seawater intake and brine outfall. Excess plant water is stored in dams.

2.4.6   *A proposed wave energy converter coupled to an RO desalination plant for Orkney, UK*

Folley *et al.* (2008) and Whittaker *et al.* (2007) reported on the potential of an independent wave-powered desalination system combined with a reverse osmosis (RO) plant utilizing a pressure exchanger-intensifier for energy recovery. A key component of the system is a wave energy converter called Oyster (Figs. 2.18 and 2.19). Oyster is essentially a wave-powered hydroelectric plant located at a nominal water depth of 12 m which in many locations is relatively close to the shoreline. The system comprises a buoyant flap, 18 m wide and 10 m high, hinged at its base to a sub-frame which is pinned to the sea bed using tensioned anchors (Fig. 2.18). The surge component in the waves forces the flap to oscillate which in turn compresses and extends two hydraulic cylinders mounted between the flap and the sub-frame which pumps water at high pressure through a pipeline back to the beach (Fig. 2.18). On the shore is a modified hydroelectric plant consisting of a Pelton wheel turbine (i.e. extracts energy from the impulse or momentum of moving water) driving a variable speed electrical generator coupled to a flywheel. Power flow is regulated using a combination of hydraulic accumulators, an adjustable spear valve, a flywheel in the mechanical power train and rectification and inversion of the electrical output. An outline schematic of the 350 kW unit to be installed at the EMEC test site off Orkney in the UK is shown in Figure 2.18.

A numerical model of the RO plant with a pressure exchanger-intensifier was developed by Folley *et al.* (2008) and shows that a specific energy consumption of less than 2.0 kWh m^{-3} over a wide range of seawater feed conditions, making it particularly suitable for use with a variable power source such as wave energy. They were able to demonstrate that it is possible

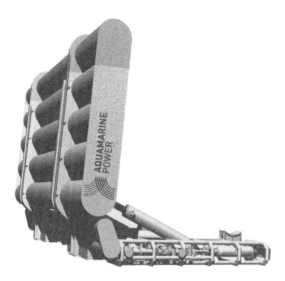

Figure 2.19.   The Oyster wave energy converter (Folley *et al.*, 2008).

to supply a desalination plant with seawater directly pressurized by the wave energy converter, eliminating the cost and energy losses associated with converting the energy into electricity and back to pressurized water. For a typical sea-state the specific hydraulic energy consumption of the desalination plant was estimated to be $1.85\,kWh\,m^{-3}$ whilst maintaining a recovery ratio of less than 25 to 35% to avoid the need for chemical pre-treatment to eliminate scaling problems. It was suggested that the economic potential for wave-powered desalination depends on these energy and cost savings more than compensating for the reduction in membrane life that occurs with variable feed conditions. Thus, the acceptability of variable feed conditions becomes an economic decision based on the relative performance and cost implications of maintaining more constant feed conditions.

Folley *et al.* (2008) went on to argue that directly-fed independent wave-powered desalination would appear to offer a promising potential for the coupling of renewable energy sources with desalination technology. Nevertheless, in order to make wave-powered desalination a reality, additional effort is necessary to design and develop the pressure exchanger-intensifier suitable for wave-powered desalination, together with tackling the challenges associated with reverse osmosis membrane durability under variable feed conditions, feed water pre-treatment and system reliability for a remote location.

2.4.7   *A proposal for combined large-scale solar power and desalination plants for the North Africa, Middle East and European region and international renewable energy alliances*

Trieb *et al.* (2002) almost a decade ago proposed a Euro-Mediterranean power pool intercon-necting the most productive sites for renewable electricity generation (Fig. 2.20). This power supply could be used for freshwater production as well as other commercial applications. Trieb *et al.* (2002) argued that the exploitation of the tremendous solar energy potential in North Africa quickly comes to its limits if it is restricted to national boundaries. Although the North African countries have vast resources of solar radiation and also land to place the necessary solar collector fields, their technological and financial resources are limited, and the local electricity demand is relatively small. The opposite is true for Europe (Fig. 2.20, *left*). In order to exploit the renewable energy potential of both regions in an efficient and economical way, an interconnection of the electricity grids was proposed to allow for the transmission of solar electricity between North

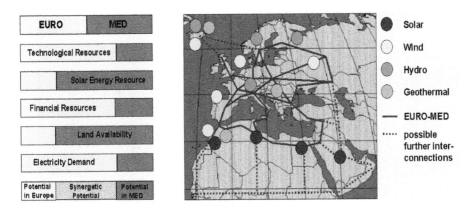

Figure 2.20. Vision of a Euro-Mediterranean power pool interconnecting the most productive sites for renewable electricity generation in the north and south of Europe. Such an international alliance will activate the large synergetic renewable energy potential of both regions, which otherwise could not be exploited to the same extend because of national limitations (Trieb *et al.*, 2002).

Africa and Europe (Fig. 2.20, *right*). The synergies of such a scheme would not only reduce considerably the cost of solar electricity in Europe, but also would create an additional income for the North African countries, enabling them to finance and develop their renewable energy resources for local use and for export. Combined solar power and desalination plants would mainly produce power and water for the increasing demand in the Mediterranean, while surplus solar electricity would be exported to the north.

2.4.8 *Solar-powered membrane distillation in Spain, Italy and Tunisia*

Several small-scale solar-powered membrane distillation (MD) demonstration pilot systems have been installed in Spain, Italy and Tunisia (Mediras project, 2011). Membrane distillation is a separation method in which a non-wetting, microporous membrane is used with a liquid feed phase (e.g. seawater) on one side of the membrane and a condensing, permeate phase (e.g. freshwater) on the other side. Separation by membrane distillation is based on the relative volatility of various components in the feed solution. The driving force for transport is the partial pressure difference across the membrane. Separation occurs when pure water vapor with its higher volatility, compared with sodium chloride, passes through the membrane pores by a convective or diffusive mechanism (Fig. 2.21). MD membranes have a pore diameter in the range of 0.1–0.4 μm. This process works at relatively low temperatures (range of 40–70°C) which can be achieved by conventional solar collectors. Once the vapor has passed through the membrane pores, it can be extracted or directly condensed in the channel on the other side of the membrane (cold side) depending on the MD system configuration. The energy required for pumping can be delivered by, for example, PV modules to make the system fully driven by solar energy.

One of the largest MD pilot plant systems established was the Mediras project (i.e. Membrane Distillation for Remote Areas) (Fig. 2.22). Several compact and multi-module two-loop systems were installed in Spain, Italy and Tunisia. Two of the compact systems, one for seawater and one for brackish water, with a capacity of 150 L day^{-1} each, were installed in Tunisia. The third compact system with a capacity of 300 L day^{-1} was installed in Tenerife, Spain. In Gran Canarias, Spain, a two-loop system powered by solar thermal field, with an expected production capacity of 3 m^3 day^{-1}, was installed (Fig. 2.22, *lower*). The second two-loop system was installed in the Italian island of Pantelleria, with a targeted production capacity of 5 m^3 day^{-1} powered simultaneously by solar energy and waste heat from the local power plant (Fig. 2.22, *upper*). About 30% of the heat source was provided from the solar panel collectors and 70% from waste heat from a diesel engine. The electrical energy was provided from the electrical grid.

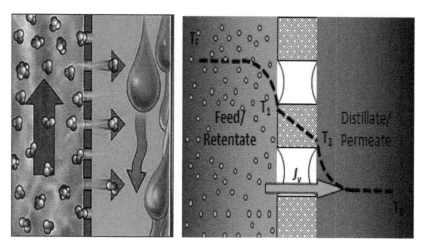

Figure 2.21.   Principle of direct contact membrane distillation (DCMD) (LHS) and temperature polarization (RHS). (Mediras project, 2011).

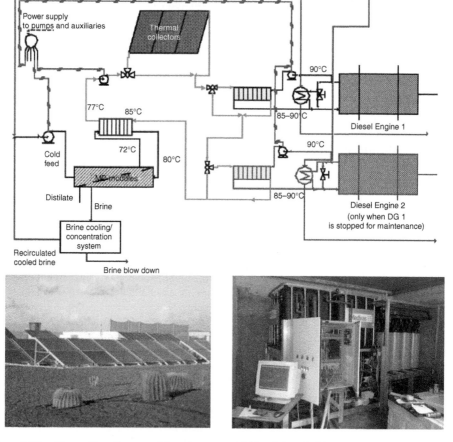

Figure 2.22.   *Upper*: Flow diagram of the solar powered MD unit installed in Pantelleria (Mediras project, 2011). *Lower LHS*: Solar still installed in Gran Canaria, Spain (Mediras project, 2011). *Lower RHS*: MD pilot plant with the modules and heat exchanger (capacity $5\,\mathrm{m}^3\,\mathrm{day}^{-1}$) (Mediras project, 2011).

2.4.9 *Solar-powered adsorption desalination prototype in Saudi Arabia*

The minimum unit cost or thermodynamic limit needed to produce a unit volume of product water by desalination is $0.78\,kWh\,m^{-3}$ at 1% salt concentration (Spiegler and El-Sayed, 2001; El-Sayed and Silver, 1980). Engineers and scientists worldwide have focused much R&D effort in trying to lower the process-based $kWh\,m^{-3}$ of thermally and mechanically driven desalination plants, namely, MSF, MED, VC and RO. The best reported $kWh\,m^{-3}$, up till now, is 3 to $5\,kWh\,m^{-3}$ by RO employing an energy recovery system from the rejected brine. In recent years, several authors (Chakraborty *et al.*, 2009; El-Sharkawy *et al.*, 2006; Ng *et al.*, 2001, 2006a,b, 2008, pers, commun. 2011; Saha *et al.*, 2006; Thu *et al.*, 2009; Wang and Ng, 2005; Wang *et al.*, 2007) have reported an emerging and yet efficient heat-driven adsorption cycle for desalination that has an unprecedented $kWh\,m^{-3}$ of twice that of the thermodynamic limit. This novel process called adsorption desalination (AD) employs a low-temperature heat source to power the sorption cycle. Such low-temperature heat sources match well with the discarded or free heat from either the exhaust of industrial processes, geothermal sources or from solar thermal energy. As opposed to higher temperature heat sources, the recovered heat is deemed as free. This is because if the renewable or exhaust heat sources are untapped, they are purged directly into the environment. Based on Ng *et al.* (2001, 2006a,b, 2008) who measured performance data, the specific energy consumption from the prototype AD plant yields a value of $1.38\,kWh\,m^{-3}$, which is unmatched by any other desalination plant.

A fully automated solar-driven AD prototype system has been commissioned at the King Abdullah University of Science and Technology (KAUST) in Saudi Arabia (Fig. 2.23) (Ng *et al.*, pers, commun. 2011). The unit is installed in the co-author's lab, WDRC (Water Desalination and Reuse Center) at KAUST. It has a nominal capacity of $3\,m^3$ per ton of adsorbent (silica gel) per day, and it is powered fully by solar energy using an array of $485\,m^2$ thermal collectors. In addition, the AD prototype can generate cooling suitable for air conditioning at a nominal capacity of 10 Rtons. Rtons stands for refrigeration tons $= 3.52\,kW$; it is a technical term used by cooling engineers. The new technology produces distillate water and cold water simultaneously, and can be seasonally adjusted to favor either cooling or potable water production which makes it particularly attractive in arid and semi-arid regions. Besides electricity efficient, the AD cycle is inherently low in maintenance by design because it has almost no major moving components. An estimation of CO_2 emission of the AD cycle yields $0.64\,kg\,m^{-3}$ which is the least polluting. In comparison, the emissions other conventional cycles are in excess of 5 to 12 folds higher than the AD cycle.

2.5 ENVIRONMENTAL CONCERNS AND SUSTAINABILITY

As mentioned in section 2.3.5 on promotion of renewable energy policy, government sponsorship of renewable energy projects in Germany is often mentioned as a model to be imitated, being based on environmental laws (e.g. Renewable Energy Sources Act) that go back nearly twenty years (Frondel *et al.*, 2010). Frondel *et al.* (2010) argued that the government's support mechanisms have essentially failed as a result of massive expenditures that showed little long-term promise for stimulating the economy, protecting the environment, or increasing energy security. Frondel *et al.* (2010) explained that it is most likely that whatever jobs are created by renewable energy promotion would vanish as soon as government support is terminated. Rather than promoting energy security or sustainability, there is still the need for backup power from fossil fuels.

Desalination of sea and brackish water requires large quantities of energy which normally results in a significant environmental impact if fossil fuels are used (e.g. CO_2 and SO_2 emissions, thermal pollution of seawater). The operating cost of different desalination techniques is also very closely linked to the price of energy as noted in section 2.3.2 on cost efficiency. This makes the use of renewable energies associated with the growth of desalination technologies very attractive. Let us take a closer look at the environmental impacts that must be considered during utilization of geothermal resources as outlined by Fridleifsson *et al.* (2008), Kagel *et al.* (2005), Lund (2007) and Rybach (2007). These include emission of harmful gases, noise pollution, water use

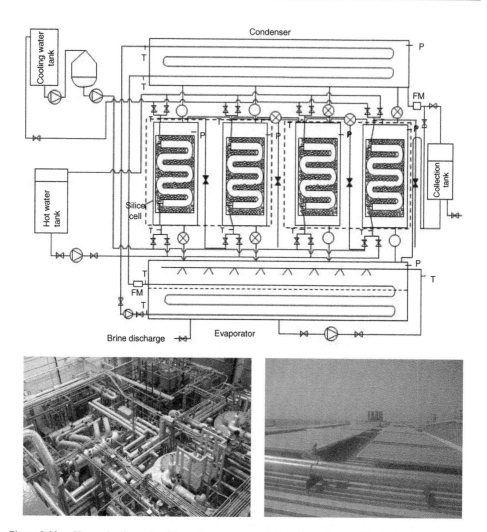

Figure 2.23.   *Upper:* A schematic of the major components of an adsorption desalination (AD) cycle. *Lower LHS*: Adsorption desalination with cooling prototype system at KAUST in Saudi Arabia. *Lower RHS*: Solar flat-plate collector system at KAUST, Saudi Arabia (Ng *et al.*, 2011).

and quality, land use, and impact on natural phenomena, as well as on wildlife and vegetation. The environmental advantages of renewable energy can be seen when comparing, for instance, a coal-fired power plant to a geothermal power plant; the former produces about 25 times as much carbon dioxide (CO_2) and sulfur dioxide (SO_2) emissions per MWh (i.e. 994 kg *vs.* up to 40 kg for CO_2, 4.71 kg *vs.* up to 0.16 kg for SO_2, respectively) (Lund, 2007; Fridleifsson *et al.*, 2008). However, in a geothermal power plant hydrogen sulfide (H_2S) also needs to be routinely treated and converted to elemental sulfur since about 0.08 kg H_2S may be produced per MWh electricity generated. We can argue that this is still much better than oil-fired power plants and natural gas fired plants which produce 814 kg and 550 kg of H_2S per MWh, respectively.

Binary power plants and direct-use projects normally do not produce any pollutants, as the water is injected back into the ground after use without exposing it to the atmosphere. We can argue that the ready availability of inexpensive oil and natural gas reserves in such areas of the world as the Arabian Gulf region may reduce the need for using renewable energy for desalination.

Table 2.6.   Environmental external costs of electricity generation in the US (year 2007 US cents kWh^{-1}) (adapted from Delucchi and Jacobson, 2011).

	Air pollution 2005			Air pollution 2030	Climate change (2005/2030)		
	5th%	Mean	95th%	Mean	Low	Mid	High
Coal	0.19	3.2	12.0	1.7	1.0/1.6	3.0/4.8	10/16
Natural gas	0.0	0.16	0.55	0.13	0.5/0.8	1.5/2.4	5.0/8.0
Wind, water and solar power	0.0	0.0	0.0	0.0	0	0	≈0

However, looking at this more closely we see that this is non-sustainable since fossil fuels are non-renewable, and with a continually growing population there is an ever increasing demand on the use of fossil fuels for desalination. Take Saudi Arabia as a specific example; in 2008 total petroleum (i.e. oil and gas) production was 10.8 million bbl day^{-1} with internal domestic oil consumption at 2.4 million bbl day^{-1} (i.e. about 25% of the total production) (U.S. Energy Information Administration, 2010a,b). Most of the internal consumption was used for electricity generation and water desalination. The population of Saudi Arabia is expected to increase from 30 million in 2010 to approximately 100 million by 2050 (U.S. Census Bureau, 2004). It has been estimated that by then 50% of the fossil fuel production will be used internally in the country for seawater desalination in order to provide freshwater for the people. This will reduce the state's income, increase pollution and is clearly non-sustainable. There are also concerns about the resulting political instability which could arise due to these effects (Cristo and Kovalcik, 2008). A possible solution to the environmental and sustainability problems is the increased use of renewable energy as well as conventional nuclear energy sources for desalination (Fridleifsson *et al.*, 2008; Lund 2006, 2007; Stock Trading, 2010). However, the former has scale-up problems and the latter has serious safety concerns.

Delucchi and Jacobson (2011) explained that there is a need to eliminate subsidies for fossil fuel energy systems or taxing fossil fuel production and use to reflect the costs of environmental damage. An example of the latter are carbon taxes that represent the expected cost of climate change due to damages as a result of CO_2 emissions (Table 2.6). Regarding environmental damages, it is estimated that the external costs of air pollution and climate change from coal and natural gas electricity generation in the US alone range from 0.0019 to 0.12 US$ kWh^{-1} for 2005 emissions, and from 0.01 to 0.16 US$ kWh^{-1} for 2030 emissions (Table 2.6). Only the upper end of the 2005 range, which is driven by assumed high climate change damages, can begin to compensate for the more than 0.10 US$ kWh^{-1} higher current private cost of solar PVs and tidal power (Table 2.6). Assuming that it is politically not feasible to add to fossil fuel generation carbon taxes that would more than double the price of electricity, eliminating subsidies and charging environmental damage taxes cannot by themselves make the currently most expensive wind, wave and solar energy options economical.

2.6   REGULATORY, POLICY AND LEGAL CONSIDERATIONS

Zhang *et al.* (2009) discussed the opportunities and challenges for renewable energy policy in one of the most rapidly developing countries in the world, China. In 2005, for example, the PRC Law of Renewable Energy was passed. Over the past three decades China's renewable energy policy has only been partially successful in allowing for the sustainable economic growth of the country. Zhang *et al.* (2009) recommended several areas for future improvement: enhanced research and development with an emphasis on domestic manufacturing of renewable energy equipment; better development of process management to ensure that policies are fully implemented; and the need to set up a market investment and financing system that will allow for bank loans to entrepreneurs. They concluded that industrialized development of renewable energy is a long-lasting and complicated process, which requires not only policy support from the state but also breakthroughs and development in techniques and markets; persistent support in terms of policy should be offered; continuous technology breakthroughs are needed, and markets need to be cultivated.

In a related development, Trieb *et al.* (2002) almost a decade ago proposed a Euro-Mediterranean power pool interconnecting the most productive sites for renewable electricity generation in the north and south of Europe (Fig. 2.20). Furthermore, a recent European Union (EU) Commission proposal for promoting the supply of power from renewable energy sources was based on a pan-European, harmonized tradable green certificate (TGC) scheme. Jacobsson *et al.* (2009) argued, on the basis of a multi-disciplinary analysis, that a pan-EU TGC system is not the way forward for Europe. They noted that it is vital that the majority of member states avoid implementation of such policy designs put forward by a coalition of vested interests. Instead they should look at the available evidence and design policies that stand a better chance of meeting the criteria of effectiveness, efficiency and equity. In particular Jacobsson *et al.* (2009) went on to explain that the policies must enable the EU to meet the innovation/industrialization challenge. Only then can the ability to implement an industrial revolution in energy systems be developed in a way that meets the needs of the member countries. The EU power oligopolies (e.g. major power companies) cannot be expected to lead this revolution. In addition to an ambitious and imaginative energy efficiency policy, they argued that a renewables policy must be designed to open up and secure attractive investment conditions for entrepreneurs in the whole value chain for a broad range of technologies. Jacobsson *et al.* (2009) noted that in a market economy, the prospect of rents is a necessary and appropriate incentive for encouraging entrepreneurship. However, rents should be channeled to risk taking innovators/entrepreneurs and should not be confused with the excess profits captured by incumbents free-riding on badly designed regulations. Only in this way, will Europe have a chance of meeting the challenge of climate change and of ensuring an economically healthy industrialization of new technologies.

Jacobsson *et al.* (2009) concluded that a trading system must be designed in a way so that it does not lead to economic inefficiencies at the EU level and endanger the ability of Europe to build up a broad capital goods industry based on electricity from renewable energy sources (RES-E). They went on to explain that the challenge for policy is to design frameworks that allow for a full use of the resource base, acknowledge the institutional diversity in the EU and, most importantly, make it possible for the member states to address the innovation/industrialization challenge.

In a related study, but this time across the Atlantic Ocean from Europe, Tonn *et al.* (2010) explored the question: is it possible for the United States to meet its energy needs sustainably without fossil fuels and corn ethanol? It can be argued that energy is one of the most pressing policy issues facing the United States today. The authors presented a scenario depicting life in the United States (US) in the year 2050. The scenario was designed to achieve energy sustainability: fossil fuels and corn ethanol were replaced by sustainable and inexhaustible energy sources. The scenario described the disappearance of the suburbs, replaced by a mix of high density urban centers and low density eco-communities. A suite of advanced technologies and significant social changes underpinned the situation. Analysis of the energy implications inherent in the scenario suggested that total US energy consumption would be around 100 quads in 2050, approximately the same as in the year 2010 despite a forecasted population increase of 130 million. A quad is a unit of energy equal to 10^{15} BTU, or 1.055×10^{18} Joules (equivalent to 8 billion gallons of gasoline).

Tonn *et al.* (2010) went on to develop a tool for constructing future energy portfolios. This tool was used to construct a US national energy portfolio for the year 2050 that is consistent with the situation offered above. The approach consisted of defining a 2050 base case, distilling the key energy-related assumptions from the scenario (Table 2.7), operationalizing the assumptions for input into the instrument and iterating among assumptions to produce a national energy portfolio.

Table 2.7.   Scenario design and supporting technological and social change assumptions (Tonn *et al.*, 2010).

Scenario design assumptions	Technology and social change assumptions
Liquid petroleum, coal, natural gas production/consumption in the US are eliminated by 2050.	Electricity transmission losses will decrease by 25% by 2050 due to advancements in high temperature superconducting lines and smart grid designs.
By 2050, approximately 50% of the population will live in super-urban high density areas and 50% will live in low-density, semi-self sufficient areas.	Transportation energy efficiency will increase by 40% by 2050 due to more efficient vehicles, and reductions in trip demand and trip length due to life style and land use changes.
Transportation energy consumption by 2050 will be 70% electricity and 30% biofuels.	Energy consumption in the residential, commercial and industrial sectors will decrease by 25% by 2050 due to improvements in energy efficiency.
Energy consumed by the US petroleum industry will fall to zero by 2050.	Energy consumption in the pulp and paper sector will decrease by another 20% by 2050 due to decreased demands for paper and packaging.
Nuclear power production to ~26 quads* by 2050.	Energy consumption in the food sector will decrease by another 20% by 2050 due to more local production.
Wind power production to ~17 quads by 2050	Energy consumption in the commercial sector will decrease by another 20% by 2050 due to decrease need for commercial space.
Geothermal power production to ~5 quads by 2050.	Energy consumption in the residential sector will decrease by another 20% by 2050 due to a decrease in the average size of homes and an increase in average household size (which will increase by 20%).
Solar power production to ~36 quads by 2050.	Requisite advances will be made in electric battery technologies, power storage technologies, and smart grid technologies.

*1 quad = 1.055 EJ

They concluded that the scenario appears plausible. Technology currently exists to support the energy system transformation. Social trends and trends in the built environment also appear favorable with respect to the scenario. It does not assume a major, voluntary de-materialization of the economy or lifestyles. This and the renewable energy aspects of the state of affairs would appear to be publicly acceptable and affordable. The authors explained that supplementary examination is needed to more accurately establish monetary costs to society and to build a well-organized and evenhanded collection of policies to encourage this future vision.

In a related regulatory and policy area, Johnstone *et al.* (2010a,b) examined the effect of environmental policies on technological innovation in the specific case of renewable energy. The analysis was conducted using patent data on a panel of 25 countries over the period 1978–2003. They found that public policy plays a significant role in determining patent applications. Different types of policy instruments were effective for different renewable energy sources. For example, broad-based policies, such as tradable energy certificates, were more likely to induce innovation on technologies that are close to competitive with fossil fuels. Johnstone *et al.* (2010a,b) argued that more targeted subsidies, such as feed-in tariffs, are needed to induce innovation on more costly energy technologies, such as solar power. Figure 2.24 shows the total number of patent applications for five renewable energy sources (solar, wind, ocean, geothermal, and biomass/waste). As expected wind and solar power had the highest counts (left axis). Solar power counts exhibited a slight U-shaped path, with growth since the early 1990s, and particularly since

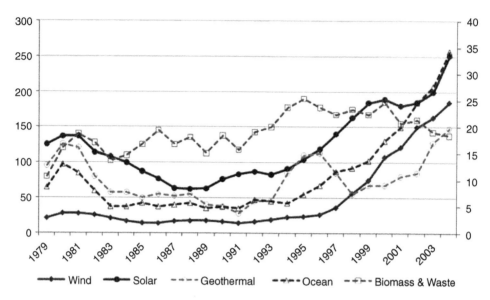

Figure 2.24.   Number of European Patent Office (EPO) patent applications for renewables by type of technology (Johnstone *et al.*, 2010a,b). *Note*: geothermal, ocean, and biomass/waste are shown
on the *right axis with dashed lines* and wind, solar are shown on the *left axis with solid lines*.

the turn of the century. In the case of wind power growth rates picked up markedly from the late
1990s. There was also high growth in ocean energy patenting recently. Finally, there appears to
have been little growth in innovation levels in the area of geothermal and biomass/waste-to-energy
since the 1970s (right axis). We can speculate that this may possibly be due to the fact that these
technologies are relatively well developed.

Empirical results generated by Johnstone *et al.* (2010a,b) indicated that public policy has had
a very significant influence on the development of new technologies in the area of renewable
energy. The passage of the Kyoto Protocol, for example, had a positive and significant impact on
patent activity with respect to wind and solar power, as well as renewable energy patents overall.
In addition, public expenditures on research and development (R&D) have had a positive and
significant effect on innovation with respect to wind and solar power in all of their models, as
well as with geothermal and ocean sources in some of their models.

2.7   CHOOSING THE MOST APPROPRIATE TECHNOLOGY FOR
FRESHWATER PRODUCTION

Hegedus and Luque (2011) have assessed the achievements and challenges of solar electricity from
PV. They argue that there is sufficient land, raw materials, safety protocols, capital, technological
knowledge and social support to allow PV to provide over 12% of the world's electrical needs by
2030. However, new ways of energy storage would need to be found. They noted that the present
PV development has been made possible by public support, driven by public opinion, which has
led to governments spending substantial money to subsidize PV. The authors predict that PV will
continue to grow at a fast pace towards the 12% goal by 2030. However, strong political support
will be necessary. Hegedus and Luque (2011) claim that the promise of significant growth in
employment, due to raw materials processing, module manufacturing, installation, and non-PV
system components, is becoming a major driving force behind that political support.

Kalogirou (2005) in a thorough study reviewed various renewable energy desalination systems and presented them together with a review of a number of pilot systems erected in various parts of the world. The selection of the appropriate renewable energy desalination technology depends on a number of factors. These include, plant size, feed water salinity, remoteness, availability of grid electricity, technical infrastructure and the type and potential of the local renewable energy resource. Among the several possible combinations of desalination and renewable energy technologies, some seemed to be more promising in terms of economic and technological feasibility than others. However, their applicability strongly depended on the local availability of renewable energy sources and the quality of water to be desalinated. Kalogirou (2005) argued that the most popular combination of technologies is multiple-effect boiling (MEB) with thermal collectors and RO with photovoltaics. PV is particularly good for small applications in sunny areas. For large units, wind energy was reported to be more attractive as it requires less space than solar collectors. With distillation processes, large sizes were more attractive due to the relatively high heat loses elements in determining water costs when water is produced from desalination plants.

Many factors enter into the capital and operating costs for desalination: capacity and type of plants, plant location, feed water, labor, energy, financing, concentrate disposal, and plant reliability (USBR, 2003). For example, the cost of desalinated seawater is about 3 to 5 times the cost of desalting brackish water from the same size plant, due primarily to the higher salt content of the former. In any state or district, the economics of using desalination is not just the number of dollars per cubic meter of freshwater produced, but the cost of desalted water *versus* the other alternatives (e.g. superior water management by reducing consumption and improving water transportation). In many arid areas, the cost of alternative sources of water (i.e. groundwater, lakes and rivers) is already very high and often above the cost of desalination (i.e. desalting). Any economic evaluation of the total cost of water delivered to a customer must include all the costs involved. This includes the costs for environmental protection (such as brine or concentrate disposal), and losses in the storage and distribution system. The capital and operating costs for desalination plants have tended to decrease over the years due primary to improvements in technical efficiency (Reif, 2008). At the same time that desalting costs have been decreasing, the price of obtaining and treating water from conventional sources has tended to increase because of the increased levels of treatment being required to meet more stringent water quality standards as well as due to continuous pollution of available water resources. This rise in cost for conventionally treated water also is the result of an increased demand for water, leading to the need to develop more expensive conventional supplies, since the readily accessible water sources have already been used up.

Bensebaa (2010) investigated the best options for thermal and PV-based large scale solar power plants. Capital along with O&M costs and other parameters from existing large scale solar farms were used to reflect actual project costs. To compete with traditional sources of power generation, he argued that solar technologies need to be able to provide electric power to respond to demand during peak hours. In spite of their high capital cost, adding energy storage is considered a better long-term solution than hybrid solar systems for large scale power plants. This is in contrast to the results of Kaldellis *et al.* (2004) who favored hybrid desalination plant configurations based on the available renewable energy sources (i.e. wind and solar) using an integrated cost-benefit analysis. In the case of Bensebaa (2010) a comparison between the two solar options was also provided that included energy storage. Although electricity storage is more expensive than thermal storage, PV power remains a competitive option. Operation and maintenance expenses in the solar thermal plant were about ten times higher than PV, an important factor resulting in higher energy cost. Based on data from proven commercial technologies, the study showed that PV holds a slight advantage even when energy storage is included.

Bensebaa (2010) went on to explain that other factors that should also be considered when comparing different solar power options include land requirements; annual power efficiency and required distance between two parallel strings of module/trough to avoid shading; larger distances are required in the case of parabolic troughs to allow trucks to move freely for regular water cleaning. The most threatening issue with solar thermal is their water requirement. A 50 MW

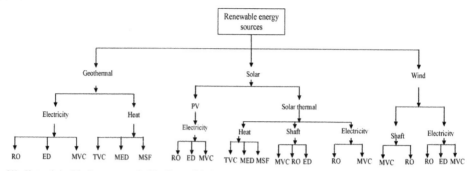

PV= Photovoltaic, RO= Reverse osmosis, ED= Electrodialysis, MVC= Mechanical vapor compression, MED= Multi effect distillation, MSF= Multi stage flash distillation, TVC= Thermal vapor compression

Figure 2.25.   Possible combinations of renewable energy sources (RES) and desalination technologies (Eltawil *et al.*, 2009).

plant for example requires around 850,000 m^3 of water annually (Hosseinia *et al.*, 2005). Given solar thermal operates only under direct irradiation (often in arid regions) there are very few areas in the world with high direct radiation is also endowed with enough renewable water resources. Solar thermal provides relatively better potential for greenhouse gas (GHG) mitigation. When compared to coal and natural gas both solar options allow significant GHG mitigation. Although CO$_2$ emission reduction is significant, Bensebaa (2010) found that its impact on the financial comparison between solar thermal and solar PV is minimal.

Forsberg (2009) discussed sustainability by combining nuclear, fossil, and renewable energy sources. The energy industries face two sustainability challenges: the need to avoid climate change and the need to replace traditional crude oil as the basis of the world's transportation system. Forsberg (2009) argues that fundamental changes in the world's energy system will be required to meet these challenges. This may require tight coupling of different energy sources (nuclear, fossil, and renewable) to produce liquid fuels for transportation, match electricity production to electricity demand, and meet other energy needs. This implies a shift in which different energy sources are integrated together, rather than being considered separate entities that compete.

Eltawil *et al.* (2009) performed a review of renewable energy technologies integrated with desalination systems. They reported that in spite of intensive research worldwide, the actual penetration of renewable energy (RE) powered desalination installations is still low. The most mature technologies for renewable energy application in desalination are wind and PV driven membrane processes and direct and indirect solar distillation. Environmental issues are associated with brine concentrate disposal, energy consumption and associated greenhouse gas production. On the other hand, social issues may include the public acceptance of using recycled water for domestic dual-pipe systems, industrial and agricultural purposes.

The suitability of a given renewable energy source for powering a desalting process depends on both the requirements of such process and the form of energy that can be obtained from a given source. Different combinations between renewable energy sources and desalination technologies are possible. Figure 2.25 shows the possible combinations. Figure 2.26 presents the design algorithm for the appropriate renewable energy source (RES)/desalination plant.

Many low-density population areas require not only freshwater availability, but in most of the cases an electrical grid connection. For these regions Mathioulakis *et al.* (2007) recommends that desalination is a solution for their needs. In using renewable energy desalination there are two separate and different technologies involved: energy conversion and desalination systems. The authors note that the real problem in these technologies is the optimum economic design and evaluation of the combined plants in order to be economically viable for remote and/or arid regions. Conversion of renewable energies requires significant investment. According to the authors, the technology is not yet mature enough to be exploited through large-scale applications.

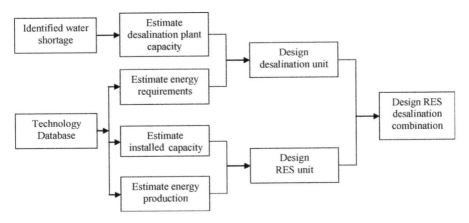

Figure 2.26. Design of the appropriate renewable energy source (RES)/desalination plant (Eltawil *et al.*, 2009).

Table 2.8. Recommended renewable energy-desalination combinations (Mathioulakis *et al.*, 2007).

Feed water quality	Product water	RE resource available	System size			Suitable combination
			Small $(1–50\,m^3d^{-1})$	Medium $(>50\,m^3d^{-1})$	Large $(>>50\,m^3d^{-1})$	
Brackish water	Distillate	Solar	×			Solar distillation
	Potable	Solar	×			PV-RO
	Potable	Solar	×			PV-ED
	Potable	Wind	×	×		Wind-RO
	Potable	Wind	×	×		Wind-ED
Seawater	Distillate	Solar	×			Solar distillation
	Distillate	Solar		×	×	Solar thermal-MED
	Distillate	Solar			×	Solar thermal-MED
	Potable	Solar	×			PV-RO
	Potable	Solar	×			PV-ED
	Potable	Wind	×	×		Wind-RO
	Potable	Wind	×	×		Wind-ED
	Potable	Wind		×	×	Wind-MVC
	Potable	Geothermal		×	×	Geothermal-MED
	Potable	Geothermal			×	Geothermal-MED

PV = photovoltaic, RO = reverse osmosis, ED = electrodialysis, MED = multi-effect distillation, MVC = mechanical vapor compressor.

Eltawil *et al.* (2009) and Mathioulakis *et al.* (2007) have reported that renewable energies and desalination plants are different technologies, which can be combined in various ways (Fig. 2.25). The energy can be provided in different forms such as thermal energy, electricity or shaft power. Table 2.8 gives an overview of recommended combinations depending on several input parameters. Photovoltaic is considered a good solution for small applications in sunny areas. For larger units, Mathioulakis *et al.* (2007) suggests that wind energy may be more attractive since it does not require much ground. This solution is often the best case for islands where flat ground is limited but there is a good wind source. In choosing the best system, the general tendency is to combine thermal energy technologies (i.e. solar thermal and geothermal energy) with thermal desalination processes and elecltromechanical energy technologies with desalination processes requiring mechanical or electrical power. The following combinations are commonly used: PV or wind-power with RO, electrodialysis or vapor compression; and solar thermal or geothermal energy with distillations processes.

Greenlee *et al.* (2009) reported that the unit operating cost for RO coupled with renewable energy is higher than for typical RO plants, due presumably to the added cost of running an additional system. Communities that would typically benefit from coupled renewable energy–RO systems are located in rural areas, where financial resources and system maintenance personnel are limited. They argued that factors including capital cost, sustainable technology, technical operation, social acceptance, and energy resource availability, have contributed to the slow growth of the renewable energy RO market. This is similar to the results of Mathioulakis *et al.* (2007). While the basic operating principles remain the same for all RO applications, individualized applications have developed, based on feed water quality. The two key types of feed water, seawater and brackish water have unique features that require explicit parameter modification and system design. Seawater RO recovery is chiefly limited by osmotic pressure increase and organic material fouling; system design typically consists of chemical and filtration pre-treatment and one RO stage. On the other hand, brackish water RO membrane systems characteristically consist of two RO stages in series; key issues include salt precipitation and concentrate management. While both seawater and brackish water RO have been sufficiently developed to be used in large-scale commercial plants, several significant challenges to the RO field remain. Further improvements are needed in membrane technology, energy use, and concentrate treatment, which will allow a wider application of RO to inland and rural communities.

As a final example of the need for choosing the appropriate combination of renewable energy and desalination technologies, consider the Greek islands in the Mediterranean Sea, where water resources are scare, limiting the economic development of the local communities. In order to try an address of this problem, Kaldellis *et al.* (2004) examined the economic viability of several hybrid desalination plant configurations based on the available renewable energy sources (i.e. wind and solar) using an integrated cost-benefit analysis. In their study all the cost parameters of the problem were taken into consideration, including the capital cost of the desalination plant, the annual maintenance and operation cost, the energy consumption cost, the local economy annual capital cost index and the corresponding inflation rate. Their results support the utilization of RES-based desalination plants as the most promising and sustainable method to satisfy the fresh, potable water demands of the small- to medium-sized Greek islands at a minimal cost. The environmental and macro-economic benefits were also recognized. The RES consisted of wind turbines (i.e. wind farm) and a photovoltaic plant which were almost exclusively used to feed the existing RO plant and only the excess of energy was sold to the grid. Kaldellis *et al.* (2004) believed that the proposed solution, in addition to its significant sustainability features, is economically competitive with any other clean water production solution for the Greek islands, thus enabling the local communities to be supplied with an adequate amount of water at a minimal cost. The authors noted that this will help to accelerate the economic development of the Aegean Archipelago, improving the quality of life of its inhabitants.

2.8   CONCLUDING REMARKS

Different combinations between renewable energy sources and desalination technologies are possible. The suitability of a given renewable energy supply for powering desalting processes depends on both the requirements of such processes and the form of energy that can be obtained. In addition, in order to aid commercialization, different types of governmental policy instruments (e.g. tax breaks; low interest loans) can be effective for different renewable energy sources. However, broad-based policies, such as tradable energy certificates, are more likely to induce innovation on technologies that are close to competitive with fossil fuels. There is also a need to eliminate subsidies for fossil fuel energy systems or taxing fossil fuel production and use to reflect the costs of environmental damage. One worrisome observation is that renewable energy sources have consistently accounted for only 13% of the total energy use over the past 40 years.

Another major concern is that in some instances government renewable energy policy may undermine and subvert market incentives, resulting in massive expenditures that show little long-term promise for stimulating the economy, protecting the environment, or increasing energy security. It is possible that whatever jobs are created by renewable energy promotion may vanish as soon as government support is terminated. Furthermore, rather than promoting energy security, the need for backup power from fossil fuels remains. The system of feed-in tariffs, for instance, may stifle competition among renewable energy producers and may create unwanted incentives to force a company to lock into existing technologies.

When considering which renewable energy desalination system to select, stand-alone electric generation hybrid systems are generally more suitable than systems that only have one energy source for the supply of electricity to off-grid applications, especially in remote areas with difficult access. Furthermore, the cost for conventional desalination can be significant because of its intensive use of energy. However, the selection of a specific process should depend on a careful study of site conditions and the application at hand. Water scarcity is an increasing problem around the world and there is a consensus that seawater desalination can help to alleviate this situation. Among the energy sources suitable to drive desalination processes, solar, wind, wave and geothermal energy, are the most promising options, due to the ability to couple the availability of energy with water demand supply requirements in many world locations.

Local circumstances will always play a significant role in determining the most appropriate process for an area. The best desalination system should be more than economically reasonable in the study stage. It should work when it is installed and continue to work and deliver suitable amounts of freshwater at the expected quantity, quality, and cost for the life of a project. Seawater desalination in itself is an expensive process, but the inclusion of renewable energy sources and the adaptation of desalination technologies to renewable energy supplies can in some cases be a particularly less expensive and economic way of providing water. The utilization of conventional energy sources and desalination technologies, notably in conjunction with cogeneration plants, is still more cost effective than solutions based on only renewable energies and, thus, is generally the first choice. While the focal point in this chapter has been on seawater desalination, the issues that were discussed also apply to desalination of saline groundwater, groundwater containing toxic substances, and residual communal/industrial effluents.

In closing, renewable energy technologies are rapidly emerging with the promise of economic and environmental viability for desalination. There is a need to accelerate the development and scale-up of novel water production systems from renewable energies. These technologies will help to minimize environmental concerns. Our investigation has shown that even though there are concerns that government policy may undermine market incentives, there is great potential for the use of renewable energy in many parts of the world. Solar, wind, wave and geothermal sources could provide a viable source of energy to power both seawater and the brackish water desalination plants. Finally, it must be noted that part of the solution to the world's water shortage is not only to produce more water, but also to do it in an environmentally sustainable way and to use less of it. It is a challenge that we are well able to meet.

REFERENCES

Adamantiades, A. & Kessides, I.: Nuclear power for sustainable development: current status and future prospects. *Energy Policy* 37 (2009), pp. 5149–5166.

Alcocer, S.M. & Hiriart G.: An applied research program on water desalination with renewable energies. *Am. J. Environ. Sci.* 4:3 (2008), pp. 204–211.

Al-Hallaj, S., Farid, M.M. & Tamimi, A.R.: Solar desalination with a humidification-dehumidification cycle: performance of the unit. *Desalination* 120 (1998), pp. 273–280.

Andritsos, N., Arvanitis, A., Papachristou, M., Fytikas, M., & Dalambakis, P.: Geothermal activities in Greece during 2005–2009. *Proceedings World Geothermal Congress 2010*, 25–29 April 2010, Bali, Indonesia, 2010.

Bensebaa, F.: Solar based large scale power plants: what is the best option? *Prog. Photovolt: Res. Appl.* (2010), published online in Wiley InterScience (www.interscience.wiley.com), DOI: 10.1002/pip.998.

Benson, S.M. & Orr, Jr., F.M.: Sustainability and energy conversions. *MRS Bulletin* 33 (2008), pp. 297–302.

Bernal-Agustin, J.L. & Dufo-Lopez, R.D.: Simulation and optimization of stand-alone hybrid renewable energy systems. *Renew. Sustain. Energy Rev.* 13 (2009), pp. 2111–2118.

BlurbWire: Kiwana desalination plant (2010). http://www.blurbwire.com/topics/Kwinana_Desalination_Plant (accessed August 2011).

Borenstein, S.: The market value and cost of solar photovoltaic electricity production. Center for the Study of Energy Markets, University of California Energy Institute, UC Berkeley, 2008, http://www.escholarship.org/uc/item/3ws6r3j4 (accessed September 2011).

Bouchekima, B.: A small solar desalination plant for the production of drinking water in remote arid areas of southern Algeria. *Desalination* 159 (2003) pp. 197–204.

Bourouni, K. & Chaibi, M.T.: Application of geothermal energy for brackish water desalination in the south of Tunisia. *Proceedings World Geothermal Congress*, 24–29 April 2005, Antalya, Turkey, 2005.

Bourouni, K., Martin, R. & Tadrist, L.: Analysis of heat transfer and evaporation in geothermal desalination units. *Desalination* 122:2–3 (1999a), pp. 301–313.

Bourouni, K., Martin, R., Tadrist, L. & Chaibi, M.T.: Heat transfer and evaporation in geothermal desalination units. *Appl. Energy* 64:1 (1999b), pp. 129–147.

Bourouni K., Chaibi, M.T. & Tadrist, L.: Water desalination by humidification and dehumidification of air: state of the art. *Desalination* 137 (2001), pp. 167–176.

Burgess, G. & Lovegrove, K.: Solar thermal powered desalination: membrane versus distillation technologies. Solar 2005, http://solar-thermal.anu.edu.au/wp-content /uploads/DesalANZSES05.pdf (accessed September 2011).

Butti, K. & Perlin, J.: *Golden thread: 2500 years of solar architecture and technology*. Cheshire Books, 1980.

Cameron, L., Doherty, R., Henry, A., Doherty, K., Van't Hoff, J., Kay, D. & Naylor, D.: Design of the next generation of the oyster wave energy converter. *3rd International Conference on Ocean Energy*, ICOE 2010, 6 October, 2010, Bilbao, Spain, 2010, pp. 1–12.

Cataldi, R., Hodgson, S. & Lund, J. (eds): *Stories from a heated earth—our geothermal heritage*. Geothermal Resources Council, Davis, CA, 1999.

Chakraborty, A., Saha, B.B., Koyama, S., Ng, K.C. & Srinivasan, K.: Adsorption thermodynamics of silica gel-water system. *J. Chem. Eng. Data* 54 (2009), pp. 448–452.

Chapa, J.: Wave energy: Aquabuoy 2.0. 2007, http://www.inhabitat.com/2007/10/08/wave-energy-aquabuoy-20-wave-power-generator/ (accessed August 2011).

Charcosset, C.: A review of membrane processes and renewable energies for desalination. *Desalination* 245 (2009), pp. 214–231.

Charlier, R.H. & Finki, C.W.: Poseidon to the rescue: mining the sea for energy—a sustainable extraction. In: *Ocean energy: tide and tidal power*. Springer, 2009, pp. 1–28.

Childs, W.D., Dabiri, A.E., Al-Hinai, H.A. & Abdullah, H.A.: VARI-RO solar powered desalting study. *Desalination* 125 (1999), pp. 155–166.

Childs, W.D. & Dabiri, A.E.: MEDRC R&D report — VARI-RO solar-powered desalting study. (www.medrc.org) research project 97-AS-005b, 2000.

Cristo, M.W. & Kovalcik, M.P.: *Population pressure and the future of Saudi state stability*. MSc Thesis, Naval Postgraduate School, Monterey, CA, 2008.

Davies, P.A.: Wave-powered desalination: resource assessment and review of technology. *Desalination* 186 (2005), pp. 97–109.

Davies, P.A. & Paton, C.: The seawater greenhouse in the United Arab Emirates: thermal modelling and evaluation of design options. *Desalination* 173 (2005), pp. 103–111.

Davies, B.V.: Low head tidal power: a major source of energy from the world's oceans. *Energy Conversion Engineering Conference*, IECEC-97, Honolulu, 1997, pp. 1982–1989.

DeCanio, S.J. & Fremstad, A.: Economic feasibility of the path to zero net carbon emissions. *Energy Policy* 39 (2011), pp. 1144–1153.

Delucchi, M.A. & Jacobson, M.Z.: Providing all global energy with wind, water, and solar power, Part II: Reliability, system and transmission costs, and policies. *Energy Policy* 39 (2011), pp. 1170–1190.

Dorn, J.G.: World geothermal power generation nearing eruption. *Plan B Updates*; Earth Policy Institute: Washington DC, USA, 14 August 2008, http://www.earthpolicy.org/Updates/2008/Update74.htm (accessed Sept. 2011).

Drude, B.C.: Submarine units for reverse osmosis. *Desalination* 2 (1967), pp. 325–328.

Dubowsky, S., Wiesman, R., Bilton, A., Kelley, L. & Heller, R.: Smart power and water for challenging environments. *MIT current projects*, 2010, http://robots.mit.edu/projects/KFUPM/index.html (accessed August 2011).

Electric Power Research Institute (EPRI): Program on Technology Innovation: Power Generation (Central Station) Technology Options-Executive Summary. Electric Power Research Institute, Palo Alto, CA, 2008.

El-Sayed, Y.M. & Silver R.S.: *Principles of desalination.* 2nd ed., Academic Press Inc. London, UK, 1980.

El-Sharkawy, I.I., Kuwahara, K., Saha, B.B., Koyama, S. & Ng, K.C.: Experimental investigation of activated carbon fibers/ethanol pairs for adsorption cooling system application. *Appl. Therm. Eng.* 26 (2006), pp. 859–865.

Eltawil, M.A., Zhengminga, Z. & Yuana, L.: A review of renewable energy technologies integrated with desalination systems. *Renew. Sustain. Energy Rev.* 13:9 (2009), pp. 2245–2262.

Energy Education: Solar ponds. www.energyeducation.tx.gov/.../index.html (accessed Sept. 2011).

European Commission (EC): *Energy sources, production costs and performance for technologies for power generation, heating, and transport.* European Commission, Brussels, Belgium, 2008.

Fahrenbruch, A. & Bube, R.H.: *Fundamentals of solar cells.* Academic Press, Orlando, FL, 1983.

Fath, M.E.S.: Solar desalination: a promising alternative for water provision with free energy, simple technology and a clean environment. *Desalination* 116 (1998) pp. 45–56.

Federation of American Scientists: Nuclear weapons. 2010, http:/www.fas.org/nuke/guide/India/nuke/ (accessed July 2011).

Fekraoui, A. & Kedaid, F.: Geothermal resources and uses in Algeria: A country update report. *Proceedings World Geothermal Congress,* 24–29 April 2005, Antalya, Turkey, 2005, pp. 24–29.

Fernández, J.L. & Chargoy, N.: Multistage, indirectly heated solar still. *Solar Energy* 44:4 (1990), pp. 215–223.

Folley, M. & Whittaker, T.: The cost of water from an autonomous wave-powered desalination plant. *Renew. Energy* 34 (2009), pp. 75–81.

Folley, M., Peñate Suárez, B. & Whittaker, T.: An autonomous wave-powered desalination system. *Desalination* 220 (2008), pp. 412–421.

Forsberg, C.W.: Sustainability by combining nuclear, fossil, and renewable energy sources. *Progr. Nuc. Energy* 51:1 (2009), pp. 192–200.

Fridleifsson, I.B., Bertani, R., Huenges, E., Lund, J.W., Ragnarsson, A. & Rybach, L.: The possible role and contribution of geothermal energy to the mitigation of climate change. In: O. Hohmeyer & T. Trittin (eds): *Proceedings IPCC Scoping Meeting on Renewable Energy Sources,* 20–25 January 2008, Lübeck, Germany, 2008, pp. 59–80.

Frondel, M., Ritter, N., Schmidt, C.M. & Vance, C.: Economic impacts from the promotion of renewable energy technologies–the German experience. *Energy Policy* 38:8 (2010), pp. 4048–4056.

Ghaffour, N.: The challenge of capacity-building strategies and perspectives for desalination for sustainable water use in MENA. *Desal. Water Treat.* 5:1–3 (2009), pp. 48–53.

Ghermandi, A. & Messalem, R.: Solar-driven desalination with reverse osmosis: the state of the art. *Desal. Water Treat.* 7 (2009), pp. 285–296.

Glueckstern, P.: Preliminary considerations of combining a large reverse osmosis plant with the Mediterranean-Dead Sea project. *Desalination* 40 (1982), pp. 143–156.

Goosen, M.F.A. & Shayya, W.: Water management, purification and conservation in arid climates. In: M.F.A. Goosen & W.H. Shayya (eds): *Water management, purification and conservation in arid climates: Volume I: Water management.* Technomic Publishing Co., Lancaster, PA, 1999, pp. 1–6.

Goosen, M.F.A., Sablani, S., Shayya, W.H., Paton, C. & Al-Hinai, H.: Thermodynamic and economic considerations in solar desalination. *Desalination* 129 (2000), pp. 63–89.

Goosen, M.F.A., Sablani, S., Paton, C., Perret, J., Al-Nuaimi, A., Haffar, J., Al-Hinai, H. & Shayya, W.: Solar energy desalination for arid coastal regions: development of a humidification-dehumidification seawater greenhouse. *Solar Energy* 75 (2003), pp. 413–419.

Goosen, M.F.A., Mahmoudi, H. & Ghaffour, N.: Water desalination using geothermal Energy. *Energies* 3 (2010), pp. 1423–1442.

Goosen, M., Mahmoudi, H., Ghaffour, N. & Sablani, S.S.: Application of renewable energies for water desalination, In: M. Schorr (ed.): *Desalination, trends and technologies,* InTech, 2011, pp. 89–118, http://www.intechopen.com/articles/show/title/ application-of-renewable-energies-for-water-desalination (accessed October 2011).

Garcia-Rodriguez, L.: Seawater desalination driven by renewable energies: a review. *Desalination* 143 (2002), pp. 103–113.

Gracia-Rodriguez, L. & Blanco-Galvez, J.: Solar-heated Rankine cycle for water and electricity production: POWERSOL project. *Desalination* 212 (2007), pp. 311–318.

Greenlee, L.F., Lawler, D.F., Freeman, B.D., Marrot, B. & Moulin, P.: Reverse osmosis desalination: water resources, technology and today's challenges. *Water Research* 43 (2009), pp. 2317–2348.

Greenpeace International, SolarPACES, and ESTELA: *Concentrating Solar Power —Global Outlook 09.* http://www.solarpaces.org/Library/docs/concentrating-solar-power-2009. pdf (accessed October 2011).

Gude, V. & Nirmalakhandan, N.: Sustainable desalination using solar energy. *Energy Conver. Manag.* 51 (2010), pp. 2245–2251.

Hegedus, S. & Luque, A.: Achievements and challenges of solar electricity from photovoltaics. In: S. Hegedus & A. Luque (eds): *Handbook of photovoltaic science and engineering.* 2nd ed. John Wiley & Sons Ltd, Hoboken, NJ, 2011, pp. 1–38.

Hosseinia, R., Soltani, M. & Valizadeh, G.: Technical and economic assessment of the integrated solar combined cycle power plants in Iran. *Renew. Energy* 30 (2005), pp. 1541–1555.

Houcine, I., Benjemaa, F., Chahbani, M.H. & Maalej, M.: Renewable energy sources for water desalting in Tunisia. *Desalination* 125:1–3 (1999), pp. 123–132.

Huang, S. & Liu, J.: Geothermal energy stuck between a rock and a hot place. *Nature* 463 (2010), p. 293.

International Desalination Association (IDA) Conference. 360 Environmental, Gran Canaria, 2007 *Environmental Management Consultants Newsletter,* Perth, Australia, 17 April 2008, http://www.360environmental.com.au/ (accessed October 2011from NEWS).

IEA/NEA (International Energy Agency/Nuclear Energy Agency): *Projected costs of generating electricity.* Paris: OECD 2005.

IEA/NEA (International Energy Agency/Nuclear Energy Agency): *Projected costs of generating electricity: 2010 Edition.* Paris: OECD 2010.

Ipsakis, D., Voutetakis, S., Seferlis, P., Stergiopoulos, F. & Elmasides, C.: Power management strategies for a stand-alone power system using renewable energy sources and hydrogen storage. *Int. J. Hydrogen Energy* 34 (2009), pp. 7081–7095.

Jacobsson, S., Bergek, A., Finon, D., Lauber, V., Mitchell, C., Toke, D. & Verbruggen, A.: EU renewable energy support policy: Faith or facts? *Energy Policy* 37 (2009), pp. 2143–2146.

Jacobson, M.Z. & Delucchi, M.A.: Providing all global energy with wind, water, and solar power, Part I: Technologies, energy resources, quantities and areas of infrastructure, and materials. *Energy Policy* 39 (2011), pp. 1154–1169.

Jayashankar, V., Anand, S., Geetha, T., Santhakumar, S., Jagadeesh Kumar, V., Ravindran, M., Setoguchi, T., Takao, M., Toyota, K. & Nagata, S.: A twin unidirectional impulse turbine topology for OWC based wave energy plants. *Renew. Energy* 34 (2009), pp. 692–698.

Johnstone, N., Hascic, I. & Popp, D.: Renewable energy policies and technological innovation: evidence based on patent counts. *Environ. Resour. Econ.* 45 (2010a), pp. 133–155.

Johnstone, N., Hašèiè, I. & Kalamova, M.: Environmental policy design characteristics and technological innovation evidence from patent data. OECD Environment Working Papers, No. 16, OECD Publishing, ENV/WKP, 2010b, http://www.oecd.org/env/workingpapers (accessed September 2011).

Kadyrzhanov, K.K., Lukashenko, S.N. & Lushchenko, V.N.: Assessment of environmental impact of reactor facilities in Khazakstan. In: J.D.B. Lambert & K.K. Kadyrzhanov (eds): *Safety related issues of spent nuclear fuel storage.* Springer, 2007, http://www.springerlink.com/content/64270j0u525166u5/ (accessed August 2011).

Kagel, A., Bates, D. & Gawell, K.: *A guide to geothermal energy and the environment.* Geothermal Energy Association, Washington, DC, 2005.

Kaldellis, J.K., Kavadias, K.A. & Kondili, E.: Renewable energy desalination plants for the Greek Islands-technical and economic considerations. *Desalination* 170 (2004), pp. 187–203.

Kalogirou, S.: Seawater desalination using renewable energy sources. *Prog. Energy Combust. Sci.* 31 (2005), pp. 242–281.

Karagiannis, I.C. & Soldatos, P.G.: Water desalination cost literature: review and assessment. *Desalination* 223 (2008), pp. 448–456.

Karytsas, C., Mendrinosa, D. & Radoglou G.: The current geothermal exploration and development of the geothermal field of Milos island in Greece. *GHC Bull.* 25, 2004, pp. 17–21.

Kedaid, F.Z.: Database on the geothermal resources of Algeria. *Geothermics* 36:3 (2007), pp. 265–275.

Kempton, W. & Dhanju, A.: Electric vehicles with V2G storage for large-scale wind power. *Windtech International* 1:2 (2006), http://www.wwindea.org/technology/ch04/en/ 4_3_4.html (accessed September 2011).

Kessides, I.N.: Nuclear power: understanding the economic risks and uncertainties. *Energy Policy* 38 (2010), pp. 3849–3864.

Khamis, I.: A global overview on nuclear desalination. *Int. J. Nuc. Desal.* 3:4 (2009), pp. 311–328.

Koschikowski, J. & Heijman, B.: Renewable energy drives desalination processes in remote or arid regions. *Membr. Technol* 8 (2008), pp. 8–9.

Laboy, E., Schaffner, F., Abdelhadi, A. & Goosen, M.F.A. (eds): *Environmental management, sustainable development & human health*. Taylor & Francis (Balkema, The Netherlands), London, UK, 2009.

Lazard Ltd.: Levelized cost of energy analysis: Version 3.0. 2009, http://blog.cleanenergy.org/files/2009/04/lazard2009_levelizedcostofenergy.pdf (accessed October 2011).

Leipoldt, E.: Alternate energy sources. http://www.alternate-energy-sources.com/history-of-solar-energy.html (accessed September 2011).

Lerner, J., Grundmeyer, M. & Garvert, M.: The role of wind forecasting in the successful intergration and management of an intermittent energy source. *Energy Central, Wind Power* 3:8 (2009), pp. 1–6.

Lu, H., Walton, J.C. & Swift, A.H.P.: Zero discharge desalination. *Int. Desal. Water Reuse Q.* 10:3 (2000), pp. 35–43.

Lund, J.W.: Chena Hot Springs. *Geo-Heat Center Quart. Bull.* 27:3 (2006), pp. 2–4.

Lund, J.W.: Characteristics, development and utilization of geothermal resources. *GHC Bulletin*, June (2007), pp. 1–9.

Magagna, D. & Muller, G.: A wave energy driven RO stand-alone desalination system: initial design and testing. *Desal. Water Treat.* 7 (2009), pp. 47–52.

Mahmoudi, H., Abdul-Wahab, S.A., Goosen, M.F.A., Sablani, S.S., Perret, J. & Ouagued, A.: Weather data and analysis of hybrid photovoltaic–wind power generation systems adapted to a seawater greenhouse desalination unit designed for arid coastal countries. *Desalination* 222 (2008), pp. 119–127.

Mahmoudi, H., Spahis, N., Goosen, M.F.A., Sablani, S., Abdul-Wahab, S., Ghaffour, N. & Drouiche, N.: Assessment of wind energy to power solar brackish water greenhouse desalination units: A case study from Algeria. *J. Renew. Sustain. Energy Rev.* 13:8 (2009a) pp. 2149–2155.

Mahmoudi, H., Ouagued, A. & Ghaffour, N.: Capacity building strategies and policy for desalination using renewable energies in Algeria. *J. Renew. Sustain. Energy Rev.* 13 (2009b), pp. 921–926.

Mahmoudi, H., Spahis, N., Goosen, M.F.A., Ghaffour, N., Drouiche N. & Ouagued, A.: Application of geothermal energy for heating and freshwater production in a brackish water greenhouse desalination unit: A case study from Algeria, *J. Renew. Sustain. Energy Rev.* 14:1 (2010), pp. 512–517.

Malik, A.A.S., Tiwari, G.N., Kumar, A. & Sodha, M.S.: *Solar distillation*. Pergamon Press, Oxford, England, 1982.

Manolakos, D., Mohamed, E.S., Karagiannis, I. & Papadakis, G.: Technical and economic comparison between PV-RO system and RO-Solar Rankine system. Case study: Thirasia island. *Desalination* 221 (2008), pp. 37–46.

Mathioulakis, E., Belessiotis, V. & Delyannis, E.: Desalination by using alternative energy: review and state-of-the-art. *Desalination* 203 (2007), pp. 346–365.

Mediras project: http://www.mediras.eu/ (accessed July 2011).

Merrick, J.H.: *An assessment of the economic, regulatory and technical implications of large-scale solar power deployment*. MSc Thesis, Massachusetts Institute of Technology, Cambridge, MA, 2010. http://18.7.29.232/handle/1721.1/62089

Miller, S.E. & Sagan, S.D.: Nuclear power without proliferation? *Daedalus*, Fall (2009) pp. 7–18.

Misra, B.M.: Sustainable desalination technologies for the future. *Int. J. Nuclear Desal.* 4:1 (2010), pp. 37–48.

Moriarty, P. & Honnery, D.: Hydrogen's role in an uncertain energy future. *Int. J. Hydrogen Energy* 34 (2009), pp. 31–39.

Narayan, G.P., Sharqawy, M.H., Summers, E.K, Lienhard, J.H., Zubair, S.M. & Antar, M.A.: The potential of solar-driven humidification–dehumidification desalination for small-scale decentralized water production. *Renew. Sustain. Energy Rev.* 14:4 (2010), pp. 1187–1201.

Ng, K.C., Chua, H.T., Chung, C.Y., Loke, C.H., Kashiwagi, T., Akisawa, A. & Saha, B.B.: Experimental investigation of the silica gel-water adsorption isotherm characteristics. *Appl. Thermal Eng.* 21 (2001), pp. 1631–1642.

Ng, K.C., Wang, X.L., Gao, L.Z., Chakraborty, A., Saha, B.B., Koyama, S., Akisawa, A. & Kashiwagi, T.: Apparatus and method for desalination. WO Patent number 121414 (2006a).

Ng, K.C., Sai, M.A., Chakraborty, A. & Saha, B.B.: The electro-adsorption chiller: performance rating of a novel miniaturized cooling cycle for electronics cooling. *ASME Trans. J. Heat Trans.* 128 (2006b), pp. 889–898.

Ng, K.C., Thu, K. & Saha, B.B.: A novel adsorption desalination method: A test programme for achieving the 1.5 kWh/m³ benchmark. *International Water Week*, 23–27 June 2008, Singapore, 2008.

Ophir, A.: Desalination plant using low grade geothermal heat. *Desalination* 40 (1982), pp. 125–32.

Pacenti, P., de Gerloni, M., Reali, M., Chiaramonti, D., Gärtner, G.O., Helm, P. & Stöhr, M.: Submarine seawater reverse osmosis desalination system. *Desalination* 126 (1999), pp. 213–218.

Palermo, R.: *Analysis of solar power generation on California turkey ranches*. MSc Thesis, College of Agriculture, Kansas State University, Manhattan, KS, 2009.

Pankratz, T.: Water desalination report (WDR). *MEDRC Workshop*, Muscat 2008, http://www.waterdesalreport.com (accessed October 2011).

Paton, C. & Davies, A.: The seawater greenhouse for arid lands. *Proc. Mediterranean Conference on Renewable Energy Sources for Water Production*, 10–12 June 1996, Santorini, Greece, 1996.

Popiel, C., Wojtkowiak, J. & Biernacka, B.: Measurements of temperature distribution in ground. *Exp. Thermal Fluid Sci.* 25 (2001), pp. 301–309.

ProDes: *Promotion of renewable energy for water production through desalination.* http://www.prodes-project.org/ (accessed September 2011).

Reali, M., de Gerloni, M. & Sampaolo, A.: Submarine and underground reverse osmosis schemes for energy-efficient seawater desalination. *Desalination* 109 (1997), pp. 269–275.

Reddy, V.K. & Ghaffour, N.: Overview of the cost of desalinated water and costing methodologies. *Desalination* 205 (2007), pp. 340–353.

Reif, T.: Profitability analysis and risk management of geothermal projects. *GHC Bulletin* January (2008), pp. 1–4.

REN21. Renewables 2010—Global Status Report. REN21 Secretariat, Paris, 2010, http://www.ren21.net/Portals/97/documents/GSR/REN21_GSR_2010_full_revised%20Sept2010.pdf (accessed July 2011).

RETI Stakeholder Steering Committee: Renewable Energy Transmission Initiative, RETI Phase 2B: Final Report. 2010. Prepared by Black & Veatch Corporation, San Francisco, http://www.energy.ca.gov/reti/documents/index.html (accessed July 2011).

Rodríguez, G., Rodríguez, M., Perez, J. & Veza, J.: A systematic approach to desalination powered by solar, wind and geothermal energy sources, In: *Proceedings of the Mediterranean conference on renewable energy sources for water production*. European Commission, EURORED Network, CRES, EDS, 10–12 June 1996, Santorini, Greece, 1996, pp. 20–25.

Rybach, L.: Geothermal sustainability. *Proceedings European Geothermal Congress*, Unterhaching, Germany, 2007.

Sablani, S., Goosen, M.F.A., Paton, C., Shayya, W.H. & Al-Hinai, H.: Simulation of freshwater production using a humidification–dehumidification seawater greenhouse. *Desalination* 159 (2003), pp. 283–288.

Sadhwani, J.J. & Veza, J.M.: Desalination and energy consumption in Canary Islands, *Desalination* 221 (2008), pp. 143–150.

Saha, B.B., Chakraborty, A., Koyama, S., Ng, K.C. & Sai, M.A.: Performance modeling of an electro-adsorption chiller. *Philos. Magazine* 86 (2006), pp. 3613-3632.

Schwarzer, K., Vieira, M.E., Faber, C. & Müller, C.: Solar thermal desalination system with heat recovery. *Desalination* 137:1–3 (2001), pp. 23–29.

Serpen, U., Aksoy, N. & Öngür, T.: 2010 present status of geothermal energy in Turkey. *Proceedings of Thirty-Fifth Workshop on Geothermal Reservoir Engineering,* 1–3 February 2010, SGP-TR-188, Stanford University, Stanford, CA, 2010.

Soerensen, B.: *Renewable energy*. Academic Press, London, UK, 1979.

Spiegler, K.S. & El-Sayed, Y.M.: The energetics of desalination processes. *Desalination* 134 (2001), pp. 109–128.

Stefansson, V.: World geothermal assessment. *Proceedings of the World Geothermal Congress*, 24–29 April 2005, Antalya, Turkey, 2005.

Stock Trading: *How quickly will Saudi Arabia turn to nuclear vitality?* (July 3rd, 2010 by admin) (http://www.profitablenicheinnovation.com/trading/day-trading/stock-trading-how-quickly-will-saudi-arabia-turn-to-nuclear-vitality) (accessed July 2010).

Szacsvay, T., Hofer-Noser, P. & Posnansky, M.: Technical and economic aspects of small-scale solar-pond-powered seawater desalination systems. *Desalination* 122 (1999), pp. 185–193.

Tester, J.W., Anderson, B.J., Batchelor, A.S., Blackwell, D.D., DiPippo, R., Drake, E.M., Garnish, J., Livesay, B., Moore, M.C., Nichols, K., Petty, S., Toksoz, M.N., Veatch, R.W., Baria, R., Augustine, C., Murphy, E., Negraru, P. & Richards, M.: Impact of enhanced geothermal systems on U.S. energy supply in the twenty-first century. *Phil. Trans. R. Soc.* A 365 (2007), pp. 1057–1094.

Thu, K., Ng, K.C., Saha, B.B., Chakraborty, A. & Koyama, S.: Operational strategy of adsorption desalination systems. *Int. J. Heat Mass Trans*. 52 (2009), pp. 1811–1816.

Tonn, B., Frymier, P., Graves, J. & Meyers, J.: A sustainable energy scenario for the United States: year 2050. *Sustainability* 2 (2010), pp. 3650–3680.

Trieb, F., Nitsch, J., Kronshage, S., Schillings, C., Brischke, L.A., Knies, G. & Czisch, G.: Combined solar power and desalination plants for the Mediterranean region-sustainable energy supply using large-scale solar thermal power plants. *Desalination* 153 (2002), pp. 39–46.

Tzen, E., Theofilloyianakos, D. & Karamanis, K.: Design and development of a hybrid autonomous system for seawater desalination. *Desalination* 166 (2004), pp. 267–274.

USBR (U.S. Bureau of Reclamation): Cost estimating procedures. In: *Desalting Handbook for Planners*, 3rd ed.; Desalination and Water Purification Research and Development Program Report No. 72; United States Department of Interior, Bureau of Reclamation, Technical Service Center, US Government Printing Office: Washington DC, USA, Chapter 9, 2003, pp. 187–231.

U.S. Census Bureau: International Population Reports WP/02, Global Population Profile: 2002, U.S. Government Printing Office, Washington, DC, 2004, http://www.census.gov/prod/2004pubs/wp-02.pdf (accessed October 2011).

U.S. Energy Information Administration: International/Country Briefs/Saudi Arabia, http://www.eia.gov/countries/country-data.cfm?fips=SA (accessed 27 July 2010a)

U.S. Energy Information Administration (EIA): *Annual Energy Outlook 2010*. Washington, DC, 2010b.

Wang, X. & Ng, K.C.: Experimental investigation of an adsorption desalination plant using low-temperature waste heat. *Appl. Therm. Eng*. 25 (2005), pp. 2780–2789.

Wang, X.L., Chakraborty, A., Ng, K.C. & Saha, B.B.: How heat and mass recovery strategies impact the performance of adsorption desalination plant: theory and experiments. *Heat Trans. Eng*. 28 (2007), pp. 147–154.

Wash Technology: Geothermal desalination: hot rocks key to producing low cost freshwater. 9 December, 2009, http://washtech.wordpress.com/2009/12/09/geothermal-desalination-hot-rocks-key-to-producing-low-cost-fresh-water/ (accessed October 2011).

Water Corporation: *Perth Metropolitan Desalination Proposal*. Perth, Australia, 2002.

White, D.E. & Williams, D.L. (eds): *Assessment of geothermal resources of the United States – 1975*. U.S. Geological Survey Circular 727, U.S., Government Printing Office, 1975.

Whittaker, T., Collier, D., Folley, M., Osterried, M., Henry, A. & Crowley, M.: The development of Oyster—A shallow water surging wave energy converter. *Proceedings of the 7th European Wave and Tidal Energy Conference*, EWTEC 2007, 11–14 September 2007, Porto, Portugal, 2007.

Wright, J.D.: Selection of a working fluid for an organic Rankine cycle coupled to a salt-gradient solar pond by direct-contact heat exchange. *J. Sol. Energy Eng*. 104:4 (1982), pp. 286 293.

Wright, M.: Nature of geothermal resources,In: J.W. Lund (ed.): *Geothermal direct-use engineering and design guidebook*. Geo-Heat Center, Klamath Falls, OR, 1998, pp. 27–69.

Zhang, P., Yang, Y., Shi, J., Zheng, Y., Wang, L. & Li, X.: Opportunities and challenges for renewable energy policy in China. *Renew. Sustain. Energy Rev*. 13 (2009), pp. 439–449.

CHAPTER 3

Use of passive solar thermal energy for freshwater production

Guillermo Zaragoza, Diego Alarcón & Julián Blanco

"The death of fire is the birth of air, and the death of air is the birth of water."
Heraclitus of Ephesus (535–475 BC)

3.1 INTRODUCTION

Distillation is one of the many processes that can be used for water purification. It requires an input of thermal energy, which can be solar radiation. Distillation is based on a phase change: water is evaporated and water vapor is separated from dissolved matter, to be condensed later as pure water. Therefore, the quality of the distillate is very high and the concentration of the impure solution is not a limiting issue, unlike for other industrial techniques based on physical separation of the solutes (reverse osmosis, electrodialysis).

Solar thermal energy is used for desalination in the most natural way by the hydrologic cycle on earth. When solar radiation is absorbed by the sea, water evaporates and rises above the surface. The vapor is moved by the natural convection currents in the atmosphere to colder areas and when it cools down to its dew point, condensation occurs and the freshwater drops down as rain. This same principle is used in man-made distillation systems, with alternative processes of heating and cooling.

The most basic technique is the direct distiller or solar still, which is frequently used for desalination. It replicates the hydrological cycle based on the passive solar heating inside a closed container with transparent cover. Impure saline feed water is contained inside the solar still, the solar radiation penetrates a transparent cover causing the water to heat up (both by the greenhouse effect from the cover, which is non-transparent to near IR radiation, and by limiting the convection losses in the contained volume) and consequently evaporate. When water evaporates, the contaminants (chemical and biological) are left in the basin. The evaporated and purified water condensates on the underside of the cover and is recovered as pure water.

Passive solar distillation in solar stills is the oldest method of desalination and can be enhanced, in active solar stills, by supplying an additional energy from external solar collectors or waste thermal energy from industry. A further improvement is obtained by separating the evaporation and condensation processes. This is the basis of the humidification-dehumidification (H-DH) distillation methods, which operate on the principle of mass diffusion using a current of air to carry water vapor between the evaporator (humidifier) and the condenser (dehumidifier). This also allows the decoupling of the solar collection and the desalination processes, which can also be fed with another source of thermal energy.

The evaporation process is endothermic, meaning that energy is required, while the condensation is an exothermic process which releases energy. By recovering this latent heat of condensation, the process can be much more energy efficient. Generally the heat recovery is used for further evaporation, turning the latent heat of condensation into latent heat of further evaporation. This drives to the concept of multi-effect or multi-stage, which can be implemented in the solar still and H-DH methods, although it is the basis of industrial processes of thermal distillation like multi-effect distillation and multi-stage flash distillation.

This chapter will concentrate only on the passive solar thermal distillation methods, therefore on solar stills.

3.1.1   A history of the solar still

Phoenician sailors which travelled along the Mediterranean Sea were already making use of the solar thermal radiation to convert seawater into fresh drinking water. The classic Greece philosopher Aristotle described the water cycle in nature and remarked that when salt water turns into vapor and the vapor condenses, it does so in the form of sweet water. In fact, boiling seawater and condensing the vapors on sponges was one of the first means of desalination (Delyannis, 2003). Another was the use of glass jars exposed directly to the sun radiation, which gave rise to the concept of the alembic. During Medieval times, Arab alchemists desalinated seawater using vessels heated with solar radiation concentrated by concave mirrors. In 1589, Della Porta mentions several methods of desalination in his book "*Magiae Naturalis*", including his solar distillation unit and a method to obtain freshwater from the air (Delyannis, 2003). But only in 1870, the first patent on solar distillation was granted in the USA to Wheeler and Evans, describing the basic operation of the solar still based on extensive experimental work.

In 1872, the first large installation of solar passive desalination was erected near Las Salinas, in northern Chile, designed by the Swedish engineer Charles Wilson. The installation was built to supply freshwater to the settlement associated to a saltpeter mine. It was fed with effluents of high salinity ($140\,\mathrm{g\,kg^{-1}}$) and produced about $22.7\,\mathrm{m^3}$ of freshwater per day, operating since 40 years. It was built of wood blackened with logwood dye and alum to absorb sunlight, using a glass cover. It consisted of 64 bays, with a total surface area of $4450\,\mathrm{m^2}$ covering $7896\,\mathrm{m^2}$ of land. Although some optimizing devices were designed in the early 20th century to increase the yield of solar distillation, basically solar concentrators (Delyannis and Delyannis, 1984), it was not until the World War II that research on solar distillation was strongly promoted to supply drinking water to troops in remote isolated places.

Maria Telkes, from the Massachusetts Institute of Technology (MIT) developed a portable air inflated plastic solar still to be used in emergency life rafts. It consisted of an inflatable device made of transparent plastic with a felt pad at the bottom and an attached container for collecting the distillate. Floating alongside the raft, the felt pad would saturate with seawater. Solar radiation passing through the transparent plastic would heat the pad creating water vapor which would condense on the inside of the plastic cover and end up sliding down to the container (Telkes, 1945). She continued working in different designs at MIT, including the multi-effect concept (Telkes, 1953).

In 1951, the Seawater Conversion Laboratory at the University of California started its investigations which led to building a solar distillation station at the Engineering Field Station in Richmond, California (Howe and Tleimat, 1974). At about the same time (1952), the United States Government set up the Office of Saline Water, which opened and financed a solar distillation program at the Daytona Beach Test Station in Florida. Many types of configuration of solar stills were analyzed in Daytona, including basin-type, multiple-effect and active solar stills (Talbert *et al.*, 1970).

In Australia, a solar distillation program was developed at the Melbourne facilities of the Commonwealth Scientific and Industrial Research Organization (CSIRO) (Wilson, 1957). A prototype of bay-type still covered with glass was designed and some were built in the desert, the largest being in Coober Pedy. In Greece, the Technical University of Athens designed some solar stills for the islands. They were asymmetric glass-covered greenhouse type with aluminum frames. One of them, in the island of Patmos, was the largest ever built so far, with capacity $8640\,\mathrm{m^3\,day^{-1}}$ (Delyannis, 1968). In other Greek islands smaller stills covered with plastic material (Tedlar) were built from designs by Frank E. Edlin tried out in Daytona Beach (Eckstrom, 1965). In the former USSR several institutions researched on solar stills and a plant was built in Ashkabad, Turkmenistan (Baum and Bairamov, 1966). There was research and experimentation on solar stills in other countries like Spain (Barasoain and Fontan, 1960), Portugal (Madeira and Cape Verde Islands), India, Cyprus, Chile, Mexico, Egypt, Algeria and Tunisia.

The information from these experiments was summarized by Eibling *et al.* (1971), analyzing the results from 27 of the largest basin-type solar stills operating up to that moment. From the

10 years of experience, they observed that the average productivities were mostly a function of total solar radiation. They estimated the productivity of a solar still to be about 3.26 L m^{-2} per day, which corresponds to about 1.2 m^3 m^{-2} per year. They stated that large durable-type, glass-covered solar stills would produce water on a consistent, dependable basis for a cost between 0.8 and 1.1 US\$ m^{-3} in most situations, this range representing the best estimates for planning purposes, at least for the next several years.

By the early 1970s, the state-of-the-art of solar stills was declared understood from the standpoints of thermodynamic and geometric effects. Some modified types of solar stills had shown to improve the productivity (tilted wicks; inclined trays; solar stills with forced convection; stills with external condensation or multiple-effect solar stills with latent heat reuse), sometimes achieving twice that of a simple-basin. However, at that time, none of those improvements could be justified on an economic basis. Also, long-term testing of the materials was needed and long life with minimum maintenance had to be demonstrated.

In the 40 years since then, a very large amount of experimental solar stills have been constructed and their performance sufficiently demonstrated. Research on solar stills has continued and only in the 21st century more than 200 scientific papers have been published so far dealing with solar stills. Although the basic knowledge and state-of-the art has not changed dramatically in the last 40 years, there are some interesting new designs and modified solar stills with increased efficiency, as well as validated models of their thermodynamics and performance.

In this chapter a description of the passive solar stills is presented. The different types of both passive and active solar stills have been already described in several publications (Tiwari and Tiwari, 2008). In this text the focus is on the performance. A simple physical model of the solar still is described, and the different designs and modifications for improving the performance are detailed.

3.2   PASSIVE SOLAR STILLS

Solar stills consist of an air-tight space in which the feed water is heated to a temperature lower than the boiling temperature, but still higher than that of the cover. Therefore, evaporation of salt water and condensation of the resulting vapor on the cover takes place. Passive solar stills are those that do not use additional energy. Active solar stills are those coupled with solar collectors and/or external devices like heat pumps or condensing units.

The most basic solar still is the basin-type. It consists of a shallow basin of salt water with a transparent cover. All joints in the cover and sides should be sealed to minimize vapor leakage and the cover must have a proper slope to allow the condensed vapor on the underside of the cover to slide into troughs and gutters to drain away the distillate. Solar energy passes through the cover and reaches the basin at the bottom. The cover can have one (Fig. 3.1) or two slopes (see Fig. 3.2).

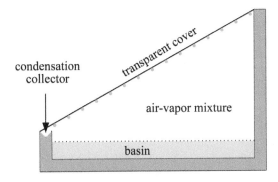

Figure 3.1.   Schematic diagram of a single-slope basin solar still.

For better absorption of the solar radiation, most basins are lined with a black, water-impervious material.

Material selection for the construction of the solar stills is very important (see ADIRA report, www.adira.gr). They must have a long life under exposed conditions, or be inexpensive enough to be replaced upon degradation. They must resist corrosion from saline and distilled water, and be strong enough to resist wind damage and slight earth movements. They must not emit volatile or soluble compounds under the temperature of operation which could give taste to the water. Selecting materials for the basin construction is one of the main problems. Metals tend to corrode, concrete and asbestos cement absorb water and can develop cracks, and plastic do not resist high temperatures easily. A good solution is to cover the basin with liner materials, which must be watertight, capable of absorbing solar radiation and withstanding high temperatures. Butyl or silicon rubber, asphalt mats and even plastic liners have been used.

The cover material must have high transmittance for the solar radiation and low transmittance for the long wavelength radiation emitted by the water. Also, it must have high stability, mechanical resistance and wettability with water to avoid condensation forming as water droplets. Glass is the best material for the cover, but transparent plastic film has been proposed as a substitute due to its possible lower cost per unit surface cover area, flexibility and the elimination of joints necessary in glass construction. However, degradation due to ultraviolet radiation and dust collection due to their high electrostatic charges must be avoided, and wettability reinforced. Cellulose acetate (after washing with a strong solution of sodium hydroxide) was tried at the Seawater Conversion Laboratory of the University of California in 1954. Polyvinyl fluoride (Tedlar, rendered wettable by mechanically roughening the surface) was tried ten years later (Tleimat and Howe, 1969). Since then, many attempts at building plastic solar stills have been carried out, struggling with the same issues. Today, 3-layer coextrusion polyethylene with inner layers of EVA (ethylene vinyl acetate) containing anti-drip additives, and wettable ETFE (ethylene tetrafluoroethylene) may represent an alternative (Zaragoza *et al.*, 2005).

Sealing must guarantee air and water tightness. Silicon or rubber materials are used. The insulation is most important under the still basin. Sometimes the least expensive option is to build the still on land which has dry soil with good drainage. Sand helps minimizing losses and may act as a sink returning heat to the still after the sun sets.

The distilled water is generally drinkable; the quality of the distillate is very high because all the salts, inorganic and organic components and microbes are left behind in the bath. Under reasonable conditions of sunlight the temperature of the water will rise sufficiently to kill all pathogenic bacteria. UV radiation is suitable to kill a reasonable part of the bacteria after some hours as well. Experiments have proved that the basin solar still can produce high quality drinking water from source water of very poor quality, as it was successful in removing non-volatile contaminants from the water (salinity, total hardness, nitrate, fluoride, etc.) and also bacteria by more than 99.9% (Hanson *et al.*, 2004).

A film or layer of sludge is likely to develop in the bottom of the basin and this should be flushed out as often as necessary. The basin must be replenished by addition of saline water and withdrawal of brine. Batch-type operation is preferable compared to continuous filling because it wastes less heat (Eibling *et al.*, 1971). The other main maintenance requirement is cleaning, since feed water treatments can be easily avoided. Glass covers do not require cleaning, but plastic covers do because of their higher electrostatic charges. Also, basin liners or wicks might require some cleaning due to scaling, and this is an expensive operation.

3.3   THERMODYNAMIC MODELLING OF THE SOLAR STILL

Heat transfer occurs across the humid air inside the enclosure of the distillation unit by free convection which is caused by the action of buoyancy force caused by density variation in the humid air. Density variation is caused by a temperature gradient in the fluid. The air close to the surface of the salt water is loaded with vapor from evaporation and hotter, therefore less dense

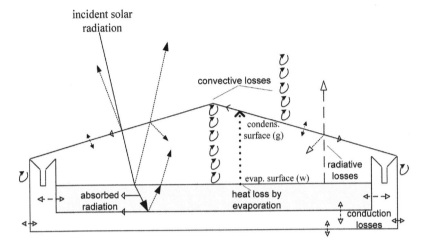

Figure 3.2.   Schematic diagram of the energy exchange processes in a double-slope solar still. Evaporation
and condensation processes (dotted line), internal and external convective exchange (looped
lines), radiation (solid line: incident solar radiation; dashed line: thermal radiation by the water;
dotted lines: reflections) and conduction exchanges (double arrows) are shown.

than that just below the cover. This difference originates convection currents which transport the
humid air upwards towards the cover, where it is cooled and therefore the vapor condenses. The
cooled and dried air falls back to the water surface restarting the cycle.

   In order to understand the energy processes that affect the performance of the solar still, we
need to consider that in the solar still there is internal and external heat transfer (Fig. 3.2). External
heat transfer mainly consists of convection and radiation processes from the cover and the sides
of the insulation to the ambient, as well as conductive heat losses from the bottom to the ground.
Internal heat transfer occurs within the confined volume, from the water to the cover, and includes
radiation, convection and evaporation heat transfer. Evaporative heat transfer is considered in fact
as the mass transfer regarding the production of condensation. In the solar still evaporative heat
transfer occurs simultaneously with convective heat transfer (transport of the vapor formed above
the water surface through an air-vapor mixture).

3.3.1   *Convective heat transfer*

The rate of convective heat transfer per unit area is given by the equation:

$$\dot{Q}_{cw} = h_{cw}\Delta T \tag{3.1}$$

where h_{cw} is the convective heat transfer coefficient [W m^{-2} K^{-1}]; and ΔT the difference of
temperature of the water vapor between the evaporation (T_w) and condensation surface (T_g),
assuming the air next to the water surface to be at the water temperature and the air next to the
condensation cover to be at the cover temperature.

   The convective heat transfer coefficient h_{cw} is a function of the operating temperature range,
the physical properties of the fluid at the operating temperature, the flow characteristics of the
fluid and the geometry of the condensing surface. Usually, heat transfer coefficients are given
by empirical relations obtained by correlating experimental results using dimensional analysis.
However, convective heat transfer is considered in terms of four dimensionless numbers.

The Nusselt number (Nu) is defined as the ratio of convective heat transfer to fluid conductive heat transfer across (normal to) the boundary under the same conditions:

$$Nu = \frac{h_{cw}}{k_a / X} \qquad (3.2)$$

where k_a is the thermal conductivity of the humid air [W m^{-1} K^{-1}]; and X the characteristic thickness (in this case the mean length between the evaporation and the condensation surface) [m]. The Nusselt number characterizes the process of heat transfer at the fluid boundary; the larger Nu, the more effective the convection is.

The Grashof number (Gr) describes the relationship between buoyancy and viscosity within a fluid; it is the ratio between the buoyancy and the viscous forces:

$$Gr = \frac{g \beta \rho_a^2 X^3 \Delta T}{\mu_a^2} \qquad (3.3)$$

where g is the acceleration due to gravity [m s^{-2}]; β the coefficient of volumetric thermal expansion [K^{-1}]; ρ_a the density of the humid air [kg m^{-3}]; and μ_a the dynamic viscosity of humid air [N s m^{-2}]. The Grashof number represents the flow regime in buoyancy driven flow (natural convection); low Gr numbers correspond to laminar flow and large values to turbulent flow.

The Prandtl number (Pr) is the ratio between momentum diffusivity (viscosity) and thermal diffusivity:

$$Pr = \frac{\mu_a C_{Pa}}{k_a} \qquad (3.4)$$

where C_{Pa} is the specific heat capacity of humid air at constant pressure [J kg^{-1} K^{-1}]. It provides a measure of the relative effectiveness of momentum transport in the velocity boundary layer with respect to thermal energy transport in the thermal boundary layer by diffusion. For gases, Pr is near unity.

The Rayleigh number (Ra) is used to characterize transition in a boundary layer during natural convection, and is defined as the product of the Grashof number (which refers to the type of flow) and the Prandtl number (the type of fluid):

$$Ra = Gr \cdot Pr \qquad (3.5)$$

When the Rayleigh number is below the critical value for a fluid, heat transfer is primarily in the form of conduction; when it exceeds the critical value, it is primarily in the form of convection.

For upwards heat flow from the horizontal water surface, Jakob (1957) suggested the following empirical relationship:

$$Nu = C (Gr \cdot Pr)^n \qquad (3.6)$$

where C and n are constants. For $Gr < 10^3$ (convection negligible): $C = 1$, $n = 0$; for $10^4 < Gr < 3.2 \cdot 10^5$ (laminar flow): $C = 0.21$, $n = 1/4$; for $3.2 \cdot 10^5 < Gr < 10^7$ (turbulent flow), $C = 0.075$, $n = 1/3$.

In solar stills, mass transfer takes place due to a differential in fluid density, which is not only a function of temperature but also composition. Since water vapor is lighter than dry air, the evaporation increases the driving buoyancy force caused by the difference in temperature. Therefore, following the approach suggested by Sharpley and Boelter (1938) it becomes necessary to use a modified Grashof number Gr' based in the following temperature difference:

$$\Delta T' = \Delta T + \frac{\Delta P}{\dfrac{M_{da} P_T}{M_{da} - M_w} - P_w}(T_w + 273.15) \qquad (3.7)$$

where M_{da} and M_w are the molecular weights of the dry air and the water vapor, respectively [kg kmol^{-1}], P_T is the total pressure of the air in the still [Pa]; and ΔP the difference of saturation vapor pressure between the evaporation (P_w) and the condensation surface (P_g) [Pa].

Dunkle (1961) was the first to present a fundamental theoretical model for the prediction of heat and mass transfer processes in solar stills, based in the previous relationships. He arrived at the following convective heat transfer coefficients:

$$h_{cw} = 0.884(\Delta T')^{1/3} \tag{3.8}$$

with $\Delta T'$ given by the simplified expression for an air water vapor system at normal atmospheric pressure:

$$\Delta T' = \Delta T + \frac{\Delta P(T_w + 273.15)}{268.9 \times 10^3 - P_w} \tag{3.9}$$

This expression was derived for operating temperatures of around 50°C, for upward heat flow in horizontal enclosed air space (i.e., parallel evaporative and condensing surfaces), and assuming saturated air and turbulent flow (expected over the normal range of operation of the still). Since then, many other researchers have studied this problem for different operating temperatures. A more general approach is to use the relation:

$$Nu = C(Gr' Pr)^n \tag{3.10}$$

where C and n are two constants which are determined from experiments. This expression can be valid for a wider range of water temperatures (i.e., of Grashof numbers), and takes into account the effects of the spacing between condensing and evaporating surfaces, the operating temperature range and the orientation of condenser cover. Therefore, the convective heat rate coefficient can be expressed as:

$$h_{cw} = C\frac{k_a}{X}(Gr' Pr)^n \tag{3.11}$$

and the values of C and n can be derived from regression analysis based on experimental data (Kumar and Tiwari, 1996).

3.3.2 Evaporative heat transfer

The rate of evaporative heat transfer per unit area is given by:

$$\dot{Q}_{ew} = h_{ew}\Delta T = \dot{m}_w L \tag{3.12}$$

where h_{ew} is the evaporative heat transfer coefficient [W m^{-2} K^{-1}]; $\dot{m}_w$ the evaporation rate per unit area of evaporation surface [kg m^{-2} s^{-1}]; and L the latent heat of vaporization [J kg^{-1}].

The evaporative heat transfer coefficient is derived from the Chilton-Colburn's analogy between heat and mass transfer, known as the Lewis relation, which can be written as (Coulson and Richardson, 1977):

$$h_{ew} = \frac{M_w L}{M_a C_{P_a}} \frac{1}{Le^{1/3}(P_a)_{LM}} \frac{\Delta P}{\Delta T} h_{cw} \tag{3.13}$$

where Le is the dimensionless Lewis number (defined as the ratio of thermal diffusivity to mass diffusivity); C_{P_a} the specific heat capacity of the humid air (water vapor-air mixture) [J kg^{-1} K^{-1}]; M_a the molecular weight of the humid air [kg kmol^{-1}], and $(P_a)_{LM}$ is the log mean of the air partial pressure [Pa]. An approximation is commonly used by assuming $Le = 1$ and $(P_a)_{LM} = P_T$ (Malik *et al.*, 1982), so that the simplified form of the Chilton-Colburn analogy is given by:

$$h_{ew} = \frac{M_w L}{M_{da} C_{P_{da}} P_T} \frac{\Delta P}{\Delta T} h_{cw} = k\frac{\Delta P}{\Delta T} h_{cw} \tag{3.14}$$

where M_{da} and $C_{P_{da}}$ correspond to the dry air; and k is a mass transfer heat transfer proportionality constant (Voropoulos *et al.*, 2000). In the above relation the effect of water vapor is assumed negligible, so it is only valid for low temperatures (Shawaqfeh and Farid, 1995). An approximation assuming saturated air at 50°C gives k the value of 0.013 (Malik *et al.*, 1982). In order to account for the effect of water vapor pressure, Dunkle (1961) used an experimental value:

$$h_{ew} = 0.016273 \, h_{cw} \frac{\Delta P}{\Delta T}$$

(3.15)

Other values of k have been reported; Clark (1990) suggested that Dunkle's value should be halved for operating temperatures greater than 55°C, while Voropoulos *et al.* (2000) found that Dunkle's model can be valid using a lower value of k in high mass-transfer rate conditions. Shawaqfeh and Farid (1995) concluded that the experimental values of T_w and T_g should be used for accurate evaluation of the evaporative heat transfer coefficient.

The theoretical production of the still is taken from the evaporation rate per unit area (Voropoulos *et al.*, 2000) from equations (3.8), (3.12) and (3.4):

$$\dot{m}_w = 0.884k \frac{\Delta P}{L}(\Delta T)^{1/3}$$

Zheng *et al.* (2002) derived from the Lewis relation another correlation:

$$\dot{m}_w = \frac{h_{cw}}{\rho_a \, C_{P_a} Le^{1-n}} \frac{M_w}{R} \left(\frac{P_w}{T_w} - \frac{P_g}{T_g} \right)$$

(3.16)

where R is the universal gas constant [kJ kmol^{-1} K^{-1}]; and the value of n is taken as 0.26 from the experiments of Chen *et al.* (1984). Another expression was obtained by Tsilingiris (2007) considering that the saturated mixture of dry air and water vapor has a different set of thermo-physical properties than dry air, especially at saturation conditions and high temperatures:

$$\dot{m}_w = \frac{h_{cw}}{C_{P_{da}}} \frac{R_{da}}{R_w} \left(\frac{P_T \Delta P}{(P_T - P_w)(P_T - P_g)} \right)$$

(3.17)

where R_{da} and R_w are the gas constants of the dry air and the water vapor, respectively [kJ kg^{-1} K^{-1}].

3.4   PERFORMANCE OF THE SOLAR STILL

The thermal instantaneous efficiency of a solar still is defined by the ratio of evaporative heat transfer to instantaneous solar radiation intensity:

$$\eta = \frac{\dot{Q}_{ew}}{I(t)}$$

(3.18)

Therefore, the performance of a solar still during a certain time Δt is given by:

$$Y = \frac{\sum \dot{m}_w \Delta t}{\sum I \Delta t}$$

(3.19)

Eibling *et al.* (1971) analyzed the results from 27 of the largest basin-type solar stills operating up to that moment. From the 10 years of experience, they observed that the average productivities were mostly a function of total solar radiation. The productivity of a large basin-type solar still was therefore approximated by the expression:

$$Y = 35.64G^{1.40}(\Delta Y \simeq \pm 25\%)$$

(3.20)

where Y is the productivity [L m^{-2} day^{-1}], ΔY its error and G the solar radiation [J m^{-2} day^{-1}]. Therefore, a fair estimation of the productivity of a solar still was found to be about 3.26 L per day and square meter of collector area, which corresponds to 1.2 m^3 per year and square meter of collector area. The effect of other atmospheric variables was found to be of minor importance in comparison.

The productivity increases slightly with ambient air temperature (~5% for each 5°C rise). The net effect of increased wind speed is a slight decrease in productivity. On one hand, wind helps decrease the cover temperature, which should increase the rate of condensation, but a lower cover temperature means larger heat losses from the brine and the balance is negative.

The highest production of a solar still occurs for high brine temperatures with minimum thermal lag. The rate of evaporation decreases exponentially with the temperature of the brine. Hirschman and Roffler (1970) found an inverse proportionality between productivity and thermal inertia of solar stills (the time lag between the maximum instantaneous solar radiation and maximum evaporation rate). Also, brine depth and insulation have an important effect. Water depth increases the volumetric heat capacity of the basin but reduces the water temperature, which decreases the evaporation rate and the productivity of the still. Cooper (1973) found that depth is inversely proportional to the productivity of the still. However, with depth the temperature and production rate are uniform and less affected by solar intensity variation in a short period of time. Also, the heat stored in the water mass is released during the absence of sunshine and production continues even during night. This thermal inertia can sometimes compensate the losses for lack of insulation to make deep basins perform better than shallow ones.

Insulation has a stronger effect on shallow basins and benefits only small stills. Bloemer *et al.* (1965) indicated that the use of insulation under the basin increases the productivity by about 15% for a two inches deep basin, and by about 20–30% for a one inch deep basin. For stills larger than a few thousands m^2 it is hard to find insulation economically viable.

The productivity obviously decreases with vapor losses, but condensate leakage back into the still can also be major factor of decline in productivity.

As reported in several works the overall efficiency of a typical basin-type solar still is only about 30% and over two decades many efforts have been devoted to increase this, as can be seen in multiple references from Löf (1961) to Murugavel *et al.* (2008) and Kaushal (2010). Another type of solar still that is designed to operate with a very low heat capacity is the tilted-wick still, first proposed by Telkes (1955). This type of still exhibits a relatively higher distillation rate per unit area and operates at relatively high temperatures. In spite of their simple construction, low operation and maintenance costs, the major shortcoming, from an energy point of view, of the basin and tilted-wick solar stills is that the latent heat of condensation is not utilized.

Multi-effect solar stills are designed to recycle some of the latent heat from the condensation by using it for preheating the feed water within the still. This can be accomplished by e.g. using the feed water duct as the condensation surface for the water vapor. The feed water is then preheated by the heat released from the condensing vapor, and the condensation surface is kept continuously cool. A multi-effect solar still can in this way produce freshwater up to 20 L per day and square meter of collector area. The increased production rate that follows by this recycling, must however be measured to the cost for the more complex construction that follows.

3.5 DESIGNS AND TECHNIQUES TO IMPROVE THE PERFORMANCE OF THE SOLAR STILL

The performance of a solar still can be enhanced by improving the radiation transmission, the evaporation and/or the condensation. Therefore, the main improvements in the basin solar still concern the cover (for light transmission) and the enhancement of the evaporation and the condensation.

3.5.1 *Enhancing the light transmission*

For stills having low-angle, symmetrical, double-sloped covers, compass orientation does not affect productivity (Eibling *et al.*, 1971).

The transmission of some of the materials used for transparent covers of solar stills is nearly constant for angles of incidence of radiation less than some critical value and then decreases to greater angles. Therefore, in designing the cover shape the goal is to maximize the area on the cover where the angle of incidence is less than the critical angle. Tent shapes transmit more energy than hemisphere or conical covers (Martens, 1966). Also, it seems obvious that double slope stills make more sense in low latitudes and single slope stills in higher latitudes. Moreover, on the basis of yearly performance data, Tiwari and Yadav (1987) concluded that a single-slope still gives better performance than a double slope for cold climatic conditions where energy loss significantly controls still performance.

By reviewing a large number of studies, Khalifa (2011) inferred a relation between the cover tilt angle and the productivity of simple solar stills in various seasons together with a relation between the optimum tilt angle and the latitude. He found that cover tilt angle should be large in winter and small in summer and increasing the tilt angle would increase the productivity through out the year. Also, the trend obtained suggests an optimum cover tilt angle that is close to the latitude angle of the site. Introducing a sun tracking system to a fixed solar still has improved the performance of the traditional fixed single slope solar system by 22% (Abdallah and Badran, 2008).

In basin solar stills the horizontal surface of the water intercepts a lesser amount of solar radiation than a surface that is tilted appropriately. For maximum radiation integrated during a year, the collecting surface should face south and be inclined to the horizontal at an angle equal to the latitude of the place. The main difficulty in designing tilted solar stills is maintaining an inclined water surface. A weir-type inclined solar still was designed by Sadineni *et al.* (2008). It consisted of an inclined absorber plate formed to make weirs, as well as a top basin and a bottom basin. The goal was to distribute the water evenly while increasing the time spent on the absorber surface, and the result was an increase in productivity of 20% with respect to conventional basin stills. Similarly, stepped trays or separating screens can be used to retain variable amounts of water in such a plane angle as to receive the highest direct solar radiation incidence (Ward, 2003).

The stepped still (Fig. 3.3) is a good approximation to the tilted still and combines the advantages of both the horizontal basin and the inclined solar still (Tleimat and Howe, 1966; Moustafa *et al.*, 1979; Radhwan, 2004). In a step (or staircase) solar still the volume of air in the enclosure is smaller and therefore heating up is faster. Also, there is more heat and mass transfer surface in the stepped basin than in the flat basin. Abdallah *et al.* (2008) report a very large increase in the production rate (180%) with respect to conventional basin-types while models by Barrera (1992) predict a more modest 10%. The steps can be equipped with weirs to force the flowing and extend the residence time of water in the still (Tabrizi *et al.*, 2010).

Further reducing the gap distance between the evaporating surface and the condensing surface increases the performance, as in the cascade-type solar still shown in Fig. 3.4a (Satcunanathan and

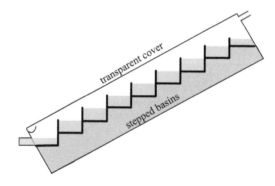

Figure 3.3.   Schematic diagram of a stepped solar still.

Hansen, 1973). Headley (1973) reports thermal efficiencies of 60–75% using a tilted double-sided solar still with cascade water trays below the glass cover, achieving a very low mean effective water depth which leads to high water temperatures.

A real inclined water surface is obtained by using a wick parallel to the cover (Fig. 3.4b), which achieves the trickling effect of the water (Frick and von Sommerfeld, 1973). Aybar *et al.* (2005) found that the wicks increased the freshwater generation by two or three times compared to a bare plate in an inclined solar still. The wick is placed on a frame which is maintained in slope, with the edge dipped in a saline water reservoir. The capillary action of the cloth sucks saline water upwards from the reservoir and after passing the maximum height, the water rolls down the cloth length under gravity. The problem with the single wick solar still is that capillary action feeds only a limited length of the cloth, so depending upon the rate of evaporation from the wet surface the cloth can be partially dry at times. Sodha *et al.* (1981) designed a multiple wick solar still in which all the surface irradiated by the sun was kept wet at all times (Fig. 3.5). By arranging a series of jute cloth pieces in layers of increasing length separated by thin polythene sheets, all the surface irradiated by the sun can be kept wet at all times; since only a small part of each cloth is exposed and the rest is covered by the polythene sheet, there is no evaporation. The concept was modified by Yadav and Tiwari (1989), who fabricated single and double slope multi-wick solar stills from fiber reinforced plastic (FRP).

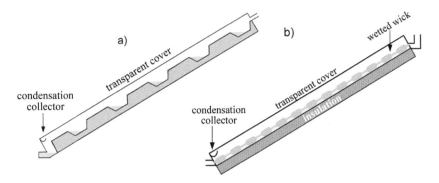

Figure 3.4.   Schematic diagram of an inclined solar still of the cascade-type (a) and the wick-type (b).

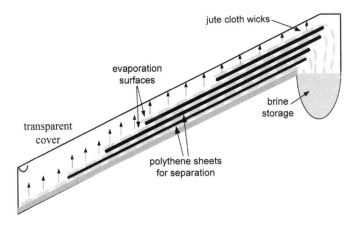

Figure 3.5.   Schematic diagram of the multi-wick solar still designed by Sodha *et al.* (1981), with jute cloth pieces (blue) separated by polythene sheets (black) so that small parts of each cloth are exposed for evaporation.

Even though the glass cover inclination is optimized to receive sun rays, part of them are received by the back and the side plates of the basin. This part of the radiation is not available to the basin water for heating. Therefore, mirrors can be used to reflect the radiation back on to the basin. Experimental results showed that the efficiency was about 15–25% higher for a basin-type still (cover tilt 45°) in Irbid (Jordan) when all its inner sides (both side and back walls) were covered with reflectors (Tamimi, 1987). Also, Al-Hayek and Badran (2004) showed that the productivity of a single-slope still with reflecting mirrors in Amman (Jordan) was 20% higher than that of a double slope still with the same cover angle (35°).

The use of an external reflector can also be a useful modification to increase the radiation into a basin solar still. Numerical analysis by Tanaka and Nakatake (2006) showed that the combination of internal and external reflectors can increase up to 48% the yearly distillate productivity of the single-slope (20°) basin still at 30°N latitude.

3.5.2   *Enhancing the evaporation*

In order to increase the evaporation, the first measure is to increase the evaporation surface. Kwatra (1996) calculated with simulations that a gain of around 20% was obtained in the yield of a basin-type solar still when the evaporation area was quadrupled, with a maximum asymptotic gain of around 30% (infinite area). In ordinary basin stills the surface increase can be achieved by different means. Using sponge cubes in the basin increase the wetted surface area while reducing the surface tension between the water molecules (Abu-Hijleh and Rababa'h, 2003). One of the advantages of the stepped solar still is the increase of the basin area exposed to solar energy and therefore the evaporating surface. A similar increase can be achieved in conventional horizontal basin solar stills by integrating fins at the basin plate. Velmurugan *et al.* (2008) report that using fins gave better result than using sponges and even an inclined wick-type still (productivity increased 45, 15 and 30%, respectively, compared to ordinary basin-type still).

Since there is a linear relationship between the latent heat of evaporation and the surface tension of the liquid mixture, increasing evaporation can be achieved by adding non-volatile surfactants to the water to be distilled (Armenta-Deu, 1997).

3.5.3   *Working in sub-atmospheric conditions*

A considerable improvement in the productivity can be obtained by reducing the operating pressure in the solar still (Yeh *et al.*, 1985). The absence of non-condensable gases, such as air, eliminates the convective heat losses from the water, so its temperature is larger and also the cover is cooler, which enhances the still productivity. According to calculations by Al-Hussaini and Smith (1995), an increase of up to 100% could be achieved in the still productivity by applying vacuum inside.

3.5.4   *Enhancing the heat absorption*

In a conventional still, most of the solar radiation is absorbed by the bottom surface, which becomes the hottest region of the still. Heat is transferred to the water by convection and lost outside by conduction through the insulating layer. Solar still basins are usually painted or lined with black materials over the insulated bottom. To avoid the need of building concrete or fiberglass basins, Madani and Zaki (1995) used soot (carbon particles 40–50 μm size), which also acts as a black porous solar absorber. Badran (2007) found that productivity was enhanced by 29% when using asphalt basin liner instead of black paint. However, Akash *et al.* (1998) obtained less increase in the daily productivity by using an absorbing black rubber mat (38%) than by using black ink-in-water and black dye-in-water solutions (45 and 60% increases, respectively).

When a dye is mixed with the water, almost all the solar radiation is directly absorbed by the water, which transfers part of the heat to the bottom. Hence the water is at a higher temperature and this accounts for the higher distillate output. Garg and Mann (1976) observed that blue and red dyes increased the absorptivity of water and therefore productivity (better the red than the blue).

Rajvanshi (1981) reported an increase of productivity as much as 29% using black naphtylamine dissolved in the water. The efficiency of dyes is larger the deeper the basin (Sodha *et al.*, 1980) and the choice of compounds are limited by their toxicity.

Dissolving certain salts in water enhances its absorptivity of solar radiation by extending the absorption to the visible region of the spectrum. This could result in increased rise in water temperature by more than 10°C (Jamal *et al.*, 1991). However, Nijmeh *et al.* (2005) observed that the addition of potassium dichromate $K_2Cr_2O_7$ (70 mg L^{-1}) and potassium permanganate $KMnO_4$ (50 mg L^{-1}) produced an enhancement in the daily efficiency of a solar still of about 17 and 26%, respectively, less than using violet dye (about 29%).

Sometimes other chemicals are used because of their heat generating photochemical properties. Al-Abbasi *et al.* (1992) used bromine and iodine in glass containers sealed and fixed inside the stills, and obtained an increase in the overall efficiency of the simple conventional basin-type solar stills of 65% with bromine (5.56 mol m^{-2}) and 42% with iodine (4.8 mol m^{-2}).

Charcoal has a high coefficient of absorption and diffuses solar energy but does not reflect it. Egarievwe *et al.* (1991) mixed the feed water with fine particles of charcoal to gain an improvement of 12.2% on the still efficiency. However, health issues jeopardize its use for drinking water production.

The addition of charcoal and coal to the water increases the energy supplied for evaporation by increasing the solar energy absorbed in the water and the basin (Akinsete and Duru, 1979). Okeke *et al.* (1990) evaluated the addition of fine particles of charcoal, charcoal pieces, coal pieces, and a coal and charcoal mixture, obtaining the highest daily efficiency when most of the charcoal pieces introduced into the still floated on the water. Egarievwe *et al.* (1991) obtained improvements in the still efficiency with the addition of charcoal pieces, coal pieces and a mixture of charcoal plus coal pieces of 27.7, 10.8 and 8.1%, respectively. Furthermore, Kumar *et al.* (2008) studied a "V"-type solar still using a layer of charcoal floating on the surface of the water as absorber and estimated an increase of 20% in the overall efficiency of the still.

The use of screens submerged into the basin water complements the enhancement in the absorption caused by the addition of dyes. Abu-Hijleh *et al.* (2001) experimented with screens made of 1 mm diameter galvanized steel wires weaved to form a square mesh. The presence of screens changed the temperature gradient in the basin water (now hotter at the top surface) and suppressed the heat transfer convection currents in the basin, increasing the evaporation rate.

Floating foils can also be used as heat absorbing elements (Szulmayer, 1973). Adding heat absorbers to the surface of the water in a basin solar still concentrates heat at the top layer, preventing the whole water mass from getting heated up by convection by reducing transmission of radiation through the water body to the bottom liner. This way, the top layer of water heats up quicker than the main bulk below, therefore: (i) evaporation starts earlier and (ii) the base of the still is kept cooler, so the heat losses to the ground are reduced. Malik *et al.* (1982) tried floating porous pads of fiber-type material on the brine surface in order to increase the temperature of the evaporation area, while allowing evaporated water particles to escape up into the air through the holes in it. This kind of floating wick solar still had already been constructed at Daytona Beach and the productivity was shown to increase. However, at the long run the porous wick material would be clogged by accumulated salt. A floating porous plate was employed by Dobrevsky and Georgieva (1983), with an increase in the evaporation of 30% at a brine depth of 18 cm. The increase in productivity by using floating absorbers is larger for deeper basins. Nafey *et al.* (2002) investigated the use of a floating perforated black aluminum plate in a single slope solar still for different brine depths and the increase in the productivity was 15% for brine depth of 3 cm and 40% for brine depth of 6 cm. Valsaraj (2002) used a floating perforated and folded aluminum absorber sheet and observed a yield more than 43% higher for brine depth of 9 cm and negligible improvement for depth of 3 cm.

Baffle plates suspended within the basin water are another device used to enhance the performance of solar stills. They are used to divide the basin into a shallow evaporating zone above the plate and a deeper zone below, leaving a small gap between them where water can flow by convection between both volumes (Fig. 3.6). The plate absorbs the energy and transfers it to the

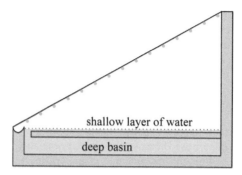

Figure 3.6.   Schematic diagram of a basin solar still with a baffle plate to separate the basin into a shallow
evaporative zone above and a deeper zone below.

upper and lower water volumes. It reduces the preheating time of the upper water and also acts
as an insulation cover for the lower water. This way, the advantages of both the shallow and the
deep basin stills are incorporated, improving the evaporation while enabling the system to store
excess heat energy in the lower volume for the continuation of desalination during off-sunshine
periods. Using an aluminum sheet painted black at the top and thermally insulated at the bottom, Rahim (2003) obtained an average yield about 30% higher than a shallow basin solar still
in the same conditions. El Sebaii *et al.* (2000) experimented with movable suspended plates in
solar stills, obtaining better performances with plates without vents and shallow depths of the
upper water volume. For metallic plates (aluminum, copper and stainless steel), the daily productivity was raised about 15–20% compared to conventional stills. Plates made of insulating,
non-corrosive materials are cheaper, and using mica the increase of daily productivity was even
higher (about 42%).

3.5.5   *Storing the incident solar energy*

The productivity of a solar still can be enhanced by the use of an energy storage medium integrated
into the still, like absorber materials that can also store and release the energy during off-sunshine
periods to enhance the productivity. Amongst these materials, black rubber or black gravel have
been investigated. Nafey *et al.* (2001) observed within a single sloped basin solar still that black
gravel absorbs and releases the incident solar energy faster than black rubber, and measured
increases in productivity of around 20% (for black rubber 10 mm thick and black gravel 20–
30 mm size). Abdallah *et al.* (2009) found that black volcanic rocks were a better absorber and
heat storage material than black coated and uncoated porous media (metallic wiry sponges made
from steel), giving around 60% gain in water yield with no corrosion problems.

   Phase change and sensible heat storage materials have also been used as heat storage. Naim
and El Kawi (2002b) used a special formulation consisting of emulsion of paraffin wax, paraffin
oil and water to which aluminum turnings were dispersed to promote heat transfer.

   Also, a sandy heat reservoir can be integrated beneath the basin liner and above the insulation
to enhance the overnight productivity of the still with 12% of the production taking place at night
(Tabrizi and Sharak, 2010).

3.5.6   *Reducing the depth of water in the basin*

As mentioned above, the productivity of the still depends on the depth of water in the basin under
normal radiation condition. The productivity of the deep basin still is limited by the presence
of the large mass of the water at the basin (Bloemer *et al.*, 1965). A reduction in the depth of
water improves the productivity due to the higher basin temperature, since the rise in the water

temperature, and hence the rate of evaporation, is inversely proportional to the mass of the water for a fixed amount of heat input (Tiwari and Tiwari, 2006). Khalifa and Hamood (2009) reviewed all the studies of the effect of brine depth on the performance of basin-type solar stills, developing and validating a correlation which showed a decreasing trend in the productivity with the increase in the brine depth. They concluded that the still productivity could be influenced by the brine depth by up to 48%.

For maintaining a thin layer of water basin, it is necessary to spread the water throughout the basin by some kinds of evaporating cloth (wick) or porous materials. Murugavel *et al.* (2008) compared the performance of a single basin double slope solar still with a layer of water approximately 2 mm depth sustained by using different wick materials like light cotton cloth, light jute cloth and sponge sheet, as well as porous materials such as washed natural rock and quartzite rock as spread materials. The results show that the still with black light cotton cloth as spread material was more productive. Naim and El Kawi (2002a) used charcoal both as a heat absorber medium and as a wick in an inclined flat solar still. A charcoal bed of particles extending the length of the still through which salt water is allowed to percolate presented an improvement in productivity over using a jute cloth of up to 15%.

Keeping the wick cloth wet entirely is complicated when the evaporation rates are very high. Al-Karaghouli and Minasian (1995) devised a floating wick consisting of a blackened jute wick floating on a 2 cm polystyrene sheet, arranged not to appear more than 0.5 cm above the water level in the still. Water suction by the capillary action of the jute fibres ensured a sufficient water flux.

On tilted-wick-type solar stills the feed water is spread as a falling film on an inclined plate. Moustafa *et al.* (1979) found that the overall efficiency for the wick-type solar still was about 60%, almost double than the basin-type solar still. Conducting an experimental study on a tilted-wick-type solar still, Tanaka *et al.* (1982) found an increase in distillate output of 20–50% against basin-types.

Unfortunately, the saline water depth in conventional basin and even in inclined stills can be only reduced to a certain limiting value. El-Zahaby *et al.* (2011) managed to control the water depth as desired by using spray feeding of the saline water to create a thin film of saline water on the absorber surface of a stepped still.

Minasian and Al-Karaghouli (1995) connected a conventional basin-type still (installed in a shadow and having an opaque cover) with a wick-type solar still in order to extract the energy contained in the lost hot drainage water, which after leaving the wick-type fed directly into the basin-type, with the basin still cover cooled. The combined stills showed higher efficiency than the two stills separately (yearly production was 85% more than the basin-type and 43% more than the wick-type).

3.5.7   *Reducing the temperature of the cover*

Cooling the cover produces also an increase in the performance of the still. Indications are that there exists an optimum cover temperature for which the yield will be a maximum (Satcunanathan and Hansen, 1973).

By shading the cover, its temperature can be decreased. An intermittent shading of the north glass cover in a simple basin solar still resulted in a 12% enhancement in the daily distillate output (Bechki *et al.*, 2010).

The most common way to reduce the cover temperature is by flowing water over the cover. The results are reduced convection and radiative energy losses to the ambient, as well as increased condensation rate on the inside of the glass cover. Theoretically, it could almost double the daily distillate production of a single slope solar still (Tiwari and Rao, 1984). Lawrence *et al.* (1990) conducted numerical simulations validated with their own experiments for a typical summer day and their results showed that the efficiency of the solar still increased as the water film flow rate increased, the effect being more significant at large heat capacity of water mass in the basin. According to numerical investigations by Abu-Hijleh and Mousa (1997), proper use of the water

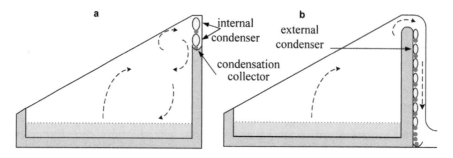

Figure 3.7.   Schematic diagram of a basin solar still with an internal (a) and an external condenser (b).

film cooling parameters may increase the efficiency of a single-basin still by up to 6%, reaching up to 20% when accounting for evaporation from the film. The regenerative solar still springs up from recovering this further evaporation. The latent heat of condensation is carried out from the cover by the cooling water and used to produce more desalinated water in a second effect. Simulations by Zurigat and Abu-Arabi (2004) showed that the productivity of the regenerative still was 20% higher than that of the conventional still.

3.5.8   *Separating evaporating and condensing zones*

Separating the evaporation and the condensation in different zones allows the temperature difference between both processes (main contributor to increasing the productivity of the system) to be increased during the day.

Internal condensers have been used for long time as an attempt to improve the performance of the basin-type solar still (Fig. 3.7a). Delyannis and Piperoglou (1965) reported an improvement of the productivity of the still up to 24% resulting from the addition of the cooler to the deep basin still. Using an internal condenser (copper tube bent to form a double-pass heat exchanger with circulation of cooling water inside) improved daily efficiency of a single basin solar still by about 10% (Ahmed, 1988). As water is used to cool the condenser, the latent heat of condensation can be recovered by preheating the water of the basin. Khalifa *et al.* (1999) observed a 34% increase in the efficiency of a single slope solar still with the addition of an eight pass internal condenser with recovery of the latent heat of condensation. They claimed that an increase in performance resulted also from the increase in the convection currents inside the still due to the local relative vacuum caused by faster condensation on the water-cooled internal condensers.

In the forced condensing technique, the water vapor is extracted from the still and forced to condense in an external condenser held at lower temperature (Fig. 3.7b). Forced convection stills were already reported by Löf (1961) and their efficiency can be very high (Rahim, 1995). However, forced condensation, either with internal or external condensers, requires additional energy input to the solar still, either pumps or fans, retracting it from the passive concept of the solar still that is the subject of this chapter.

Enhancing of condensation can also be achieved in a passive way. In a double slope solar still, if one side of the cover receives the sun rays close to normal the other side is in shadow, allowing for the use of passively cooled condensers on the roof (Fatani *et al.*, 1994). El Bassouni (1993) used a finned galvanized steel plate in the unexposed side of the cover of a double sloped basin solar still to enhance the condensing surface, as well as open glass tubes crossing the basin so that the current of outside cooler air circulating inside them helped condensing the vapor accumulated in the still.

Passive condensers can also be added in the shaded region of a single-slope solar still, acting as a heat and mass sink which continuously sucks water vapor from the still, condensing it and maintaining the still at low pressure. Theoretical studies showed that the mass transfer of water vapor was better by natural circulation (density difference of air between still and condenser) than by purging (relative pressure difference) and diffusion (difference in water vapor concentration). Experiments showed a 70% increase in the yield for the first case (Fath and Elsherbiny, 1993).

Another way of separating the condensation and the evaporation is in the inverted trickle solar still (Badran *et al.*, 2004). The devise is based on the flow of a thin layer of water (trickle) on the back of an absorber plate by means of a wire screen welded to the plate. The layer was maintained attached to the plate by the surface tension forces, and as water evaporated, vapor was transferred to another compartment to be condensed. By eliminating condensation on the cover, solar radiation reaching the absorber plate increased and so did the temperature difference between evaporator and condenser.

3.5.9   *Reusing the latent heat of condensation in two or more stages*

By reusing the latent heat of condensation for further evaporation, multiple-effect stills have greater productivity than conventional single-effect stills such as basin-type still and tilted-wick-still.

Progressing from the solar still with internal condenser, Fath (1996) proposed a two-effect basin-type still in which vapor from the first effect was partially purged to a second effect, where it condensed releasing its heat to the basin water of the second effect for additional evaporation. Purging the vapor maintained the first effect still at relatively low pressure, enhancing evaporation. In addition, the evaporative and convective heat load on the first effect glass cover was reduced, and therefore its temperature decreased.

The simplest two-effect solar still is the double basin solar still, in which the upper basin has a transparent floor and reuses the latent heat of condensation from the lower basin. Single slope (Sodha *et al.*, 1980) and double slope (Kumar *et al.*, 1991) double-basin solar stills have been studied. The daily average still production for the double-basin still is around 40% higher than the production of the single-basin still (Al-Karaghouli and Alnaser, 2004).

A similar concept can be developed by using a wick-type solar still as the lower basin. Singh and Tiwari (1992) proposed this double-effect solar still by adding a further transparent surface on top of a two-slope multi-wick solar still to recover the latent heat of condensation with further evaporation of running water. Similarly, Yeh and Ma (1990) added between the absorbing plate and the cover of a tilted wick solar still a further transparent glass plate with water running slowly through a system of weirs.

Multiple basins solar stills are limited by the need of transmitting solar radiation through all the stages before reaching the absorber, so the number of effects is limited. Al Mahdi (1992) found the optimum number of basins to be three. This number can be extended by adding heat from below. This must be done indirectly, either from an external heat source, in which case the solar still is active and therefore escapes the subject of this chapter, or by using an inverted absorber solar still, where the solar radiation is reflected to be absorbed at the bottom of the still, allowing for more effects on the top (Suneja and Tiwari, 1999).

Direct use of solar energy in multi-effect solar stills can be done by keeping the condensing surface below the absorber to create a downward diffusion of thermal energy. Selçuk (1964) designed a multiple-effect tilted solar still with direct solar heating of an absorber plate with a glass cover on top (Fig. 3.8). Below the collecting surface and in contact with it, a series of cascading evaporator troughs like in a stepped solar still constituted the first effect. The sensible heat transmitted trough the absorber evaporated the water and the vapor condensed on the underlying plate, which transmitted the latent heat to the second series of troughs below for a second distillation effect and so on. The last underlying plate (condenser plate of the last effect) was cooled by ambient air. Cooper and Appleyard (1967) constructed a similar device but in this case soaked wicks were used in contact with the bottom surfaces of the solar absorber plate and parallel partitions. Solar energy absorbed by the plate caused evaporation of saline water in the

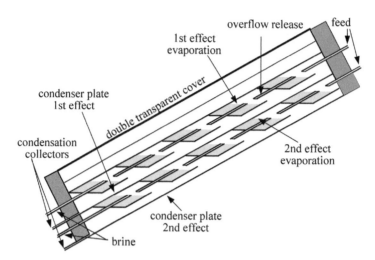

Figure 3.8.   Schematic diagram of a multiple-effect tilted solar still with stepped basins as designed by Selçuk (1966).

brine-soaked wick attached to its bottom surface. Evaporated water vapor was diffused across the gap downwards and condensed on the uncovered top surface of the facing partition. Latent heat of condensation contributed to further evaporation of saline water in the wick attached underneath the partition. The process continued until the last partition, which could be cooled by water or air. According to the authors, the spongy fabric used as wick adhered poorly to the overhanging plate. Ouahes et al. (1987) used a very thin fabric comprising a single, finely woven layer. This fabric was held in contact with the overhanging plate through the interfacial tension, much greater than the force due to gravity, and a capillary film formed at the plate-fabric interface.

Yeh and Chen (1987) constructed a very simple version of the multi-stage wick solar still with only two stages, while Toyama et al. (1987) simulated a laboratory test plant of five stages and a field test plant of ten stages. Much simpler was the double-effect achieved by Tiwari et al. (1984) in a multi-wick solar still by adding a further condensing surface (a galvanized iron sheet) below the blackened wet jute cloth with a slight spacing around the absorbing surface, such that the excess of vapor can also be condensed below the sheet.

When the gap between the evaporating water layer and the condensing surface is reduced so much that heat transfer by convection is suppressed, we talk about diffusion solar stills. Multi-effect diffusion stills have been modeled and tested experimentally (Elsayed et al., 1984). Distillate productivity can be greatly increased by reducing the diffusion gaps between partitions (Ohshiro et al., 1996). However, the small gaps may cause contamination of condensate with saline water or re-absorption of condensate by the evaporating wick, since the deformation of partitions due to gravity and thermal expansion would cause contact between the evaporating wicks and the condensing surfaces. The vertical multiple-effect diffusion still (Dunkle, 1961), in which the partitions are vertically arranged, can decrease the deformation due to gravity and reduce the diffusion gaps between partitions compared to the other types of multiple-effect diffusion stills in which the partitions are inclined. However, the vertical still has a disadvantage in that additional equipment may be required to absorb the solar energy effectively on the first partition. Dunkle (1961) and Kiatsiriroat et al. (1987) proposed a vertical still coupled with a flat plate solar collector. Tanaka et al. (2002) proposed to couple it with a one-slope basin-type still acting as a condensing vertical wall. The basin-type distillation section performed distillation and supplied energy (i.e., the latent heat of condensation) to the multiple-effect diffusion still section. A much simpler solution was proposed by Tanaka and Nakatake (2005) with a newly designed vertical multi-effect diffusion

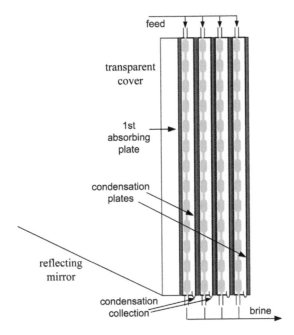

Figure 3.9.   Schematic diagram of a vertical multiple-effect diffusion still with a reflecting mirror as designed by Tanaka and Nakatake (2005).

still coupled with a flat plate reflector (Fig. 3.9). The angle of the fiat-plate mirror can be adjusted to absorb solar radiation on the first partition effectively.

REFERENCES

Abdallah, S. & Badran, O.O.: Sun tracking system for productivity enhancement of solar still. *Desalination* 220 (2008), pp. 669–676.

Abdallah, S., Badran, O. & Abu-Khader, M.M.: Performance evaluation of a modified design of a single slope solar still. *Desalination* 219 (2008), pp. 222–230.

Abdallah, S., Abu-Khader, M.M. & Badran, O.: Effect of various absorbing materials on the thermal performance of solar stills. *Desalination* 242 (2009), pp. 128–137.

Abu-Hijleh, B.A.K., Abu-Qudias, M. & Al-Khateeb, S.: Experimental study of a solar still with screens in basin. *Int. J. Solar Energy* 21 (2001), pp. 257–266.

Abu-Hijleh, B.A.K. & Mousa, H.A.: Water film cooling over the glass cover of a solar still including evaporation effects. *Energy* 22:1 (1997), pp. 43–48.

Abu-Hijleh, B.A.K. & Rababa'h, H.M.: Experimental study of a solar still with sponge cubes in basin. *Energy Convers. Manage.* 44 (2003), pp. 1411–1418.

ADIRA (Autonomous Desalination system concepts for seawater and brackish water In Rural Areas with Renewable Energies – Potential, Technologies, Field Experience, Socio-Technical and Socio-Economic Impacts). Internet site: www.adira.gr (accessed October 2011).

Ahmed, S.T.: Study of single-effect solar still with an internal condenser. *Solar Wind Technol.* 5:6 (1988), pp. 637–643.

Akash, B.A., Mohsen, M.S., Osta, O. & Elayan, V.: Experimental evaluation of a single-basin solar still using different absorbing materials. *Renew. Energy* 14: 1–4 (1998), pp. 307–310.

Akinsete, V.A. & Duru, C.U.: A cheap method of improving the performance of roof type solar stills. *Solar Energy* 23 (1979), pp. 271–272.

Al-Abbasi, M.A., Al-Karaghouli, A.A. & Minasian, A.M.: Photochemically assisted solar desalination of saline water. *Desalination* 86 (1992), pp. 317–324.

Al-Hayek, I. & Badran, O.O.: The effect of using different designs of solar stills on water distillation. *Desalination* 169 (2004), pp. 121–127.

Al-Hussaini, H. & Smith, I.K.: Enhancing of solar still productivity using vacuum technology. *Energy Convers. Manage.* 36:11 (1995), pp. 1047–1051.

Al-Karaghouli, A.A. & Alnaser, W.E.: Performances of single and double basin solar-stills. *Appl. Energy* 78 (2004), pp. 347–354.

Al-Karaghouli, A.A. & Minasian, A.N.: A floating-wick type solar still. *Renew. Energy* 6:1 (1995), pp. 77–79.

Al Mahdi, N.: Performance prediction of a multi-basin solar still. *Energy* 17:1 (1992), pp. 87–93.

Armenta-Deu, C.: Increasing the evaporation rate for freshwater production – application to energy saving in renewable energy sources. *Renew. Energy* 11:2 (1997), pp. 197–209.

Aybar, H., Egeliofglu, F. & Atikol, U.: An experimental study on an inclined solar water distillation system. *Desalination* 180 (2005), pp. 285–289.

Badran, A.A., Assaf, L.M., Kayed, K.S., Ghaith, F.A. & Hammash, M.I.: Simulation and experimental study for an inverted trickle solar still. *Desalination* 164 (2004), pp. 77–85.

Badran, O.O.: Experimental study of the enhancement parameters on a single slope solar still productivity. *Desalination* 209 (2007), pp. 136–143.

Barasoain, J.A. & Fontan, L.: First experiment in the solar distillation of water in Spain. *Revista de Ciencia Aplicada* 72 (1960), pp. 7–17.

Barrera, E.: A technical and economical analysis of a solar water still in Mexico. *Renew. Energy* 2:4/5 (1992), pp. 489–495.

Baum, V.A. & Bairamov, R.: Prospects of solar stills in Turkmenia. *Solar Energy* 10:1 (1966), pp. 38–40.

Bechki, D., Bouguettaia, H., Blanco-Galvez, J., Babay, S., Bouchekima, B., Boughali, S. & Mahcene, H.: Effect of partial intermittent shading on the performance of a simple basin solar still in south Algeria. *Desalination* 260 (2010), pp. 65–69.

Bloemer, J.W., Eibling, J.A., Irwin, J.R. & Löf, G.O.G.: A practical basin-type solar still. *Solar Energy* 9: 4 (1965), pp. 197–200.

Chen, Z., Ge X., Sun, X., Bar, L. & Miao Y.X.: Natural convection heat transfer across air layers at various angles of inclination. *Eng. Thermophy.* (1984), pp. 211–220.

Clark, J.A.: The steady-state performance of a solar still. *Solar Energy* 44:1 (1990), pp. 43–49.

Cooper, P.I.: The maximum efficiency of single-effect solar stills. *Solar Energy* 15 (1973), pp. 205–217.

Cooper, P.I. & Appleyard, J.A.: The construction and performance of a three effect wick type, tilted solar still. *Sun at Work* 12:1 (1967), pp. 4–8.

Coulson, J.M. & Richarson, J.F.: *Chemical Engineering*. 3rd ed., Pergamon Press, New York, 1977.

Delyannis, A.: The Patmos solar distillation plant. *Solar Energy* 11 (1968), pp. 113–115.

Delyannis, A.A. & Delyannis, E.: Solar desalination. *Desalination* 50 (1984), pp. 71–81.

Delyannis, A. & Piperoglou, E.: Solar distillation in Greece. *1st Int. Symp. Water Desalination*, Washington, DC, 1965.

Delyannis, E.: Historic background of desalination and renewable energies. *Solar Energy* 75 (2003), pp. 357–366.

Dobrevsky, I. & Georgieva, M. Some possibilities for solar evaporation intensification. *Desalination* 45 (1983), pp. 93–99.

Dunkle, R.V.: Solar water distillation: the roof type still and a multiple effect diffusion still, international development in heat transfer. In: *ASME Proceedings. International Heat Transfer Conference, Part V*, University of Colorado, Boulder, CO, 1961, pp. 895–902.

Eckstrom, R.M.: Design and construction of the Symi still. *Sun at Work* 10:1 (1965), p. 7.

Egarievwe, S.U., Animalu, A.O.E. & Okeke, C.E.: Harmattan performance of a solar still in the Guinea Savannah. *Renew. Energy* 1:5-6 (1991), pp. 799–801.

Eibling, J.A., Talbert, S.G. & Loef, G.O.G.: Solar stills for community use – digest of technology. *Solar Energy* 13 (1971), pp. 263–276.

El-Bassouni, A-M.A.: Factors influencing the performance of basin-type solar desalination units. *Desalination* 93 (1993), pp. 625–632.

El Sebaii, A.A., Aboul-Enein, S., Ramadan, M.R.I. & El-Bialy, E.: Year-round performance of a modified single-basin solar still with mica plate as a suspended absorber. *Energy* 25 (2000), pp. 35–49.

El-Zahaby, A.M., Kabeel, A.E., Bakry, A.I., El-Agouz, S.A. & Hawam, O.M.: Enhancement of solar still performance using a reciprocating spray feeding system—An experimental approach. *Desalination* 267 (2011), pp. 209–216.

Elsayed, M.M., Fathalah, K.K., Shams, J. & Sabbagh, J.: Performance of multiple effect diffusion still. *Desalination* 51 (1984), pp. 183–199.

Fatani, A.A., Zaki, G.G. & Al-Turki, A.: Improving the yield of simple basin solar stills as assisted by passively cooled condensers. *Renew. Energy* 4:4 (1994), pp. 377–386.

Fath, H.E.S.: High performance of a simple design, two effect solar distillation unit. *Desalination* 107 (1996), pp. 223–233.

Fath, H.E.S. & Elsherbiny, S.M.: Effect of adding a passive condenser on solar still performance. *Energy Convers. Manage.* 34:1 (1993), pp. 63–72.

Frick, G. & von Sommerfeld, J.: Solar stills of inclined evaporating cloth. *Solar Energy* 14 (1973), pp. 427–431.

Garg, H.P. & Mann, H.S.: Effect of climatic, operational and design parameters on the year round performance of single-sloped and double-sloped solar still under Indian arid zone condition. *Solar Energy* 18 (1976), pp. 159–164.

Hanson, A., Zachritz, W., Stevens, K., Mimbela, L., Polka, R. & Cisneros, L.: Distillate water quality of a single-basin solar still: laboratory and field studies. *Solar Energy* 76 (2004), pp. 635–645.

Headley, O.St.C.: Cascade solar still for distilled water production. *Solar Energy* 15 (1973), pp. 245–258.

Hirschman, J.R. & Roffler, S.K.: Thermal inertia of solar stills and its influence on performance. Paper 5/26, *International SES Conference*, Melbourne, Australia, 1970.

Howe, E.D. & Tleimat, B.W.: Twenty years of work on solar distillation at the University of California. *Solar Energy* 16 (1974), pp. 97–105.

Jakob, M.: *Heat Transfer*, Vol. 2. Wiley, New York, NY, 1957.

Jamal, M., Junaidi, T.& Muaddi, J.: A step forward towards an ideal absorber for solar energy. *Int. J. Energy Res.* 15:5 (1991), pp. 367–375.

Kaushal, V.A.: Solar stills: A review. *Renewable and Sustainable Energy Reviews* 14 (2010), pp. 446–453.

Khalifa, A.J.N.: On the effect of cover tilt angle of the simple solar still on its productivity in different seasons and latitudes. *Energy Convers. Manage.* 52 (2011), pp. 431–436.

Khalifa, A.J.N. & Hamood, A.M.: On the verification of the effect of water depth on the performance of basin type solar stills. *Solar Energy* 83 (2009), pp. 1312–1321.

Khalifa, A.J.N., Al-Jubouri, A.S. & Abed, M.K.: An experimental study on modified simple solar stills. *Energy Convers. Manage.* 40 (1999), pp. 1835–1847.

Kiatsiriroat, T., Bhattacharya. S.C. & Wibulswas, P.: Performance analysis of multiple effect vertical still with a flat plate solar collector. *Solar & Wind Technology* 4:4 (1987), pp. 451–457.

Kumar, A., Anand, J.D. & Tiwari, G.N.: Transient analysis of a double slope-double basin solar distiller. *Energy Convers. Manage.* 31:2 (1991), pp. 129–139.

Kumar, B.S., Kumar, S. & Jayaprakash, R.: Performance analysis of a "V" type solar still using a charcoal absorber and a boosting mirror. *Desalination* 229 (2008), pp. 217–230.

Kumar, S. & Tiwari, G.N.: Estimation of convective mass transfer in solar distillation systems. *Solar Energy* 57 (1996), pp. 459–464.

Kwatra, H.S.: Performance of a solar still: predicted effect of enhanced evaporation area on yield and evaporation temperature. *Solar Energy* 56:3 (1996), pp. 261–266.

Lawrence, S.A., Gupta, S.P. & Tiwari, G.N.: Effect of heat capacity on the performance of solar still with water flow over the glass cover. *Energy Convers. Manage.* 30:3 (1990) pp. 277–285.

Löf, G.O.G.: Fundamental problems in solar distillation. *Proc. Symposium on Research Frontiers in Solar Energy Utilization* 47, National Academy of Sciences, Washington, DC, 1961, pp. 1279–1290.

Madani, A.A. & Zaki, G.M.: Yield of solar stills with porous basins. *Appl. Energy* 52 (1995), pp. 273–281.

Malik, M.A.S., Tiwari, N., Kumar, A. & Sodha, M.S.: Active and passive solar distillation: a review. In: *Solar distillation*, Pergamon Press, UK, 1982.

Martens, C.P.: Theoretical determination of the flux entering solar stills. *Solar Energy* 10:2 (1966), pp. 77–80.

Minasian, A.N. & Al-Karaghouli, A.A.: An improved solar still: the wick-basin type. *Energy Convers. Manage.* 36:3 (1995), pp. 213–217.

Moustafa, S.M.A., Brusewitz, G.H. & Farmer, D.M.: Direct use of solar energy for water desalination. *Solar Energy* 22 (1979), pp. 22–141.

Murugavel, K.K, Chockalingam, Kn.K.S.K. & Srithar, K.: Progresses in improving the effectiveness of the single basin passive solar still. *Desalination* 220 (2008), pp. 677–686.

Nafey, A.S., Abdelkader, M., Abdelmotalip, A. & Mabrouk, A.A.: Solar still productivity enhancement. *Energy Convers. Manage.* 42 (2001), pp. 1401–1408.

Nafey, A.S., Abdelkader, M., Abdelmotalip, A. & Mabrouk, A.A.: Enhancement of solar still productivity using floating perforated black plate. *Energy Convers. Manage.* 43 (2002), pp. 937–946.

Naim, M.M. & El Kawi, M.A.A.: Non-conventional solar stills. Part 1. Non-conventional solar stills with charcoal particles as absorber medium. *Desalination* 153 (2002a), pp. 55–64.

Naim, M.M. & El Kawi, M.A.A.: Non-conventional solar stills. Part 2. Non-conventional solar stills with energy storage element. *Desalination* 153 (2002b), pp. 71–82.

Nijmeh, S., Odeh, S. & Akash, B.: Experimental and theoretical study of a single-basin solar still in Jordan. *International Communications in Heat and Mass Transfer* 32 (2005), pp. 565–572.

Ohshiro, K., Nosoko, T. & Nagata, T.: A compact solar still utilizing hydrophobic poly(tetrafluoroethylene) nets for separating neighboring wicks. *Desalination* 105 (1996), pp. 207–217.

Okeke, C.E., Egarievwe, S.U. & Animalu, A.O.E.: Effects of coal and charcoal on solar-still performance. *Energy* 15:11 (1990), pp. 1071–1073.

Ouahes, R., Ouahes, C., Le Goff, P. & Le Goff, J.: A hardy, high-yield solar distiller of brackish water. *Desalination* 67 (1987), pp. 43–52.

Radhwan, A.M.: Transient analysis of a stepped solar still for heating and humidifying greenhouses. *Desalination* 161 (2004), pp. 89–97.

Rahim, N.H.A.: Utilization of a forced condensing technique in a moving film inclined solar desalination still. *Desalination* 101 (1995), pp. 255–262.

Rahim, N.H.A.: New method to store heat energy in horizontal solar desalination still. *Renew. Energy* 28 (2003), pp. 419–433.

Rajvanshi, A.K.: Effect of various dyes on solar distillation. *Solar Energy* 27 (1981), pp. 51–65.

Sadineni, S.B., Hurt, R., Halford, C.K. & Boehm, R.F.: Theory and experimental investigation of a weir-type inclined solar still. *Energy* 33 (2008), pp. 71–80.

Satcunanathan, S. & Hansen, H.P.: An investigation of some of the parameters involved in solar distillation. *Solar Energy* 14 (1973), pp. 353–363.

Selçuk, M.K.: Design and performance evaluation of a multiple-effect, tilted solar distillation unit. *Solar Energy* 8:1 (1964), pp. 23–30.

Sharpley, B.F. & Boelter, L.M.K.: Evaporation of water into quiet air from a one-foot diameter surface. *Ind. Eng. Chem.* 30 (1938), pp. 1125–1131.

Shawaqfeh, A.T. & Farid, M.M.: New development in the theory of heat and mass transfer in solar stills. *Solar Energy* 55:6 (1995), pp. 527–535.

Singh, A.K. & Tiwari, G.N.: Performance study of double effect distillation in a multiwick solar still. *Energy Convers. Manage.* 33:3 (1992), pp. 207–214.

Sodha, M.S., Kumar, A. & Tiwari, G.N.: Effects of dye on the performance of a solar still. *Appl. Energy* 7 (1980), pp. 147–162.

Sodha, M.S., Kumar, A., Tiwari, G.N. & Tyagi, R.C.: Simple multiple wick solar still: analysis and performance. *Solar Energy* 26 (1981), pp. 127–131.

Suneja, S. & Tiwari, G.N.: Parametric study of an inverted absorber triple effect solar still. *Energy Convers. Manage.* 40 (1999), pp. 1871–1884.

Szulmayer, W.: Solar stills with low thermal inertia. *Solar Energy* 14 (1973), pp. 415–421.

Tabrizi, F.F. & Sharak, A.Z.: Experimental study of an integrated basin solar still with a sandy heat reservoir. *Desalination* 253 (2010), pp. 195–199.

Tabrizi, F.F., Dashtban, M., Moghaddam, H. & Razzaghi, K.: Effect of water flow rate on internal heat and mass transfer and daily productivity of a weir-type cascade solar still. *Desalination* 260 (2010), pp. 239–247.

Talbert, S.G., Eibling J.A. & Loef G.O.G.: *Manual on solar desalination of saline water*. US office of saline water research and development progress report 546, US Dept. of the Interior, USA, 1970.

Tamimi, A.: Performance of a solar still with reflectors and black dye. *Solar & Wind Technology* 4:4 (1987), pp. 443–446.

Tanaka, H. & Nakatake, Y.: A simple and highly productive solar still: a vertical multiple-effect diffusion-type solar still coupled with a flat-plate mirror. *Desalination* 173 (2005), pp. 287–300.

Tanaka, H. & Nakatake, Y.: Theoretical analysis of a basin type solar still with internal and external reflectors. *Desalination* 197 (2006), pp. 205–216.

Tanaka, H., Nosoko, T. & Nagata, T.: Experimental study of basin-type, multiple-effect, diffusion coupled solar still. *Desalination* 150 (2002), pp. 131–144.

Tanaka, K., Yamashita, A. & Watanabe, K.: Experimental and analytical study of the tilted wick type solar still. In: *Solar World Forum*, Vol. 2. Pergamon Press, Oxford, UK, 1982.

Telkes, M.: *Solar distiller for life raft*. Office of Technical Services Research and Development Report 5225, PB 21120, Washington, Development of Commerce, DC, 1945.

Telkes, M.: Freshwater from seawater by solar distillation. *Industrial and Engineering Chemistry* 45:5 (1953), pp. 1080–1114.

Telkes, M.: Improved solar stills, In: *Transactions of the Conference on the Use of Solar Energy*, Tucson, AZ, 1955, pp. 145–153.

Tiwari, A.K. & Tiwari, G.N.: Effect of water depths on heat and mass transfer in a passive solar still: in summer climatic condition. *Desalination* 195 (2006), pp. 78–94.

Tiwari, G.N. & Rao, V.S.V.B.: Transient performance of a single basin solar still with water flowing over the glass cover. *Desalination* 49 (1984), pp. 231–241.

Tiwari, G.N. & Tiwari, A.K.: *Solar distillation practice for water desalination systems*. Anshan,Tunbridge Wells, UK, 2008.

Tiwari, G.N. & Yadav, Y.P.: Comparative designs and long term performance of various designs of solar distillery. *Energy Convers. Manage.* 27: 3 (1987), pp. 327–333.

Tiwari, G.N., Sharma, S.B. & Sodha, M.S.: Performance of a double condensing multiple wick solar still. *Energy Convers. Manage.* 24:2 (1984), pp. 155–159.

Tleimat, B.W. & Howe, E.D.: Nocturnal production of solar distillers. *Solar Energy* 10:2 (1966), pp. 61–66.

Tleimat, B.W. & Howe, E.D.: Comparison of plastic and glass condensing covers for solar distillers. *Solar Energy* 12 (1969), pp. 293–304.

Toyama, S., Aragaki, T., Salah, H.M. & Murase, K.: Dynamic characteristics of a multistage thermal diffusion type solar distillator. *Desalination* 67 (1987), pp. 21–32.

Tsilingiris, P.T.: The influence of binary mixture thermophysical properties in the analysis of heat and mass transfer processes in solar distillation systems. *Solar Energy* 81 (2007), pp. 1482–1491.

Valsaraj, P.: An experimental study on solar distillation in a single slope basin still by surface heating the water mass. *Renew. Energy* 25 (2002), pp. 607–612.

Velmurugan, V., Gopalakrishnan, M., Raghu, R. & Srithar, K.: Single basin solar still with fin for enhancing productivity. *Energy Conversion and Management* 49 (2008), pp. 2602–2608.

Voropoulos, K., Mathioulakis, E. & Belessiotis, V.: Transport phenomena and dynamic modelling in greenhouse-type solar stills. *Desalination* 129 (2000), pp. 273–281.

Ward, J.: A plastic solar water purifier with high output. *Solar Energy* 75 (2003), pp. 433–437.

Wilson, B.W.: Solar distillation research and its application in Australia. In: *Proceedings of symposium on saline water conversion*, Washington, DC, 1957, No. 568, pp. 123–130.

Yadav, Y.P. & Tiwari, G.N.: Demonstration plant of FRP multiwick solar still: an experimental study. *Solar Wind Technol.* 7 (1989), pp. 653–666.

Yeh, H.M. & Chen, L.C.: Experimental studies on double-effect solar distillers. *Energy* 12:12 (1987), pp. 1251–1256.

Yeh, H.M. & Ma, N.T.: Energy balances for upward-type, double-effect solar stills. *Energy* 15:12 (1990), pp. 1161–1169.

Yeh, H.M., Ten, L.W. & Chen, L.C.: Basin-type solar distillers with operating pressure reduced for improved performance. *Energy* 10:6 (1985), pp. 683–688.

Zaragoza del Águila, G., Agüera Zurano, J.M., López, J.C. & Pérez-Parra, J.: Tecnologías para la desalación de agua: Funcionamiento y caracterización de una desaladora solar pasiva con cubierta de plástico. *Horticultura* 185 (2005), pp. 24–29.

Zheng, H., Zhang, X., Zhang, J. & Wu, Y.: A group of improved heat and mass transfer correlations in solar stills. *Energy Convers. Manage.* 43 (2002), pp. 2469–2478.

Zurigat, Y.H. & Abu-Arabi, M.K.: Modelling and performance analysis of a regenerative solar desalination unit. *Appl. Therm. Eng.* 24 (2004), pp. 1061–1072.

CHAPTER 4

Solar desalination with humidification-dehumidification process: design and analysis

Habib Ben Bacha

> *"When the well's dry, we know the worth of water."*
> Benjamin Franklin (1706–1790), Poor Richard's Almanac, 1746

> *"Water is the most critical resource issue of our lifetime and our children's lifetime. The health of our waters is the principal measure of how we live on the land."*
> Luna Leopold

4.1 INTRODUCTION

Drinking water is a more and more scarce resource. The increment of population, environmental impacts and climate change are reducing the water availability per person. Unfortunately, about 97% of the earth's water is seawater, and only 3% is freshwater (glaciers and polar ice caps 2.4%, and other land surface water such as aquifers, rivers and lakes 0.6%). Bar charts presented in Figure 4.1 detail the distribution of earth's water. To cope with this issue and to respond to increasing water demand, desalination of brackish water and seawater is being considered. However, the energy requirements for that process are high and can be a problem, mainly in isolated areas, where connection to the public electricity grid is either not cost effective or not feasible. Renewable energies such as solar energy are the best way to supply the energy needs, because of their availability near the desalination plants and avoiding environmental impacts and availability problems associated with fossil fuels. Furthermore, increase in energy consumption makes the cost of renewable solar energy highly competitive against fossil fuels. Solar energy has been receiving greater attention in recent years for various applications using different desalination techniques. Solar desalination systems can be devised in two main types, i.e. direct and indirect collection systems. The direct collection systems use solar energy to produce distillate directly, i.e. in the solar still, which has a single structure and does not require sophisticated technical construction or operation procedure. The solar still distillers do not compete with the conventional methods such as multi-stage-flash (MSF), multi-effect distillation (MED) and reverse osmosis (RO), in particular not for the production of large quantities of water. However, they seem to be useful in providing small communities with freshwater with low efficiency. The low efficiency stems from the fact that only one basin is used in which the seawater or brackish water is heated, evaporated and then condensed. In fact, the saline water acts as heat absorber and evaporator, while the glass cover acts as condenser (Ben Bacha *et al.*, 1999a).

Solar thermal desalination units using indirect collection of solar energy can be classified into the flowing categories: multi-stage flash (MSF), multi-effect distillation (MED), mechanical vapor compression (MVC), membrane distillation (MD) and atmospheric humidification/dehumidification process (H-DH). The atmospheric humidification-dehumidification process is constituted essentially of three main parts: the heater, the humidifier, and the condenser. A water solar collector, air solar collector or other forms of low-grade energy can be used as a heater. The humidifier or evaporation tower is used to increase the air humidity. The humidified air is cooled and condensed in the condenser. All three stages are set-up in three independent compartments. This makes the overall system efficiency higher than the still distiller described above since we

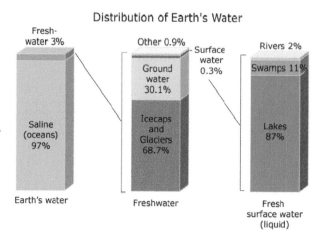

Figure 4.1.   Distribution of earth's water (source: U.S. Geological Survey; http://ga.water.usgs.gov/edu/earthwherewater.html).

can act directly on each compartment and independently of the others to optimize its output parameters. So, mainly this permits the increasing of the temperature difference between the hot humidified air and the cooler system in the condenser which allows the increase of the distilled water production.

Many designs and achievements using H-DH principle have been developed and tested. The solar powered water desalination unit based on solar multiple condensation evaporation cycle principle (SMCEC) can be considered as one of the performed techniques developed in the humidification/dehumidification process. It is an innovative technology with promising diffusion and application due to its flexibility, simplified design, low maintenance, extended life time for over twenty years, quasi null energy consumption, low capital cost, construction and adaptation for use in rural areas to produce freshwater for drinking and irrigation (Bohner, 1989; Ben Bacha, 1996; Ben Bacha et al., 1999b).

In this chapter, six major tasks are accomplished. A technical review of the state-of-the-art in the H-DH technologies is provided as a first task. The second task is the detailed description of design and operation principle of the SMCEC desalination unit. The third task is to present the developed models of the different sections of the unit using thermal energy and mass balances. The fourth and fifth tasks deal with the numeric simulation and the experimental validation of the established models, respectively. The last tasks of this document are dedicated to present the economic evaluation and the environmental impact of the desalination procedures.

4.2   STATE-OF-THE-ART

The H-DH cycle was developed over the years and many researchers have investigated many different variations on these systems. The principle of functioning of the H-DH process has been reviewed by many investigators such as Bourouni et al. (2001), Parekh et al. (2004), and Narayan et al. (2010). There are fundamentally two designs of H-DH process: the open-water/closed-air cycle and closed-water/open-air cycle.

In the open-water/closed-air configuration, heated air circulates in a closed loop between the evaporator (evaporation tower or humidifier) and the condenser (condensation tower or dehumidifier). In the closed-water/open-air configuration, the air is heated directly in an air solar collector or heated and humidified in the evaporator by the hot water received from a water solar collector. Then, air is dehumidified in the condenser cooled with relatively cold saline feed and the partially dehumidified air leaves the unit. A low-grade thermal energy such as solar, geothermal, recovered

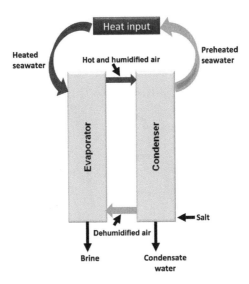

Figure 4.2.   Open-water/closed-air cycle (water-heated system).

energy or cogeneration can be used for H-DH systems. Air circulates in these systems by either natural convection due to the difference in the density and temperature of the humid air in the evaporation tower or by forced convention using mechanical means.

4.2.1   *Open-water/closed-air cycle*

Basically, three principal designs are developed in this classification: (i) water-heated systems, (ii) air-heated systems, and (iii) water- and air-heated systems.

4.2.1.1   *Water-heated systems*
Water is heated by a heat source such as solar collectors, by heat from a thermal storage device, or by waste heat. In this design there are two configurations represented by Figures 4.2 and 4.3, respectively. The first one includes water heater device (for example solar collector) and a separate evaporator and condenser as studied by Farid *et al.* (1996), Al-Hallaj *et al.* (1998) and Ben Bacha *et al.* (1999a). Air is heated by direct contact with water inside the evaporation tower. In the second one, which is known as multi-effect H-DH process, the evaporation tower and the condensation tower are integrated in the same module as developed by Müller-Holst *et al.* (1998). "Multi-effect" expression is used here not in reference to the number of constructed stages, but to the ratio of heat input to heat utilized for distillate production. Air is removed from the evaporator at various points and supplied to the condenser at corresponding points. This allows continuous temperature stratification resulting in small temperature gap to keep the process running. This in turn results in a higher heat recovery from the condenser.

Based on this concept, two pilot plants with direct flow through the collectors have been installed and tested on the island of Fuerteventura and in the laboratory at ZAE Bayern (Müller-Holst *et al.*, 1998). For the first pilot, the daily averaged heat recovery factor (GOR: gain output ratio, defined as the ratio of the energy consumed in the production of the condensate to the energy input), was between 3 and 4.5. However, for the second one, a GOR more than 8 at steady-state conditions was achieved (Müller-Holst *et al.*, 1998). A new concept which enabled continuous (24 h per day) distillate production was developed and implemented in Sfax, Tunisia, which includes solar collectors, conventional heat storage tank and heat exchanger between the collector circuit (desalted water) and the distillation circuit. Ulber *et al.* (1998) and Ben Bacha *et al.* (1999b) investigated this new concept.

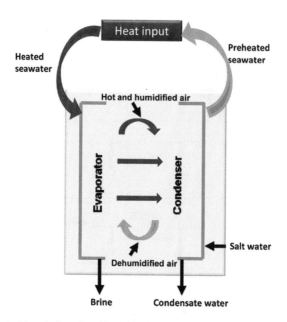

Figure 4.3.   Open-water/closed-air cycle with multi-effect H-DH process (water-heated system).

A lot of works are cited in the literature on this kind of cycle. Based on the work developed by Parekh *et al.* (2004), and Narayan *et al.* (2010), we present in Table 4.1 a review of the most important technical results detailed by different researchers.

4.2.1.2   *Air-heated systems*
The schematic diagram presented in Figure 4.4 describes the basic design of the H-DH systems using an air heated system. Air is heated in a solar collector and forwarded to the evaporator to be cooled and saturated. It is then dehumidified and cooled in the condenser. The main weakness of this process is that the humidity ratio of air that can be achieved is very low which slows down the water production of the process. To address this problem, Chafik (2001, 2003, and 2004) suggested a multi-stage system where the air goes through a series of air heating-air humidification processes to reach a higher value of absolute humidity before it is cooled by one cooler for dehumidification. Or, as discussed by Narayan *et al.* (2010), from an energy efficiency point of view, air-heated systems have higher energy consumption than water-heated systems. This is because air heats up the water in the evaporator and this energy is not subsequently recovered from the water, unlike in the water-heated cycle in which the water stream is cooled in the evaporator. A summary of main technical results reported by Narayan *et al.* (2010) is shown in Table 4.2.

4.2.1.3   *Water- and air-heated systems*
The main results of the model simulation and experimental validation of the H-DH plant developed by Ben Bacha (1996) and Ben Bacha *et al.* (1999a, 2000, 2003a) suggest that the optimal operation and production for this type of system need essentially a:

- high water temperature and flow rate at the entrance of the evaporator,
- high air temperature, humidity and flow rate at the entrance of the condenser.

Consequently, to improve its production, a flat plate solar air collector and a humidifier are added to the unit. Figure 4.5 presents the schematic diagram of the new design of the H-DH process. This system differs from the previously explored published works by using a humidifier

Table 4.1. Summary of main studies on open-water/closed-air cycle (water-heated systems).

Reference	Unit characteristics & description	Main results
Younis et al. (1993)	Solar pond H-DH with 1700 m² acts as heat storage tank and provides heated seawater to be purified. • Forced air circulation. • Packing material-meshed curtains. • Latent heat recovered in the condenser to pre-heat seawater going to the humidifier.	• Air flow rate has a significant effect on the production of water. • Study of different packing materials is required for improving performance.
Farid et al. (1995, 1996)	• Natural and forced circulation. • 1.9 m² solar collector to heat the water. • Air was in forced circulation. • Wooden shaving packing was used for the humidifier. • Multi-pass shell and tube heat exchanger used for dehumidification. • Forced circulation. • Humidifier packing-wooden shavings.	• The effect of water flow rate on the heat and mass transfer coefficients in the condenser and humidifier is significant. • The effect of air flow rate is small and natural circulation is preferred over forced convection. • A simulation program is required to conduct a detailed study. • Production of 12 L m⁻² day⁻¹ of desalinated water was achieved. • The authors report the effect of air velocity on the production is complicated and cannot be stated simply. • The water flow rate was observed to have an optimum value. • Decreasing the water flow rate causes more efficient evaporation and condensation. • Decreasing the water flow rate below 70 kg h⁻¹ reduces the performance factor due to an expected decrease in efficiency of collector at elevated temperatures.
Al-Hallaj et al. (1998)	Two units (bench scale and pilot) were built and tested in Jordan: • Closed-air cycle. • Air circulated by both natural and forced convection to compare the performance of both of these modes. • Solar collector (tubeless flat-plate type of 2 m² area) has been used to heat the water to 50–70°C. • Humidifier, a cooling tower with wooden surface (surface area 87 m² m⁻³ for the bench unit and 14 m² m⁻³ for the pilot unit). • Humidifier inlet temperature was between 60 and 63°C. • Condenser area of the bench unit: 0.6 m² with single condenser and 8.0 m² in pilot unit with double condenser. • Condenser area of the pilot unit: 8.0 m² with double condenser.	• Production of the pilot MEH unit increased with night operation using rejected water from humidifier. Productivity around 8 L m⁻² day⁻¹ was estimated. • The authors noted that results show that the water flow rate has an optimum value at which the performance of the plant peaks. • They found that at low top temperatures forced circulation of air was advantageous and at higher top temperatures natural circulation gives better performance. • No significant improvement in performance with forced circulation was noticed. • A significant effect was observed only at higher temperatures of >50°C. • Large mass of the outdoor unit (~300 kg) was a negative factor. • Use of a lighter construction material for the unit is proposed. • Night-time operation of the unit is recommended.

(Continued)

Table 4.1. Continued

Reference	Unit characteristics & description	Main results
Muller-Holst *et al.* (1998, 1999, 2007)	• Closed-air open-water cycle with natural draft circulation for the air. • Thermal storage tank of 2 m³ size to facilitate 24 h operation. • 38 m² collector field size heats water up to 80–90°C. • Latent heat recovered to heat the water to 75°C. Pilot solar-MEH unit at Canary Islands and desalination unit at Sfax, Tunisia, were studied with: • Natural circulation. • Evaporator and condenser were located in same unit. • Evaporator packing material-fleeces made of polypropylene. Simulation of solar thermal seawater desalination system using TRNSYS.	• The authors report a GOR of 3–4.5 and daily water production of 500 L for a pilot plant in Tunisia. • There is a 50% reduction in cost of water produced because of the continuous operation provided by the thermal storage device. • A simulation tool for optimizing collector field and storage was developed at ZAE Bayern. • The cost of water was determined using laboratory-based simulation criteria. • Laboratory unit yielded reduced specific energy demand due to better evaporation surfaces and thinner flat-plate heat exchangers on the condensation side of the unit. • The simulation results for optimization of 24 h operation are to be verified in field tests. • Simulation results were compared with data from pilot plant in Fuerteventura, which were in good agreement. • Increasing evaporator inlet temperature causes rising distillate volume flow. • A higher distillate volume flow was achieved by increasing brine volume flow (load flow > 600 L h⁻¹). • The optimum condenser inlet temperature was found to be 40°C. • A simulation tool for the desalination unit would be used in a unit configuration with storage tank.
Nawayseh *et al.* (1999)	• Two units in Jordan and one in Malaysia were studied. • Different configurations of condenser and humidifier were studied and mass and heat-transfer coefficients were developed. • Solar collector heats up the water to 70–80°C. • Air circulated by both natural and forced draft. • Humidifier with vertical/inclined wooden slates packing. • Heat recovered in condenser by pre-heating the feed water.	• The authors observed that the water flow rate has a major effect on the wetting area of the packing. • They also note that natural circulation yields better results than forced circulation. • The heat/mass transfer coefficients calculated were used to simulate performance and the authors report that the water production was up to 5 kg h⁻¹. • Due to a larger contact area of humidifier and condenser, productivity improved significantly in the unit constructed in Malaysia compared with that in Jordan. • The increase in humidifier area was 136%, while the increase in the single condenser area was 122.5% and in the double condenser area was 11.25%. • The humidifier cross-sectional area of the unit constructed in Malaysia was reduced by 39% to that of the Jordan unit; but complete wetting was not achieved. • A computer simulation for design of the unit components is required to achieve optimization of unit. • The effect of water is significant on heat and mass transfer coefficients in the humidifier and condenser. • The effect of air flow rate on performance is small. • A simulation program was developed to correlate with experimental results. • The effect of air flow rate is found to be insignificant.

Reference		
		• Increasing water flow rate decreases production due to lower evaporation. However, reducing water flow rate to extreme values lowers production due to drop in collector efficiency.
		• Simulation shows fast convergence of temperatures and production rate to final values.
Ben Bacha (1996) Ben Bacha et al. (2000, 2003b, 2007)	• Solar collector used for heating water (6 m² area). • There is a water storage tank which runs with a minimum temperature constraint. • Cooling water provided using brackish water from a well. • Thorn trees constitute the packed bed. • Dehumidifier made of polypropylene plates.	• A daily water production of 19 L was reported. • Without thermal storage 16% more solar collector area was reported to be required to produce the same amount of distillate. • The authors also stated that the water temperature at inlet of humidifier, the air and water flow rate along with the humidifier packing material play a vital role in the performance of the plant. • The developed models and the numerical simulation of the desalination unit in the dynamic mode helped to predict the transient behavior and the evolution of the output signals as a function of the unit input parameters variations and meteorological fluctuations. • Experimental validation of the developed models. • Development of software for sizing this kind of unit. • The main results of the simulation suggest that for the optimal operation of the unit, the water debit (L) to air debit (G) ratio must be as low as possible with L and G being at their maximum limits.
Garg et al. (2002)	• System has a thermal storage of 5 L capacity and hence has longer hours of operations. • Solar collector area (used to heat water) is about 2 m². • Air moves around due to natural convection only. • The latent is recovered partially.	• The authors conclude that the water temperature at the inlet of the humidifier is very important to the performance of the cycle. • They also observe that the heat loss from the distillation column (containing both the humidifier and the dehumidifier) is important in assessing the performance accurately.
Nafey et al. (2004)	• This system is unique in that it uses a dual heating scheme with separate heaters for both air and water. • Humidifier is a packed bed type with canvas as the packing material. • Air cooled dehumidifier is used and hence there is no latent heat recovery in this system.	• The authors reported a maximum production of $1.2\,L\,h^{-1}$ and about $9\,L\,day^{-1}$. • Higher air mass flow gave less productivity because increasing air flow reduced the inlet temperature to humidifier.
Klausner (2003)	• A unique H-DH cycle with a direct contact packed bed dehumidifier was used in this study. • The system uses waste heat to heat water to 60°C. • Uses a part of the water produced in the dehumidifier as coolant and recovers the heat from this coolant in a separate heat exchanger.	• The authors demonstrated that this process can yield a freshwater production efficiency of 8% with an energy consumption of $0.56\,kWh\,kg^{-1}$ of freshwater production based on a feed water temperature of only 60°C. • It should be noted that the efficiency is the same as the recovery ratio. • Also the energy consumption does not include the solar energy consumed.

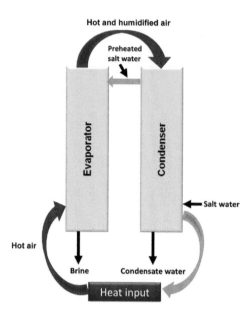

Figure 4.4.   Open-water/closed-air cycle (Air-heated systems).

coupled with flat-plate air solar collectors on the one hand, and an evaporator coupled with a field of flat-plate water solar collectors on the other, which makes the system more flexible and increases the freshwater production. The advantage of this design is that condensation process occurs in a higher temperature range than the evaporation process with a high air flow rate in the condenser. The design and performance simulation of these units is discussed in detail in previous publications (Zhani and Ben Bacha, 2010a,b).

4.2.2   Closed-water/open-air cycle

Figure 4.6 represents a schematic diagram for a typical closed-water/open-air cycle of H-DH process. In this design water circulates in a closed loop between the condenser, heat source, and evaporator. The closed-water circulation is in contact with a continuous flow of cold outside air in the evaporator. The air is heated and loaded with moisture as it passes upwards through the falling hot water in the evaporator. After passing through a condenser cooled with non-evaporated water from the evaporator, the partially dehumidified air is rejected to ambient.

The advantage of the closed-water/open-air cycle is that generally air enters the evaporator at ambient conditions with a low relative humidity. Therefore, it has a high capacity to load water vapor compared with the saturated air at the outlet of the condenser. On the other hand, there are two disadvantages of this cycle: (i) the heat of the rejected air is not recovered in the unit; and (ii) as discussed by Narayan et al. (2010), when the humidification process does not cool the water sufficiently, the coolant water temperature to the inlet of the condenser increases. This limits the dehumidification of the humid air resulting in a reduced water production compared to the open-water cycle. However, when an efficient evaporator at optimal operating conditions is used, the water may be potentially cooled to a temperature below the ambient temperature (up to the limit of the ambient wet-bulb temperature). Under those conditions, the closed-water system is more productive than the open-water system.

Based on the work developed by Parekh et al. (2004), and Narayan et al. (2010), we present a review of the important features of the unit studied and main observations from these studies (Table 4.3).

Table 4.2. Summary of main studies on open-water/closed-air cycle (air-heated systems), Narayan et al. (2010).

Reference	Unit characteristics & description	Main results
Chafik (2003, 2004)	• Solar collectors (four-fold-web-plate, or FFWP, design) of 2.08 m² area heat air to 50–80°C. • Multi-stage system that breaks up the humidification and heating in multiple stages. • Pad humidifier with corrugated cellulose material. • 3 separate heat recovery stages. • Forced circulation of air.	• The author reported that the built system is too costly and the solar air heaters constitute 40% of the total cost. • Also he observed that the system can be further improved by minimizing the pressure drop through the evaporator and the dehumidifiers.
Ben-Amara et al. (2004)	• FFWP collectors (with top air temperature of 90°C) were studied. • Polycarbonate covers and the blackened aluminum strips make up the solar collector. • Aluminum foil and polyurethane for insulation.	• Variation of performance with respect to variation in wind velocity, inlet air temperature and humidity, solar irradiation and air mass flow rate was studied. • Endurance test of the polycarbonate material showed it could not withstand the peak temperatures of summer and it melted. Hence a blower is necessary. • Minimum wind velocity gave maximum collector efficiency.
Orfi et al. (2004)	• The experimental setup used in this work uses a solar heater for both air and water (has 2 m² collector surface area). • There is a heat recovery unit to pre-heat seawater. • The authors have used an evaporator with the heated water wetting the horizontal surface and the capillaries wetting the vertical plates and air moving in from different directions and spongy material used as the packing.	• The authors report that there is an optimum mass flow rate of air to mass flow rate of water that gives the maximum humidification. • This ratio varies for different ambient conditions.
Houcine et al. (2006)	• 5 heating and humidification stages. • First two stages are made of 9 FFWP type collectors of each 4.98 m² area. The other collectors are all classical commercial ones with a 45 m² area for the third and fourth stage and 27 m² area for the final stage. Air temperature reaches a maximum of 90°C. • Air is forced-circulated. • All other equipment is the same as used by Chafik (2001, 2003, 2004).	• Maximum production of water was 516 L day⁻¹. • Plant tested for a period of 6 months. • Major dilation is reported to have occurred on the polycarbonate solar collectors. • The water production cost for this system is (for a 450–500 L day⁻¹ production capacity) 28.65 € m⁻³ which is high. • 37% of the cost is that of the solar collector field.
Yamali et al. (2008)	• A single stage double-pass flat-plate solar collector heats the water. • A pad humidifier is used and the dehumidifier used is a finned tube heat exchanger. • Also a tubular solar water heater was used for some cases. • The authors also used a 0.5 m³ water storage tank. • Heat is not recovered.	• The plant produced 4 kg day⁻¹ maximum. • Increase in air flow rate had no effect on performance. • An increase in mass flow rate of water increased the productivity. • When the solar water heater was turned on the production went up to 10 kg day⁻¹ maximum primarily because of the ability to operate it for more time.

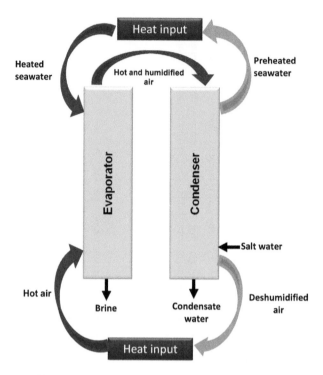

Figure 4.5.   Open-water/closed-air cycle (water- and air-heated systems).

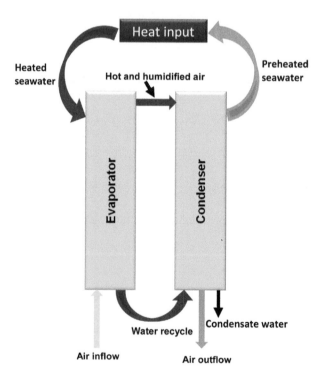

Figure 4.6.   Closed-water/open-air cycle (water-heated systems).

Table 4.3. Summary of main studies on closed-water/open-air cycle (water-heated systems).

Reference	Unit characteristics & description	Main results
Bohner (1989)	• The closed-water circulation is in contact with a continuous flow of cold outside air in the evaporation chamber. • This system with a closed salt water cycle ensures a high utilization of the salt water for freshwater production. • Polypropylene solar collector designed. • Forced or natural circulation of air. • Water is recycled or recirculated. • Incoming cold air provides a cooling source for the circulating water before it re-enters the condenser. • Polypropylene plate was used in the condenser.	• In the closed water cycle, the salt water is continuously evaporated in the evaporation chamber. For example, 1 m^3 saltwater with 1% salt results in 330 L distillate water and a brine concentration as high as ~15%.
Khedr (1993)	• The system has a packed tower (with 50 mm ceramic Raschig rings) dehumidifier. • The first system to use a direct contact condenser in a H-DH technology. • The performances are calculated numerically.	• The author reports the GOR for their system as 0.8 which shows that heat recovery is limited. • Based on an economic analysis, he concludes that the H-DH process has significant potential for small capacity desalination plants as low as 10 m^{3}day^{-1}. • Studies covered: water to air mass ratio ranging from 60 to 80 and concentration ratio between 2 and 5 at temperature difference between 10 and 16°C along the liquid for dehumidification. • 76% of the energy consumed in humidifier was recovered by condensation. • An increase of concentration ratio to 5 can reduce make-up water and rejected brine by approximately 58% and 24%, respectively. • The H-DH process has a significant potential for small-capacity desalination plants as small as 10 m^{3}day^{-1}.

(Continued)

Table 4.3. Continued

Reference	Unit characteristics & description	Main results
Al-Enezi et al. (2006)	• Solar collector designed to heat air to 90°C. • Forced circulation of air. • Cooling water circuit for the condenser. • Heater for preheating water to 35–45°C. • Plastic packing was used in the humidifier.	• The authors have studied the variation of production in $kg\,day^{-1}$ and heat and mass transfer coefficients with respect to variation in cooling water temperature, hot water supply temperature, air flow rate and water flow rate. • They conclude that the highest production rates are obtained at high hot water temperature, low cooling water temperature, high air flow rate and low hot water flow rate. • The authors have considered that the variation in parameters is very limited and hence these conclusions are true only in that range.
Dai et al. (2000, 2002)	• Forced convention for the air circulation. • Packing material-honey comb paper used as humidifier packing material. • Condenser is fin tube type which also helps to recover the latent heat by pre-heating seawater.	• The performance of the system was strongly dependent on the temperature of inlet salt water to the humidifier, the mass flow rate of salt water, and the mass flow rate of the process air. • The top water temperature has a strong parameters effect on the production of freshwater. • The thermal efficiency and water production increase with increase of mass flow rate of water to the humidifier. • The effect of rotation speed of fan on thermal efficiency was studied. It was observed that the lower the temperatures of inlet water to humidifier, the smaller the optimum rotation speed required. • An optimum mass flow rate of air exists. A higher or lower mass flow rate of air is not recommended for increasing both water production and thermal efficiency. • The thermal efficiency and water production increase significantly with temperature of inlet water into humidifier. • Productivity achieved was around $6.2\,kg\,m^{-2}day^{-1}$.

Due to the flexible designs of the plant developed by Ben Bacha *et al.*, it allows the plant to work in the presented configurations. More details about the design and performance simulation of these designs are discussed in detail in this chapter.

4.3 DESIGN AND WORKING PRINCIPLE OF THE SMCEC DESALINATION UNIT

Figure 4.7 presents a schematic diagram of the SMCEC desalination unit. The developed unit operates at atmospheric pressure using air as a carrier for vapor. Three compartments constitute the installation: solar collectors, evaporation tower and condensation tower.

The solar collector is the major component of the unit. In fact, its characteristics have an important influence on the operation and efficiency of the unit. It has an absorber plate containing uniformly spaced parallel copper pipes with its upper surface painted matt black to increase the absorptivity of the system. The absorber can be also made from a polypropylene material with very thin and tightly spaced capillary tubes. The fluid circulates in a forced convection and in one direction in the absorber, which is covered with a single glass of high transmissivity to solar radiation. All the parts are fitted inside an external case. The undersides of the absorber and the side casing are well insulated to reduce the heat losses to ambient air. More details about solar collectors design and predicted operations are presented in Ben Bacha *et al.* (2003b, 2007).

The evaporation tower produces the water vapor. It is equipped with packed bed to increase the contact surface and therefore improve the humidification rate. Due to the nature of the water

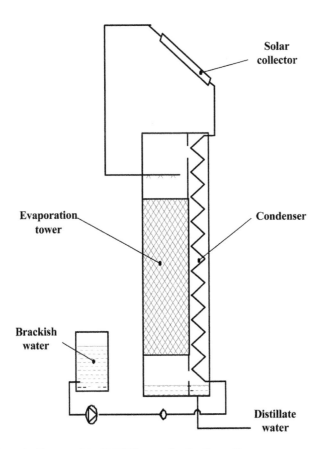

Figure 4.7. Schematic diagram of the SMCEC water desalination unit.

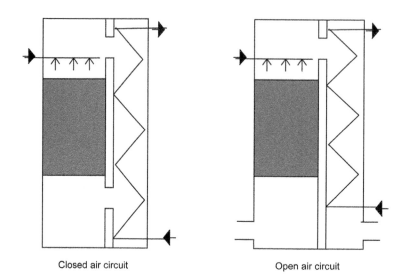

Closed air circuit   Open air circuit

Figure 4.8.   Natural convection.

(high salinity level, chalky, solid residuals and corrosion problems, etc.), the packed bed used in the unit must be carefully chosen. Thorn trees or palm tree leaves are well suited for this application. They are abundant, free and have a high resistance against the forth-mentioned problems.

The condensation chamber contains polypropylene condensation plates through which the cold salty water circulates for preheating. The structure of the evaporation tower and the condensation tower is made of aluminum. The thermal insulation is achieved by Styrofoam layers with a thickness of 40 to 50 mm. To insure isolation against vapor and water circulating in the unit and to protect the styrofoam plates against corrosion caused by brackish water, the inside of the unit is covered with polypropylene plates. The distilled water is collected in a basin at the bottom of the condensation tower. Water not evaporated is collected in a basin at the bottom of the evaporation tower and then recycled or discarded in case that the salt concentration is too high. The two basins are made of polypropylene material. Pumps are installed to facilitate water circulation in the different parts of the desalination unit. Polypropylene ducts are used due to their resistance to corrosion.

First, brackish water or seawater is heated by the solar collectors as much as possible. Then, the hot water is injected to the top of the evaporation tower. A pulverizer with a special shape is used to assure a uniform pulverization of the hot water in all the sections of the tower. Hot and saturated air mixes with the rising air current toward the condensation tower. Then it condenses in contact with the cold condensation plates. As presented earlier in section 4.2, the circulation of the air in the evaporation tower may occur in natural convection (Fig. 4.8) or in forced convection (Fig. 4.9) with two functioning modes: closed-air circuit or open-air circuit. In the course of working in free convection, the displacement of humid air in the ascending direction of the tower is caused by the difference of the density and temperature of the humid air in the tower. The forced convection is insured by a helical fan fixed at the bottom or at the top of the tower (Fig. 4.9a,b). The choice of the air circulation mode in the tower depends on the shape and the dimension of the packed bed.

The brackish water is preheated in the condenser by exchange of heat with the vapor at condensation. It allows the reduction of the consumption of thermal energy necessary for heating the water in the solar collector.

Theoretical studies and experimental investigation of the distillation module (evaporation tower and condensation tower) have been developed by Ben Bacha et al. (1999a, 2003a, 2007).

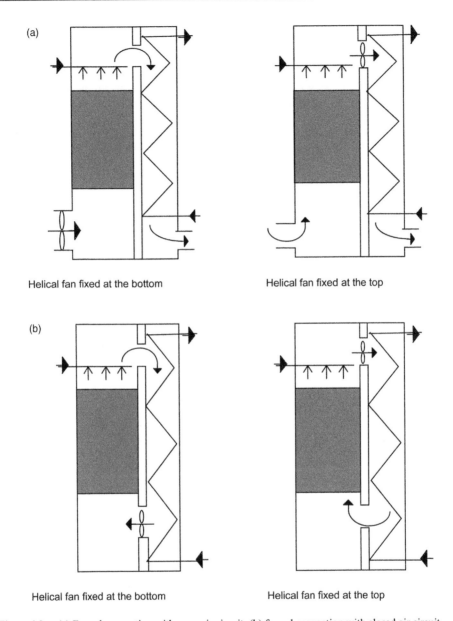

(a)

Helical fan fixed at the bottom   Helical fan fixed at the top

(b)

Helical fan fixed at the bottom   Helical fan fixed at the top

Figure 4.9.   (a) Forced convection with open air circuit; (b) forced convection with closed air circuit.

From the results obtained during the previous works developed by Ben Bacha *et al.*, (1999a, 2000, 2007), it was possible to identify potential relevant improvements that could be implemented in the SMCEC unit to boost its production. This analysis concluded that it is essential to elevate the capacity of air to load water vapor by heating and subsequent humidification of air at the exit of the condensation tower instead of rejecting or recycling it. The vapor carrying capability of air increases with temperature: 1 kg of dry air can carry 0.5 kg of vapor and about 670 kcal (2804 kJ) when its temperature increases from 30°C to 80°C (Parekh *et al.*, 2004). Therefore, to attend this objective it is necessary to incorporate into the SMCEC unit a flat plate solar air collector for heating air and a humidifier for its humidification.

Figure 4.10 presents a schematic of the improved SMCEC water desalination unit. The newly designed process differs from the basic design by using a humidifier and an evaporation tower, on the one hand, and a field of flat-plate air solar collectors and a field of flat-plate water solar collectors on the other. The injected air inside the condensation tower is loaded by moisture due to heat and mass transfers between the hot water and the heated air stream in the humidifier in the case of working in closed air loop and between the hot water and the ambient air stream in the evaporation tower in the case of working in open air loop.

The air solar collector device used by the present desalination prototype is formed by a single glass cover and an absorber. The absorbing aluminum material that captures the solar energy is constituted by separated rectangular channels. Polyurethane is utilized for rear and sides insulations. A silicon sealant is applied between the different components of the air solar collector to ensure insulation from the environment.

A pad humidifier is used in the desalination prototype. At the top, there is a liquid distributor, which can feed the pad with hot brackish water coming from water solar collectors, while at the bottom there is a liquid collector, where brine is collected as it drains down the pad. Thus, the hot brackish water flows downward, while the air passes in a cross-flow direction. Textile (viscose) is used as packing to increase the interface area between the air and water, which form the wetted surface. On the outside, the humidifier is covered with a polyethylene sheet and insulated with a layer of armaflex.

The preliminary results show that the improvement implemented in the SMCEC unit increased the production by 44% for hot water temperature at the entrance of the evaporation tower thereabouts 55°C. Despite this increase, simplification in unit design and optimization of the distillate water production are in progress to reduce the cost of the distilled water due to the high cost of

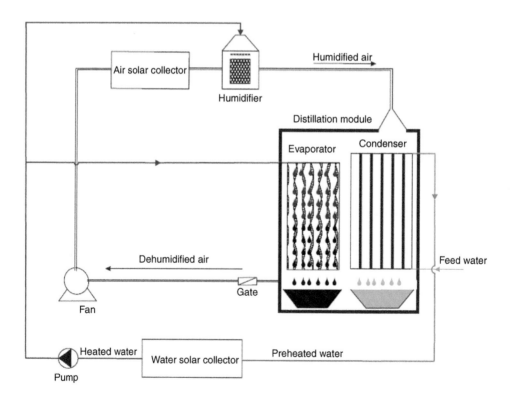

Figure 4.10.   Improved SMCEC water desalination unit.

air and water solar collectors. More details about the improved SMCEC water desalination unit are presented in Zhani and Ben Bacha (2010a,b).

4.4 COMPONENTS MATHEMATICAL MODELING

Multiple phenomena such as heating, evaporation and condensation are all involved in the desalination unit and many of its variables are time and space dependent. In this chapter we are interested in the sizing of the SMCEC water desalination unit and determining the optimal operation conditions. Therefore, steady state models of the different sections of the desalination unit are developed from the governing heat and mass transfer equations. The air circulation mode used for the installation is a natural convection with a closed air circuit.

4.4.1 *Solar collector modeling*

The plane collector has an absorber with parallel and narrow channels as presented in Figure 4.11. In this case, the fluid circulates in a forced convection and in one direction. To establish the model for the system, the following assumptions are made (Ben Bacha, 1996):

- The speed of the fluid is uniform, therefore the local state of the fluid and of the absorber depends only on one side x that varies from 0 to L.
- The absorber and fluid have the same temperature at any point.
- The fluid temperature remains under the 100°C point.

The thermal balances for the system formed by the absorber and the fluid for a slice of the collector with a width of l and a length of dx, yield the following local equation for the fluid behavior:

$$\frac{dT_f}{dx} = \frac{l}{m_f c_f}(BI - U_1(T_f - T_a)) \tag{4.1}$$

The mathematical model for the solar collector, allows the study for constant insolation level and constant ambient temperature, the fluid temperature variation at the outlet of the collector as a function of the control, geometrical, and physical parameters (flow rate, collector area, material, and inclination angle of the collector). In addition, the fluid temperature and flow rate at the collector outlet are the two parameters with the most significant impact on the unit production as they are among the input parameters of the evaporation tower.

4.4.2 *Evaporation tower modeling*

The model of the evaporation tower is a set of equations developed using thermal and mass balances for the water phase, the air phase, and the air-water interface on an element of volume

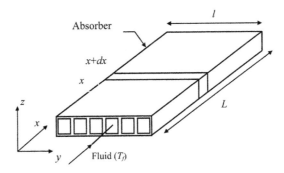

Figure 4.11. Absorber of the solar collector.

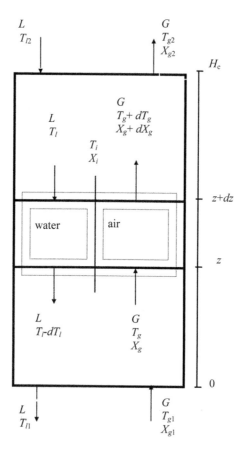

Figure 4.12.   An element of volume of the evaporation chamber of height dz (Ben Bacha et al., 1999a).

of height dz (Fig. 4.12). The approach is similar to that used in the cooling tower study (Younis et al., 1987; Kaguel et al., 1988).

To establish the model for the system, the following assumptions are made:

• The tower is adiabatic.
• Head losses in the tower column are low and it can be assumed that the density of the air mass flux is constant.
• Since the evaporation rate is small compared to the water flow rate, the water mass flux remains constant throughout the column.
• The water distribution over the packed bed is uniform.

The mathematical model for the evaporation tower is formed by the following system of equations (Ben Bacha et al., 1999a; Ben Bacha, 1996):

for the water phase:

$$\frac{dT_l}{dz} = \frac{h_l a(T_l - T_i)}{LC_l} \tag{4.2}$$

for the air phase:

$$\frac{dT_g}{dz} = \frac{h_g a(T_i - T_g)}{GC_g} \tag{4.3}$$

and

$$\frac{dX_g}{dz} = \frac{k_g a (X_i - X_g)}{G} \tag{4.4}$$

at the air-water interface:

$$X_i = X_g + \frac{h_l a (T_l - T_i) + h_g a (T_g - T_i)}{L_v k_g a} \tag{4.5}$$

also, it can be expressed as (Cretinon, 1995; Sonntag, 1990; Wexler, 1976, 1977):

$$X_i = 0.62198 \frac{P_{ws}}{1 - P_{ws}} \tag{4.6}$$

where P_{ws} is the saturation pressure,

$$\ln(P_{ws}) = -6096.938 \frac{1}{T_i} + 21.240964 - 2.71119 \times 10^{-2} T_i + 1.67395 \times 10^{-5} T_i^2 + 2.43350 \ln(T_i) \tag{4.7}$$

The steady state model of the evaporation tower using the equations presented above could be used to monitor the influence of the command parameters on the tower operation. In particular, it allows to determine the temperature and the amount of water in the air at the outlet of the evaporation tower as a function of the water temperature at the entrance (top of the tower) for a given tower height. The obtained model can be used to size this type of evaporation tower for desired water and air temperatures.

Due to the nature of water (high salinity, deposits of calcite, solid residues, corrosion problems, etc.), the garnishing to be used in this application must be chosen carefully. For instance, the thorn trees, abundant and free are appropriate under these operating conditions and their characteristics need to be specified beforehand. To do so, a set of experiments were carried out on a test bench for a cooling tower to determine the different specific heat and mass exchange coefficients at the contact surface between the garnishing and the humid air (Ben Bacha *et al.*, 1999a; Ben Bacha, 1996). The relationships giving the mass exchange coefficients (k_g) and the heat exchange coefficients h_l as a function of the air flow rate (G) and water flow rate (L) are:

$$k_g = \frac{2.09 G^{0.11515} L^{0.45}}{a} \tag{4.8}$$

$$h_l = \frac{5900 G^{0.5894} L^{0.169}}{a} \tag{4.9}$$

The air-film heat transfer coefficient and the mass transfer coefficient on the air-water interface are coupled by the Lewis relation (Ben Bacha *et al.*, 2007; Younis *et al.*, 1987) as follows: $h_g = C_g k_g$.

4.4.3 *Condensation tower modeling*

The formulation of the mathematical model of the condensation tower in the steady state regime is based on thermal and mass balances. It allows developing the coupling equations between the temperature of the cooling water of the condenser and the humid air temperature and water content. The condensation rate is determined using an algebraic equation that relates the variation of the water content with the height of the tower. The balances are done on an element of volume of the tower of height dz (Fig. 4.13).

To establish the model for the system, the following assumptions are made (Ben Bacha *et al.*, 1999a; Ben Bacha, 1996):

- The tower is adiabatic.
- Each stage is completely homogeneous as to avoid thermal gradients in the direction normal to the condenser plates.

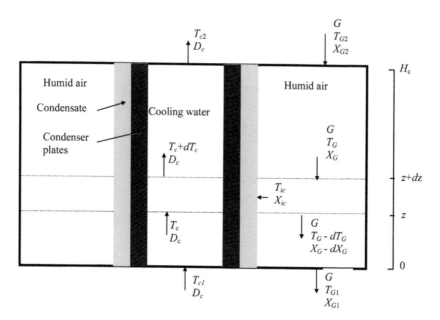

Figure 4.13.   An element of volume of the condensation chamber of height dz (Ben Bacha et al., 1999a).

- The length dz of the tube element is small; therefore, the temperature in each stage changes linearly throughout the length of the element.
- Colburn and Hougen method developed by Votta, which states that the sensitive heat transfer of the condensed water droplets is negligible, was used (Grehier and Rojey, 1989).
- The air heat and mass transfer coefficients at the air-water interface are related by the Lewis equation.
- The condensate film flow is normally streamlined.
- The heat flows through the film by conduction.

The mathematical model for the condensation tower is formed by the following system of equations (Ben Bacha, 1996; Ben Bacha et al.,1999a):

for the water phase:

$$\frac{dT_c}{dz} = \frac{UA(T_{ic} - T_c)}{D_c C_c} \tag{4.10}$$

for the air phase:

$$\frac{dT_G}{dz} = A\frac{(h_G(T_G - T_{ic}) + L_v k_G(X_G - X_{ic}))}{GC_G} \tag{4.11}$$

$$\frac{dX_G}{dz} = k_G A\frac{(X_{ic} - X_G)}{G} \tag{4.12}$$

at the air-condensate interface:

$$X_{ic} = X_G + \frac{(h_G(T_G - T_{ic}) + U(T_c - T_{ic}))}{L_v k_G} \tag{4.13}$$

also, it can be expressed as (Cretinon, 1995; Sonntag, 1990; Wexler, 1976, 1977):

$$X_{ic} = 0.62198\frac{P_i}{1 - P_i} \tag{4.14}$$

where P_i is the saturation pressure.

Following the water balance:

$$dW_c = G dX_G \tag{4.15}$$

then the flow rate of the condensed water is:

$$dW_c = k_G A (X_{ic} - X_G) dz \tag{4.16}$$

For a flow between two vertical plates, the heat transfer coefficient h_G is given by (Sacadura, 1993):

$$h_G = \frac{\lambda_G}{D_{h1}} 0.479 (Gr^{1/4}) \tag{4.17}$$

The coefficient k_G is determined using the Lewis relation: $k_G = h_G / C_G$.

The overall heat transfer coefficient from the air-condensate interface to the cooling water inside the condenser is approximated by:

$$U = \frac{1}{(1/h_c) + (e/\lambda_p) + (1/h_e)} \tag{4.18}$$

where h_e is given by (Sacadura, 1993):

$$h_e = \frac{\lambda_e}{D_{h2}} \left(0.023 Re_G^{0.8} Pr_G^{0.33} \right) \tag{4.19}$$

and h_c is expressed using the Nusselt relation (Sacadura, 1993):

$$h_c = \sqrt[4]{\frac{\rho_c^2 g L_v \lambda_c^3}{4 \mu_c z (T_{ic} - T_p)}} \tag{4.20}$$

The steady state model developed allows finding the influence of the command parameters on the tower operation and the sizing of this type of condensation tower for desired amount of distilled water W_c produced by the unit. More details about modeling, design and predicted operation of the solar collectors, evaporation tower, and condensation tower are presented in Ben Bacha *et al.* (1999, 1996, 2003b, 2007).

4.5 NUMERICAL RESULTS

A series of simulation scenarios in the steady state mode are carried out to study the influence of the different command signals on the output parameters of each unit section and the entire desalination unit. Since the evaporation chamber is fed from the solar collectors, its performance depends on the meteorological conditions. For this reason and for simulation purposes, the following two assumptions are made:

- Constant outlet temperature of the field collectors.
- A storage tank connected to the solar collector.

In addition, the optimal condensation tower operation depends on the functioning of the evapora-tion tower since its outputs parameters (temperature and humidity of air at the top) are considered as inputs to the condensation chamber.

4.5.1 *Solar collector*

Several simulations of the solar collector steady state model are made using the model developed above. The effect of solar radiation (I), fluid temperature at the collector entrance (T_{fe}) and fluid flow rate variations (m_f) on the solar collector outlet temperature (T_{fs}) are presented.

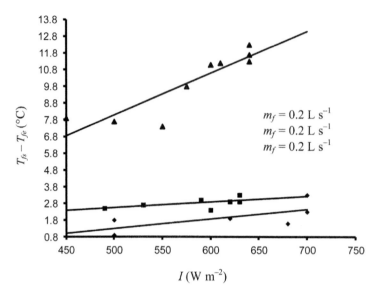

Figure 4.14.   Impact of fluid flow rate and solar radiation on the temperature difference ($T_{fs} - T_{fe}$).

The fluid temperature at the entrance of the solar collector depends on the ambient temperature (T_a). So, we study the variation of the temperature difference ($T_{fs} - T_{fe}$) according to the fluid flow rate and the solar radiation instead of the exit temperature (Fig. 4.14). It can be seen that for a constant fluid flow rate the temperature varies linearly as a function of solar radiation. In addition, as m_f decreases, the temperature variation ($T_{fs} - T_{fe}$) increases and becomes more significant.

4.5.2   Evaporation tower

The water temperature (T_{12}) and the water flow rate (L) at the top entrance and the air flow rate (G) at the bottom entrance of the evaporation tower are its command parameters. The water temperature (T_{12}) is considered as a key variable in this parametric study. Several cases of numerical simulation of the evaporation tower operation conducted to monitor the evolution of the evaporation tower operation as a function of its command parameters (L) and (G) are presented in Ben Bacha (1996), and Ben Bacha et al. (2000, 2003a).

Discussions of the experimental results are included in the experimental validation section. Figure 4.15 indicates that the amount of water X_{g2} in the humid air at the top of the evaporation tower increases significantly as the water temperature (T_{12}) at the evaporation tower inlet increases.

Figure 4.16 shows the effect of water temperature (T_{12}) at the entrance of the evaporation tower on humid air temperature (T_{g2}) at the outlet of the latter. From this figure we can note that increasing the water temperature provokes an elevation of the temperature of humid air at the outlet of the evaporation tower. The increase of the hot water temperature at the entrance of the evaporation tower produces more heat inside the installation.

4.5.3   Condensation tower

The main objective of the numerical simulation of the condensation tower operation is to investigate the influence of the main command parameter, the air temperature (T_{g2}), on the amount of condensate flow rate W_c. The impact of the other input signals of the condensation tower such as the cooling water flow (D_c) and the cooling water temperature (T_c) on the amount of condensate flow rate W_c are studied in Ben Bacha (1996) and Ben Bacha et al. (2000, 2003a).

Figure 4.17 illustrates the effect of air temperature (T_{g2}) at the condensation tower inlet on the amount of condensate flow rate W_c. From this figure we remark that increasing the air temperature

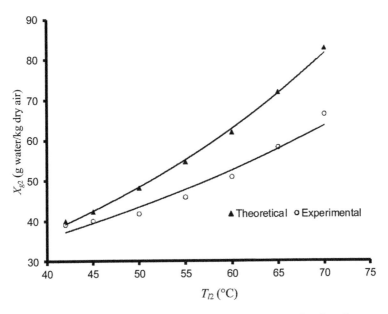

Figure 4.15. Air humidity variation at the top of the evaporation tower as a function of water temperature at the evaporation tower inlet.

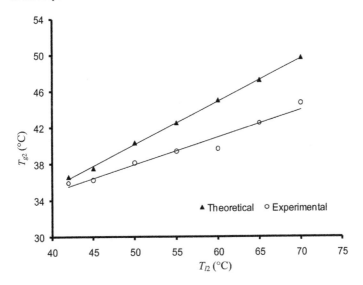

Figure 4.16. Air temperature variation at the top of the evaporation tower as a function of water temperature at the evaporation tower inlet.

generates an elevation of the distillated water flow rate. The increase of the air temperature at the entrance of the condensation chamber produces more distillate water.

4.5.4 *Entire desalination unit*

Since the modules of the desalination unit (solar collectors, evaporation tower and condensation tower) are coupled and the output of each module is the input to the next, it is necessary to

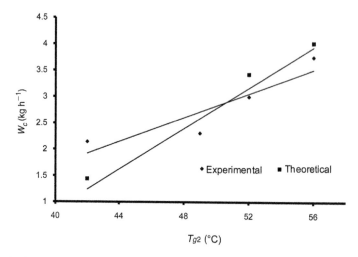

Figure 4.17.   Variation of the condensation flow rate as a function of air temperature at the condensation tower inlet.

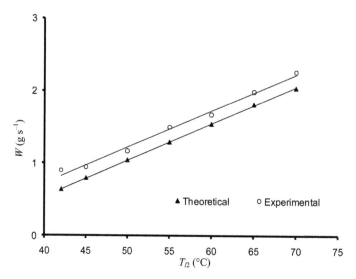

Figure 4.18.   Variation of the condensation flow rate as a function of water temperature at the evaporation tower inlet, $T_{c1} = 26°C$ and $D_c = 0.08$ kg s^{-1}.

couple the different models and obtain a global model that provides a good description of the entire system. Several cases of numerical simulation of the entire desalination unit operation have been conducted to monitor the evolution of the distillate water production as a function of water temperature at the evaporation tower inlet, water flow rate at the evaporation tower inlet, and water temperature at the condenser entrance.

Figure 4.18 shows that the flow rate of the distilled water varies linearly as a function of the water temperature at the evaporator inlet (T_{l2}). The increase of the hot water temperature at the entrance of the evaporation tower produces more heat inside the installation. Similarly, the temperature of the cooling water at the outlet of the condenser is proportional to the water temperature at the evaporator inlet (Fig. 4.19). Useful amount of heat in the vapor could be recovered

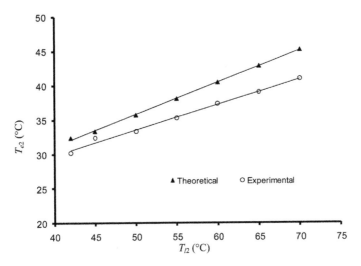

Figure 4.19.   Variation of cooling water temperature as a function of water temperature at the evaporation tower inlet, $T_{c1} = 26°C$ and $D_c = 0.08 \, kg \, s^{-1}$.

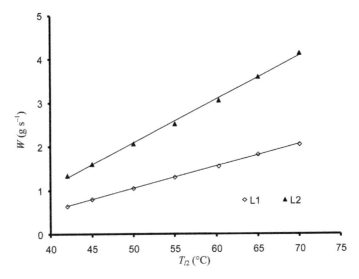

Figure 4.20.   Variation of the condensation flow rate as a function of water flow rate at the evaporation tower inlet, $T_{c1} = 26°C$, $L_1 = 0.08 \, kg \, s^{-1}$ and $L_2 = 0.16 \, kg \, s^{-1}$.

to preheat the brackish water in the plates before its injection to the solar collector. The cooling water in the condensation tower has a constant flow rate (D_c) and a constant temperature (T_c).

Figure 4.20 demonstrates the impact of the circulating water flow rate in the evaporation tower (L) on the condensation rate (W) as a function of water temperature at the evaporator inlet (T_{l2}). It can be seen that doubling the flow rate yields twice the condensation rate. This is true for any temperature T_{l2} shown in Figure 4.20. Moreover, as T_{l2} increases, the condensation rate difference for two different water flow rates becomes more significant.

Figure 4.21 shows that as the water temperature at the entrance of the condenser (T_c) increases, the condensation flow rate (W) decreases and *vice versa*. That is, low temperature at the condenser entrance is desired.

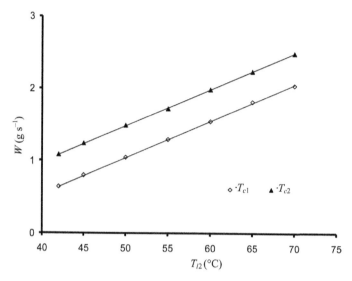

Figure 4.21.   Variation of the condensation flow rate as a function of water temperature at the condenser entrance, $T_{c1} = 26°C$, $T_{c2} = 15°C$ and $D_c = 0.08$ kg s^{-1}.

Table 4.4.   Characteristics of the pilot desalination unit.

Solar Collector	
Area	$7.20\,m^2$
Effective transmission absorption	0.83
Riser tube material	copper
Absorber surface	paint mat black
Loss coefficient	$3.73\,W\,m^{-2}\,K^{-1}$
Back insulation, thickness	Fibre glass, 50 mm
Evaporation chamber	
Size	$1.20\,m \times 0.50\,m \times 2.55\,m$
Packed bed type	Solid packing: thorn trees
Mass transfer coefficient	$k_g = (2.09\,G^{0.11515}L^{0.45})/a$
Heat transfer coefficient	$h_l = (5900\,G^{0.5894}L^{0.169})/a$
	$h_g = C_g\,k_g$
Condensation tower	
Size	$1.2\,m \times 0.36\,m \times 3.0\,m$
Plates type	polypropylene

4.6   EXPERIMENTAL VALIDATION

To validate the developed models of the desalination unit, a series of experiments was conducted using the pilot desalination unit located at the National Engineering School of Sfax (E.N.I.S) – Sfax University, Tunisia. Experimental measurements were taken on the solar collector and the distillation module (evaporation and condensation chambers). The characteristics of the pilot desalination unit are listed in Table 4.4.

4.6.1   *Solar collector*

Figure 4.22 shows the variations of the experimental and the simulation output variable of the solar collector (T_{fs}). According to this figure, the experimental (E1) and simulation (S1) variations have the same trends.

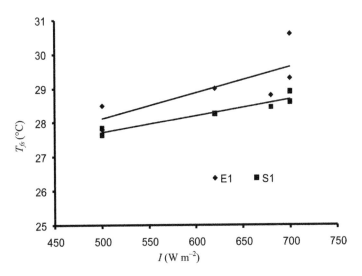

Figure 4.22.   Impact of solar radiation on the solar collector outlet temperature, $m_f = 0.2 \, \text{L s}^{-1}$.

4.6.2   *Distillation module*

The measurements of relative humidity show that the air circulating in the installation is saturated at the top and at the bottom of the distillation module in the steady state regime. Indeed, the theoretical curve of water content X_{g2} as a function of the temperature T_{l2} exceeds the experimental curve, as seen in Figure 4.15. The gap between the theoretical and experimental values increases with T_{l2}. Similarly, the air temperature at the top of the evaporation tower (T_{g2}) is lower than the theoretical value (Fig. 4.16), in particular for high T_{l2}. The difference is essentially due to the energy losses that increase with the temperature, even though the process is assumed to be theoretically adiabatic. This is what would explain the increase in the gap between the theoretical and experimental values. These losses would influence the water temperature at the outlet of the condenser since the flux of the heat generated during the condensation is supposed to be completely absorbed by the cooling water. However, this is not the case. This is why the experimental values of T_{e2} shown in 4.19 are lower than the theoretical values.

For a given water temperature at the entrance of the evaporation tower (T_{l2}), and knowing the experimental values for $W_{cond.}$, X_{G1} and X_{G2}, the humid air mass velocity is (Ben Bacha *et al.*, 1999a):

$$G_{exp} = \frac{W_{cond}}{(X_{G2} - X_{G1})} \qquad (4.21)$$

(with the air being saturated at the top and at the bottom of the tower).

Using this equation, the air circulates in the unit with a speed lower than the theoretical one. Tests show that $G_{exp} \cong 0.60 G_{th}$. The decrease in the air flow rate is compensated by an increase in the difference between the air temperatures at the top and that at the bottom of the condensation tower. This is why the actual water distillation rate is close to or sometimes higher than the theoretical rate, as seen in Figure 4.17.

4.7   COST ANALYSIS

Desalination is becoming a less expensive form of water treatment. For large desalination units, the solar desalination units using humidification-dehumidification technology would not compete with the desalination plants based on the RO, MSF and VC methods. However, for a medium and small desalination unit the earlier is a viable solution with high energy savings due to the fact that solar energy is free (of course without considering the investment cost of the solar collectors) and that the solar distillation systems do not need complex technology. Compared to water price produced by the local water authority, the cost per cubic meter of solar distillation systems can be justified by the autonomy of this kind of desalination unit and by its setup in arid zones where other options for water supply such as conventional desalination units or water conveyance from remote sources are difficult, or even impossible to develop.

The solar desalination units using humidification-dehumidification process usually consist of three sections: solar collectors to heat brackish or seawater, an evaporation tower and a condensation tower. The solar collectors have a high cost that represents 10 to 35% of the total desalination unit cost. The amount of distillate water produced by the unit depends on the solar collector size. Projections of fuel costs (coal, oil, natural gas and uranium) show an increase of about 50% every 10 years, however, solar collectors offer a low energy cost (Ben Bacha et al., 1999b).

Simplification in conception and optimization of the distillate water production are necessary to reduce the cost. It is therefore necessary to develop a methodology that guarantees lower desalination cost. The task is to maximize the utilization rate, i.e., a 24 hour operation using an energetic storage device to extend the unit operation beyond daylight hours and cloudy days. This is possible when using insulated storage tanks and solar ponds. The use of geothermal water exploited from wells with temperatures exceeding 50°C can significantly reduce the distillate water price by 20 to 30% since it becomes unnecessary to have solar collectors to preheat the water (Ben Bacha et al., 1999a). Additionally, wide-scale implementation would push the costs further down.

Based on economic evaluation of a solar powered prototype unit using humidification-dehumidification process developed by Zhani and Ben Bacha (2010), we study in this chapter the cost of the water produced and the payback period of the experimental setup. The investment cost of each component constituting the experimental setup is presented in Table 4.5. The payback period of the experimental setup depends on overall cost of fabrication, maintenance cost and operating cost. The investment cost of the entire prototype unit studied is 6697 €. The life time for a solar desalination unit was assumed to be 20 years with annual production of 120,000 L (20 L day^{-1} × 300 day year^{-1} × 20 years = 120,000 L).

Cost of water per liter = investment cost of the entire unit/number of liters that would be produced = 6697/120,000 = 0.056 € L^{-1}.

Cost of water produced per day = cost of water per liter × productivity = 0.056 € L^{-1} × 20 L day^{-1} = 1.12 € day^{-1}. The maintenance cost is estimated at 0.01 € day^{-1}.

Net earning = Cost of water produced − Maintenance cost = 1.12 − 0.01 = 1.11 €.

Payback period = Investment/Net earning = 6697/1.11 = 6033 days.

On the basis of the economic evaluation presented, the solar collectors have a high cost that represents 10 to 35% of the total desalination unit cost. Consequently, the difficulty for the deployment of this technology in developing countries is to obtain enough capital to purchase and

Table 4.5.   Investment cost of each component constituting the experimental prototype.

Unit components	Quantity	Cost of components
Water solar collector	12 m^2	12 m^2 × 235 € m^{-2} = 2820 €
Distillation module	1	2490 €
Immersed pump	2	2 × 111 € = 222 €
Fan	1	277 €
Ducts	−	888 €
	Total cost	6697 €

install solar collectors. Therefore, if one does not have the necessary capital to start with, solar distillers are then the good solution as they are installed by module. Their modular feature enables the installation of a small number of units every year or every time it is financially possible to purchase some of the needed equipment. It is therefore possible to reduce the cost of freshwater since solar collectors are the main component of a solar desalination unit and any improvement in their efficiency will have a direct bearing on the water production rate and the product cost. Also, the amount of distillate water produced by the unit depends on the solar collector size.

NOTATION

A	air-water exchanger area in the condensation tower	$[\text{m}^2\,\text{m}^{-3}]$
a	air-water exchanger area in the evaporation tower	$[\text{m}^2\,\text{m}^{-3}]$
B	effective transmission absorption product ($B = \alpha\tau$)	
C	water specific heat	$[\text{J}\,\text{kg}^{-1}\,\text{K}^{-1}]$
C_g	humid air specific heat in the evaporation tower	$[\text{J}\,\text{kg}^{-1}\,\text{K}^{-1}]$
C_G	humid air specific heat in the condensation tower	$[\text{J}\,\text{kg}^{-1}\,\text{K}^{-1}]$
D_c	water mass velocity in the condensation tower	$[\text{kg}\,\text{m}^{-2}\,\text{s}^{-1}]$
D_{h1}	hydraulic diameter of air flow	$[\text{m}]$
D_{h2}	hydraulic diameter of water flow	$[\text{m}]$
e	thickness of the condenser plate	$[\text{m}]$
g	gravitational acceleration	$[\text{m}\,\text{s}^{-2}]$
G	mass velocity of humid air	$[\text{kg}\,\text{m}^{-2}\,\text{s}^{-1}]$
Gr	Grashof number	$[\text{ad}]$
h_c	heat transfer coefficient for the condensate film	$[\text{W}\,\text{m}^{-2}\,\text{K}^{-1}]$
h_e	heat transfer coefficient at the water-condenser inside wall interface	$[\text{W}\,\text{m}^{-2}\,\text{K}^{-1}]$
h_l	water heat transfer coefficient at the air-water interface in the evaporation tower	$[\text{W}\,\text{m}^{-2}\,\text{K}^{-1}]$
h_g	air heat transfer coefficient at the air-water interface in the evaporation tower	$[\text{W}\,\text{m}^{-2}\,\text{K}^{-1}]$
h_G	air film heat transfer coefficient in the condensation tower	$[\text{W}\,\text{m}^{-2}\,\text{K}^{-1}]$
I	flux of incident radiation	$[\text{W}\,\text{m}^{-2}]$
k	water vapor mass transfer coefficient at the air-water interface	$[\text{kg}\,\text{m}^{-2}\,\text{s}^{-1}]$
L	water mass velocity in the evaporation tower	$[\text{kg}\,\text{m}^{-2}\,\text{s}^{-1}]$
L_v	latent heat of water vaporization	$[\text{J}\,\text{kg}^{-1}]$
l	Solar collector width	$[\text{m}]$
m_f	fluid flow rate in the absorber	$[\text{kg}\,\text{s}^{-1}]$
Nu	Nusselt number	$[\text{ad}]$
Pr	Prandtl number	$[\text{ad}]$
Ra	Rayleigh number	$[\text{ad}]$
T_a	ambient temperature	$[\text{K}]$
T	temperature	$[\text{K}]$
T_i	temperature at the air-water interface in the evaporation tower	$[\text{K}]$
T_{ic}	temperature at the air-water interface in the condensation tower	$[\text{K}]$
T_p	wall temperature	$[\text{K}]$
U_1	overall heat loss coefficient from the absorber to the surroundings	$[\text{W}\,\text{m}^{-2}\,\text{K}^{-1}]$
U	overall heat transfer coefficient in the condenser	$[\text{W}\,\text{m}^{-2}\,\text{K}^{-1}]$
X	air humidity	$[\text{kg}_{\text{water}}\,\text{kg}_{\text{dry air}}^{-1}]$
X_{ic}	saturation humidity in the condensation tower	$[\text{kg}_{\text{water}}\,\text{kg}_{\text{dry air}}^{-1}]$
X_i	saturation humidity in the evaporation tower	$[\text{kg}_{\text{water}}\,\text{kg}_{\text{dry air}}^{-1}]$
Z	axial distance variable	$[\text{m}]$

GREEK SYMBOLS

α absorptance of the collector absorber surface
λ_e water thermal conductivity $[\mathrm{W\,m^{-1}\,K^{-1}}]$
λ_c condensed water thermal conductivity $[\mathrm{W\,m^{-1}\,K^{-1}}]$
λ_G humid air thermal conductivity $[\mathrm{W\,m^{-1}\,K^{-1}}]$
λ_p wall thermal conductivity $[\mathrm{W\,m^{-1}\,K^-1}]$
ρ_c water density water $[\mathrm{kg\,m^{-3}}]$
μ_c dynamic viscosity of condensed water $[\mathrm{N\,s\,m^{-2}}]$
τ transmittance

SUBSCRIPTS

1   tower bottom
2   tower top
c condensation tower
f absorber
g evaporation tower
G condensation tower
l evaporation tower

REFERENCES

Abdel-Salam, M.S., Hilal, M.M., El-Dib, A.F. & Abdel Monem, M.: Experimental study of humidification-dehumidification desalination system. *Energy Sources* 15 (1993), pp. 475–490.
Al-Enezi, G., Ettouney, H.M. & Fawzi, N.: Low temperature humidification-dehumidification desalination process. *Energy Convers. Manage.* 47 (2006), pp. 470–484.
Al-Hallaj, S., Farid, M.M. & Tamimi, A.R.: Solar desalination with humidification-dehumidification cycle: performance of the unit. *Desalination* 120 (1998), pp. 273–280.
Ben-Amara, M., Houcine, I., Guizani, A. & Maalej, M.: Experimental study of a multiple-effect humidification solar desalination technique. *Desalination* 170 (2004), pp. 209–221.
Ben Bacha, H.: *Modélisation et simulation en vue de la commande d'une unité de dessalement d'eau par l'énergie solaire.* PhD Thesis, Tunis University, E.S.S.T.T, Tunisia, 1996.
Ben Bacha, H., Bouzguenda, M., Abid, M.S.A. & Maalej, Y.: Modelling and simulation of a water desalination station with solar multiple condensation evaporation cycle technique. *Renew. Energy* 18 (1999a), pp. 349–365.
Ben Bacha, H., Maalej, A.Y., Ben Dhia, H., Ulber, I., Uchtmann, H., Engelhardt, M. & Krelle, J.: Perspectives of solar powered desalination with "SMCEC" technique. *Desalination* 122 (1999b), pp. 177–183.
Ben Bacha, H., Bouzguenda, M., Damak, T., Abid, M.S. & Maalej A.Y.: A study of a water desalination station using the SMCEC technique: production optimisation. *Renew. Energy* 21 (2000) pp. 523–536.
Ben Bacha, H., Damak, T., Bouzguenda, M. & Maalej, A.Y.: Experimental validation of the distillation module of a desalination station using the SMCEC principle. *Renew. Energy* 28 (2003a), pp. 2335–2354.
Ben Bacha, H., Damak, T., Bouzguenda, M., Maalej, A.Y. & Ben Dhia, H.: A methodology to design and predict operation of a solar collector for a solar-powered desalination unit using the SMCEC principle. *Desalination* 156 (2003b), pp. 305–313.
Ben Bacha, H., Damak, T., Ben Abdalah, A.A., Maalej, A.Y. & Ben Dhia, H.: Desalination unit coupled with solar collectors and a storage tank: modelling and simulation. *Desalination* 206 (2007), pp. 341–352.
Ben Bacha, H., Maalej, A.Y. & Ben Dhia, H.: A methodology to predict operation of a solar powered desalination unit. Advanced research workshop. In L. Rizzuti, H.M. Ettouney & A. Cipollina (eds): *Solar Desalination for the 21st Century.* Springer, Dordrecht, The Netherlands, 2007, pp. 69–82.
Bohner, A.: Solar desalination with a high efficiency multi effect process offers new facilities. *Desalination* 73 (1989), pp. 197–203.
Bourouni, K., Chaibi, M., Martin, R. & Tadrist, L.: Heat transfer and evaporation in geothermal desalination units. *Appl. Energy* 64 (1999), pp. 129–147.

Bourouni, K., Chaibi, M.T. & Tadrist, L.: Water desalination by humidification and dehumidification of air: state of the art. *Desalination* 137 (2001), pp. 167–176.

Chafik, E.: A new seawater desalination process using solar energy. *Proceeding 6th AQUA-TECH Conference*, Cairo, Egypt, 2001.

Chafik, E.: A new type of seawater desalination plants using solar energy. *Desalination* 156 (2003), pp. 333–348.

Chafik, E.: Design of plants for solar desalination using the multi-stage heating/humidifying technique. *Desalination* 168 (2004), pp. 55–71.

Coulson, J.M. & Richardson, J.F. (eds): *Chemical engineering*, Volume 1. 3rd edition, Pergamon Press, 1977, pp. 216–219.

Cretinon, B.: Mesure des paramètres de l'air humide. In: *Techniques de l'Ingénieur* R3045 1995, pp. 1–17.

Dai, Y.J. & Zhang, H.F.: Experimental investigation of a solar desalination unit with humidification and dehumidification. *Desalination* 130 (2000), pp. 169–175.

Dai, Y.J., Wang, R.Z. & Zhang, H.F.: Parametric analysis to improve the performance of a solar desalination unit with humidification and dehumidification. *Desalination* 142 (2002), pp. 107–118.

Farid, M.M. & Al-Hajaj, A.W.: Solar desalination with humidification–dehumidification cycle. *Desalination* 106 (1996), pp. 427–429.

Farid, M.M., Nawayseh, N.K., Al-Hallaj, S. & Tamimi, A.R.: Solar desalination with humidification dehumidification process: studies of heat and mass transfer. *Proceeding SOLAR 95*, Hobart, Tasmania, 1995, pp. 293–306.

Farid, M.M., Parekh, S., Selman, J.R. & Al-Hallaj, S.: Solar desalination with humidification–dehumidification cycle: mathematical modeling of the unit. *Desalination* 151 (2002), pp. 153–164.

Fritz, S.C., Metcalfe, S.E. & Dean, W.: Holocene climate patterns in the Americas inferred from paleolimnological records. In: V. Markgraf (ed.): *Interhemispheric climate linkages*. Academic Press, San Diego, CA, 2001, pp. 241–263.

Garg, H.P., Adhikari, R.S. & Kumar, R.: Experimental design and computer simulation of multi-effect humidification (MEH)–dehumidification solar distillation. *Desalination* 153 (2002), pp. 81–86.

Grehier, A. & Rojey, A.: Echangeurs à condensation en matériau polymère. *Revue de l'Institut Français du Pétrole* 1 (1989), pp. 77–89.

Houcine, I., Amara, M.B., Guizani, A. & Maalej, M.: Pilot plant testing of a new solar desalination process by a multiple-effect-humidification technique. *Desalination* 196 (2006), pp. 105–124.

Kaguel, S., Nishio, M. & Wakao, N.: Parameter estimation for packed cooling tower operation using a heat input-response technique. *Int. J. Heat Mass Transf.* 31(1988), pp. 2579–2585.

Khedr, M.: Techno-economic investigation of an air humidification-dehumidification desalination process. *Chem. Eng. Technol.* 16 (1993), pp. 270–274.

Klausner, J.F., Mei, R. & Li, Y.: Innovative freshwater production process for fossil fuel plants. U.S. DOE—Energy Information Administration annual report, 2003.

McQuiston, F.C.: Heat, mass and momentum transfer data for five plate-fin tube heat transfer surfaces. *ASHRAE Transactions* 84 (1978) pp. 266–293.

Müller-Holst, H.: Solar thermal desalination using the multiple effect humidification (MEH) method. Advanced research workshop. In: L. Rizzuti, H.M. Ettouney & A. Cipollina (eds): *Solar desalination for the 21st Century*. Springer, Dordrecht, The Netherlands, 2007, pp. 215–225.

Müller-Holst, H., Engelhardt, M., Herve, M. & Scholkopf, W.: Solar thermal seawater desalination systems for decentralized use. *Renew. Energy* 14 (1998), pp. 311–318.

Müller-Holst, H., Engelhardt, M. & Scholkopf, W.: Small-scale thermal seawater desalination simulation and optimisation of system design. *Desalination* 122 (1999), pp. 255–262.

Nafey, A.S., Fath, H.E.S., El-Helaby, S.O. & Soliman, A.M.: Solar desalination using humidification-dehumidification processes. Part II. An experimental investigation. *Energy Convers. Manage.* 45 (2004), pp. 1263–1277.

Narayan, G.P., Sharqawy, M.H., Summers, E.K., Lienhard, J.H., Zubair, S.M. & Antar M.A.: The potential of solar-driven humidification–dehumidification desalination for small-scale decentralized water production. *Renew. Sustain. Energy Rev.* 14 (2010), pp. 1187–1201.

Nawayseh, N.K., Farid, M.M., Omar, A.Z., Al-Hallaj S. & Tamimi, A.R.: A simulation study to improve the performance of a desalination unit constructed in Jordan. *Desalination* 109 (1997), pp. 277–284.

Nawayseh, N.K., Farid, M.M., Al-Hallaj, S. & Tamimi, A.R.: Solar desalination based on humidification process. Part I. Evaluating the heat and mass transfer coefficients. *Energy Convers. Manage.* 40 (1999), pp. 1423–1439.

Orfi, J., Laplante, M., Marmouch, H., Galanis, N., Benhamou, B., Nasrallah S.B. & Nguyen, C.T.: Experimental and theoretical study of a humidification–dehumidification water desalination system using solar energy. *Desalination* 168 (2004), pp. 151–159.

Parekh, S., Farid, M.M., Selman, J.R. & Al-Hallaj, S.: Solar desalination with a humidification dehumidification technique a comprehensive technical review. *Desalination* 160 (2004), pp. 167–186.

Sacadura, J.F.: Initiation aux transferts thermiques. In: *Techniques et Documentation*. Fourth printing, 1993, pp. 251–260.

Sonntag, D.: Vapor pressure formulations based on the ITS-90 and psychometer formulate-Important new values of the physical constants of 1986. *Z. Meteorology* 70 (1990), pp. 5/340–5/344.

Ulber, I., Vajen, K., Uchtmann, H., Engelhardt, M., Krelle, J., Baier, W., Uecker, M., Orths, R. & Ben Bacha, H.: A new concept for solar thermal desalination—results of in-situ measurements. *Proceeding EuroSun '98*, 14–17 September 1998, Portoroz (Slovenia),Volume 2, III.2.51–1.7.

US Geological Survey: Internet site: http://ga.water.usgs.gov/edu/ earthwherewater.html (accessed November 2007).

Wexler, A.: Vapor pressure formulation for water in range 0 to 100°C. *Journal of Research of the National Bureau of Standards A., Physics and Chemistry* 80:A (1976), pp. 775–785.

Wexler, A.: Vapor pressure formulation for ice. *Journal of Research of the National Bureau of Standards A., Physics and Chemistry* 81:A (1977), pp. 5–20.

Yamali, C. & Solmus, I.: A solar desalination system using humidification–dehumidification process: experimental study and comparison with the theoretical results. *Desalination* 220 (2008), pp. 538–551.

Younis, M.A., Fahim, M.A. & Wakao, N.: Heat input-response in cooling tower-zeroth moments of temperature variations. *J. Chem. Eng. Jpn.* 20 (1987) pp. 614–618.

Younis, M.A., Darwish, M.A. & Juwayhel, F.: Experimental and theoretical study of a humidification–dehumidification desalting system. *Desalination* 94 (1993), pp. 11–24.

Zhani, K. & Ben Bacha, H.: Experimental investigation of a new solar desalination prototype using the humidification dehumidification principle. *Renew. Energy* 35 (2010a), pp. 2610–2617.

Zhani, K. & Ben Bacha, H.: Modeling and simulation of a new design of the SMCEC desalination unit using solar energy. *Desalin. Water Treat.* 21 (2010b), pp. 346–356.

CHAPTER 5

Solar PV powered RO systems

Vicente J. Subiela, Baltasar Peñate, Fernando Castellano &
F. Julián Domínguez

> *"Upholding the human right to water is an end in itself and a means for giving substance to the wider rights in the Universal Declaration of Human Rights and other legally binding instruments—including the right to life, to education, to health and to adequate housing. Ensuring that every person has access to at least 20 litres of clean water each day to meet basic needs is a minimum requirement for respecting the right to water—and a minimum target for governments."*
>
> United Nations Development Programme, UNDP (2006)

5.1 INTRODUCTION

As it was presented in Chapter 1, demographic and economic growth is resulting in a continuous increase of freshwater demand for irrigation, industrial and drinking purposes. On the other hand freshwater resources are naturally limited and further on undergo a continuous deterioration by anthropogenic contamination. These developments, in combination with impacts due to climate change, poverty, inequality and unequal power relationships, as well as flawed water management policies, resulting in the fact that the number of regions and countries living under water stress, i.e. a water availability of less than $1700 \, m^3 \, person^{-1} \, year^{-1}$, have increased especially during the last few decades and will further increase significantly in the future (Rogers, 2008; United Nations, 2006).

The situation is particularly severe in developing countries of North Africa and Middle East. The minimum drinking water requirement is much lower than defined by water stress: 30–40 liters per capita and day have been estimated for dry zones in developing countries (Gleick, 1996).The most dramatic effect behind these large numbers is the child mortality; the lack of water quality or quantity is related to more than 1.4 millions of deaths per year (WHO, Water Sanitation and Health, 2010). Nevertheless, international awareness about water is coming up; despite water is not specifically mentioned as a human right, the United Nations have been aware of the importance of water by declaring the period 2005–2015 as the International Decade for Action: Water for Life (United Nations, 2005).

This increasing water demand has caused a strong development of desalination technologies, which have experienced an exponential increment during the last years; according to the IDA (International Desalination Association) reports, the cumulative contracted capacity of desalination plants was 62.8 millions of daily cubic meters at the end of June 2008 (Global Water Intelligence, 2008).

However, desalination requires large quantities of energy, in terms of heat for distillation processes (like multi-effect distillation, or multi-stage flash), or electricity for membrane processes (like reverse osmosis or electrodialysis). Energy is mostly provided from combustion of fossil fuels; thus, water desalination implies CO_2 emissions. As a rough reference, each cubic meter of desalted water from a seawater RO plant means a generation of 3 kg of CO_2. Moreover, desalination increases energy demand, which means a higher external dependence and the consequent economic expense in those countries with low energy resources.

Desalination by using renewable energies overcomes these disadvantages: on one hand, it is a pollution-free system and on the other hand, it uses a local energy source such as solar radiation or wind energy. There is a wide variety of renewable energy powered desalination systems, depending on the type of energy source and water treatment process (for details see Chapter 2). This chapter focuses on a particular combination of autonomous desalination: photovoltaic energy coupled to reverse osmosis; the purpose is to present a theoretical and a practical approach and the experiences obtained from the systems installed in North Africa (Morocco and Tunisia). This coupling has already been tested for a significant period (Herold *et al.*, 1998) and is the most popular option; about 32% of autonomous desalination systems are based on PV powered RO units (Tzen, 2005).

The Canary Islands Institute of Technology (ITC) started to test this combination in 1998 (Subiela *et al.*, 2009). This long experience of ITC in this field has allowed passing from the laboratory to the real world scale, offering a solution to water supply needs in remote inland areas already several years ago. For example a brackish water system, with a capacity of 2.08 m³ h⁻¹, was installed in the Tunisian village of Ksar Ghilène, and commissioned in May 2006 (Peñate *et al.*, 2007). More recently, and thanks to the EU initiative MEDA-Water Program, six units were installed and commissioned in Morocco during 2008 (ADIRA project, 2003–2008); two by the Moroccan NGO FM21 and the other four, by the ITC. The experience has demonstrated the useful role of autonomous desalination to solve the water supply situation in rural areas (Outzourhit *et al.*, 2006, 2009).

This chapter presents a general description of the technical combination concept of photovoltaic driven reverse osmosis units, and provides real data of the technical characteristics of the systems installed in Tunisia and Morocco. The technical issues are complemented with a section on the economic and social aspects. Finally, the chapter concludes with a selection of the most relevant technical recommendations and final conclusions.

5.2   REVIEW OF THE STATE-OF-THE-ART

The connection of desalination with renewable energies has implied an increasing number of publications describing generalities and presenting the state-of-the-art of the technologies (Hanafi, 1994; Bellesiotis *et al.*, 2000; García-Rodríguez, 2002; Delyannis, 2003; Kalogirou, 2005; Mathioulakis, 2007; Eltawil, *et al.*, 2009). In all these references, specific mentions are given to PV powered RO systems.

As mentioned above, the PV-RO combination is the most applied autonomous desalination concept. Table 5.1 presents a list of the installed systems using this technology (Ghermandi and Mesalem, 2009); some data have been completed from other list. The table indicates the location of the systems, their capacity, the type of energy generation system and commissioning year.

5.3   HOW TO IMPLEMENT A PV-RO SYSTEM

5.3.1   *Generalities*

A PV-RO system is a renewable energy driven desalination combination consisting of a photovoltaic field that powers a reverse osmosis unit. The implementation of this type of systems must fulfill a set of conditions:

- *Planning of the project must take the local conditions into account*: A set of data of the particularities of each location must be collected. Not only the elemental data of water demand, feed water composition and solar radiation, but also other related technical, economic, environmental and social aspects have to be considered from the initial steps of the project on; for instance the existence and state of the infrastructures, uses of water, and cultural aspects related to water.

Table 5.1. List of installed PV-RO units.

Location[1]	Capacity [m³ h⁻¹][2]	Energy system type	Year	Ref[3]
Concepción del Oro, Mexico (BW)	0.06	2.5 kWp PV	1978	(a)
Giza, Egypt (BW)	0.25	7 kWp PV	1980	(b)
Jeddah, Saudi Arabia (SW)	0.13	8 kWp PV	1981	(b)
Cituis West Jawa, Indonesia (BW)	1.5	25 kWp PV	1982	(a)
Perth, Australia (BW)	0.5–0.1	1.2 kWp PV	1982	(a)
Vancouver, Canada (SW)	0.02–0.04	0.48 kWp PV	1983–84	(a), (b)
Doha, Qatar (SW)	0.24	11.2 kWp PV	1984	(a), (b)
Sant Nicola, Italy (BW)	0.5	65 kWp PV	1984	(b)
Hassi-Khebi, Algeria (BW)	0.95	2.59 kWp PV	1987–1988	(a)
University of Almeria, Spain (BW)	2.5	23.5 kWp PV	1988	(a)
Lipari Island, Italy (SW)	2	63 kWp PV	1991	(a)
Sadous Riyadh, Saudi Arabia (BW)	0.24	10 kWp PV	1994	(b)
St. Lucie Inlet State Park, Florida, U.S.A. (SW)	0.02	2.7 kWp PV + Diesel Gen.	1995	(a)
Gillen Bore, Australia (BW)	0.05	4.16 kWp PV	1996	(b)
Pozo Izquierdo[5], Gran Canaria, Spain (SW)	0.4	4.8 kWp PV	1998–2002	(c)
Coite-Pedreiras, Brazil (BW)	0.25	1.1 kWp PV	2000	(b)
CREST, Loughborough University, U.K. (SW)	0.06	1.54 kWp PV	2001	(b)
Lavrio, Attiki, Greece (SW)	0.13	3.96 kWp PV + 900 W WEC[4]	2001	(d)
Agricultural Univ. of Athens (AUA), Greece (BW)	0.042	850 Wp PV + 1 kW WEC[4]	2000	(b)
Mesquite, Nevada (BW)	0.05	540 Wp PVs	2003	(b)
Pozo Izquierdo[6], Gran Canaria, Spain (SW)	1.25	5.6 kWp PV	2005	(d)
Ksar Ghiléne, Tunisia	2.1	10.5 kWp PV	2006	(c)
Tamaguerte, Alhaouz Province, Morocco (BW)	1.0	4.8 kWp PV	2008	(e)
Had Dra Commune, Essaouira province, Morocco (BW)	1.0	3.9 kWp PV	2008	(e)
Amellou, Province of Tiznit, Morocco (BW)	1.0	4 kWp PV	2008	(c)
Tangarfa, Province of Tiznit, Morocco (BW)	0.5	2.5 kWp PV	2008	(c)
Azla, Ida ou Azza, Province of Essaouira, Morocco (BW)	1.0	4 kWp PV	2008	(c)
Tazekra, Sidi Ahmed Esayeh, Province of Essaouira, Morocco (BW)	1.0	4 kWp PV	2008	(c)

[1]BW: Brackish Water, SW: Seawater; [2]Nominal capacity; [3](a) García-Rodríguez (2002); (b) Ghermandi and Messalem (2009); (c) Subiela et al. (2009); (d) Tzen et al. (2004); (e) Outzurhuit et al. (2009). [4]Wind energy converter; [5]1998 (1st version); 2000 (2nd version); 2003 (3rd version); [6]With solar trackingenergy

- *Tailor made and tough design*: Despite the concept of a PV-RO system is common to all the projects, there are many options for the design at each part of the system (control, RO unit, power supply system, civil works, etc.).
- *The implementation does not end the day after the commissioning*: A maintenance and operation procedure should be included. This is part of the self-management of the project, which means the local people must have received a previous training that guarantees that they are able to operate & maintain the system.

Thus, the implementation of a PV-RO system, particularly when the installation is in developing countries, implies a complex process in which technical and non-technical aspects must be considered.

5.3.2   *The implementation process*

This section describes, in chronological order, the stages for a general case to implement a PV-RO system.

5.3.2.1   *Stage 1: Analysis of the local conditions*
The first step is to learn the characteristics of the location or the potential locations wherein the system should be installed. In some cases, the decision of the final location is not fixed and requires a previous analysis to select the most appropriate one. A good starting point is to have a basic idea of the potential areas or regions of the country; fortunately basic data about water situation and availability of solar resources are known for many countries. The United Nations Water and Sanitation Program describes the water situation in a specific report (United Nations, 2006). Among others, the initiative from NASA allows access to local solar radiation data (CETC-Varennes, RET Screen).

Secondly it is recommended to elaborate a questionnaire with all the relevant issues to evaluate the potential locations (technical, economic, and social aspects) as a reference the reader can consult Banat *et al.* (2007). After collecting all the data, the information should be analyzed and then, according to a set of objective or quantifiable criteria, initiate a selection process to find the most suitable location.

5.3.2.2   *Stage 2: Design*
A good design necessarily requires the information from stage 1; in other words, the design must be subjected to the local conditions and not the opposite. As a reference, the characteristics to be considered within the design phase are the following:

- *Selection of high quality equipment & materials*: Although it means a higher investment cost, it will reduce the possibilities of operation failures in the future and will facilitate the maintenance tasks.
- *Control strategy*: The control system must be conceived to minimize the energy consumption, maximize the water production and increase the lifetime of the components.
- *Local training*: A specific training phase, addressed to the local technicians who will assume the local maintenance and follow-up actions, is one of the most important stages. The training should take into account the basic operation aspects of the system, and how to carry out the preventive and corrective maintenance actions.
- *Physical-chemical analysis of the raw water*: This is key information for the most appropriate selection of RO elements, mainly affecting the membranes, high pressure pumps and pre-treatment.
- *The particular location of the system*: The selection of the location must find a compromise between the feed water point and the freshwater source; if the system is far from the raw water well, then the pumping power will increase and a larger PV field will be needed. Nevertheless, the system is normally positioned close to the water demand point, from where water is distributed by gravity in order to avoid possible leakages along the freshwater pipe.

- *Use of batteries*: The inclusion of this energy storage system will allow operating the system for more hours (periods with low solar radiation).
- *Use of an energy recovery system*: This is a strongly recommended decision when the feed flow is seawater. The installation of this device makes possible to exploit the pressure energy of brine flow to pressure up part of the feed flow, leading to a significant reduction on the specific energy consumption, and consequently on the size of the photovoltaic energy system.
- *Simple design*: The simpler the design the easier are the manufacturing and installation processes. On the other hand, for the particular question of the building and related civil works, a simple design will be helpful for the construction companies, normally assumed by local entities with low to medium qualification workers.

5.3.2.3 *Stage 3: Manufacturing*

Despite most of the components can be obtained at commercial level (batteries, PV panels, pumps or converters), as the system has been tailor-made designed, in some cases it is necessary to manufacture parts of it. The reverse osmosis plant can be purchased but it makes a lot of sense to design and manufacture this element since there are many possible options according to the amount of water to be produced and the quality of the raw water. There are several companies, expert in RO, able to manufacture specific designs for low capacity units (under $200\,\mathrm{m}^3\,\mathrm{day}^{-1}$).

The elements of the RO unit (filters, chemical products tanks, high pressure pump, pressure vessels, pipes, valves and hydraulic accessories, measurement devices, transmitters, etc.) should be assembled in a compact way, in order to take into account that the unit will be lately transported inside containers either of 6 or 12 m.

5.3.2.4 *Stage 4: Transport*

After all the equipment is ready it must be transported to the final location. This is not a direct process at all and a strong logistics is behind it. On one hand, the equipment must be prepared in such a way that it can be kept in a container, so the different elements should be grouped into numbered or referenced lots. On the other hand it is necessary to establish a transport plan, considering the packaging and preparation of lots for container(s), shipping by road and sea, proceedings at the customs agent of the destination port and finally moving by truck(s) to the final location.

Additionally, this stage implies significant cost and time, since the process at the local customs is not easy. The use of a letter from a local authority certifying that the equipment is part of the supplies of an international cooperation project and is not for commercial/business purposes (if that is the case) could be very helpful to avoid or reduce the related expenses and taxes.

Due to the specific difficulties of the local actions of the transport stage, it is necessary to include the participation of a subcontracted company to arrange this process directly; otherwise, the unexpected delays, costs and problems will make it difficult to guarantee an appropriate delivery of the equipment.

5.3.2.5 *Stage 5: Installation*

The installation stage can start as soon as the equipment has been transported to the final location and the building of the required civil works has been completed. The installation stage includes actually several activities to be performed for the two main parts of the system: the hydraulic set and the electric set.

- *Hydraulic part-low pressure circuit*: This section includes the raw water intake and feed pumping system to the inlet of the high pressure pump, and the outgoing brine and desalinated water pipes. At inland locations the feed pump is located inside a well, several meters underwater, but always with a free space over the bottom point; a one way valve and a filter are installed at the water intake. The commonly used material is high-density polyethylene.

- *Hydraulic part-high pressure circuit*: The high pressure circuit includes the feed pipe to the pressure vessels wherein the reverse osmosis membranes extract freshwater from the salty water. In case of seawater, the rejected water (brine) flows out with high pressure (close to 60 bar (6 MPa) for a seawater salinity of 35 g L^{-1}) that can be used to pressurize part of the feed flow by an energy recovery device, and then a booster pump supplies the additional pressure required to achieve the operation pressure at the inlet of the RO modules. If the feed water salinity is low (under 3–4 g L^{-1}) then it is possible to increase the desalinated flow by mixing a small amount of feed flow with the product water flow (blending). In case of seawater, the high operation pressure (60–70 bar) implies the selection of high quality stainless steel (AISI 316L or higher) for the material used in the pipes; in case of brackish water, the lower operation pressure allows the use of less resistant materials, as polypropylene.
- *Electric installation-DC circuit*: This circuit starts in the photovoltaic field, wherein the solar radiation is converted into direct current (DC). Photovoltaic panels must be connected in series in order to get the required DC operation voltage and also in parallel to obtain the appropriate output current according to the power demand. This circuit is connected to the batteries by a charge controller; this device operates as a switch to control when the DC electricity is stored in the batteries and when used directly to satisfy the load.
- *Electric installation-AC circuit*: The alternating current (AC) circuit starts in the inverter, a DC/AC converter that supplies the power to the loads of the whole system: feed pump, high pressure pump, lighting and auxiliary loads.
- *Monitoring installation*: At least a minimum set of sensors and transmitters must be installed to monitor the system. The key signals to be monitored are the following: feed and product water flows, product water conductivity, and high pressure (before and after the membranes), and low pressure (filters inlet and outlet points). Therefore, a set of wires will connect the transmitters with the data acquisition system, which will collect and store the analog signals to be subsequently downloaded to a PC, or transmitted online to a remote server.
- *Control system installation*: The main mission of the control system is to avoid failures and operate the unit to produce the maximum or the desired amount of water per day. Thus, the control system, usually implemented by a programming logic controller (PLC), will receive a set of input signals (digital and analog information) to decide when to stop/start the unit and to inform and manage the alarms (output signals).

5.3.2.6   *Stage 6: Commissioning*
Once the installation is completed, it is necessary to carry out a commissioning procedure to check that all components of the system are able to operate correctly. As in the case of the installation stage, the commissioning considers all the parts of the system; the most relevant actions are the following:

- *Hydraulic part*: A set of pressure tests is made in order to identify possible leakages in the low and high pressure circuits. It is required to run the high pressure pump and check the quality and quantity of produced water. It is also indicated to check that all the alarms and security systems operate correctly and to guarantee that all of them can be activated. The auxiliary circuits, flushing and chemical dosing must also be verified. Finally the commissioning must include the definition of the set-up operation points such as pressure, flows dosing and blending (this last item only for low salinity raw waters, i.e., under 10 g L^{-1}).
- *Electric part*: There is a number of parameters to be set up in the charge controller and the inverter; the operator will have to follow the indications given in the specific handbook of both devices. There are no specific recommendations since this issue depends on each manufacturer. The DC voltage and current must be checked in all the points in order to make sure that all the elements are ready to operate correctly and generate the foreseen power according to the technical characteristics of each device. The batteries require a specific attention since they are one of the most fragile elements, by checking the electrolyte density, and the individual voltage, of each element.

- *Control system*: As a previous action to regular operation, the control unit has to be proved to confirm that the internal software runs properly; in some cases this step is used to adjust the values of some internal programming control variables or parameters.

5.3.2.7 Stage 7: Operation & Maintenance

The project does not finish after the commissioning. The operation and maintenance (O&M) stage is in fact more critical than it seems. Despite the system has been designed and developed from the low maintenance point of view, there is always a set of preventive and corrective actions related to O&M. The system will be considered as a sustainable project provided that there is a concept of self-management; this includes two major aspects to be taken into account:

- *Technical sustainability*: The availability of a local staff with the capacity to carry out these actions and the possibility of obtaining the consumables (mainly chemical products for the pre- and post-treatment of the RO unit and distilled water for the batteries) and finding local suppliers for possible equipment replacement.
- *Economic sustainability*: A very common mistake in the economic assessment of an autonomous desalination system is to consider just the investment, and exclude the associated O&M costs. A minimum expense is required for the suitable operation of the system, mainly personal and consumable costs. If the project takes place in a developing country with strong economic restrictions, it is essential on the one hand to include the local contribution to co-funding the capital expenses, and on the other hand, to implement a strategy to cover the O&M costs.

Each system is different and needs its specific O&M activities; nevertheless, as a reference, a set of the most important tasks to be completed are the following:

- *Hydraulic part-basic maintenance*: General cleaning, daily follow-up by filling a form with the main operation data, refilling of the chemical products tanks, general inspection of the RO unit to identify leakages, unexpected vibrations or noises.
- *Hydraulic part-advanced maintenance*: Appropriate comprehension and corrective actions during alarms, analysis of the operation parameters, general checking of equipment and replacement of damaged and old elements (case of cartridge filters).
- *Electric part-basic maintenance*: Washing and cleaning of PV panels, checking the electrolyte density of batteries and refilling (with distilled water) if necessary, tighten the connection screws of all the electric cabinets/panels.
- *Electric part-advanced maintenance*: Review the operation parameters, checking the batteries voltage, general checking of equipment and substitution of damaged and old elements.

5.4   DESCRIPTION OF THE TECHNOLOGICAL CONCEPT

5.4.1 Approach to the solution

The choice of the technological solution will be mainly determined by the characteristics of the raw water and the energy source existing at the target site. Other local conditions such as level and type of qualifications of operators that could carry out further O&M tasks will be also taken into account in the selection.

From the technical point of view, the parameters to evaluate with respect to the choice of the desalination technology are the specific energy consumption, the need for chemicals and durability of components, the modularity options, the flexibility in terms of power demand, as well as the simplicity of the operation and maintenance.

On the other hand, solar radiation (at least monthly mean values) are the main data to be evaluated; if available, wind speed data can be helpful in order to consider a possible hybrid power supply. Finally, another point to include is the number, type and commercial activities of the local companies related with water and electricity installations; the identification of these providers will be useful when there is a breakdown to be repaired.

After this analysis, it is probable that there is more than one possible technical solution; then an economic study will allow finding the most favorable option.

5.4.2   *Description of the PV technology*

Photovoltaic energy is a method of generating electrical power by converting solar radiation into direct current electricity using semiconductors that exhibit the photovoltaic effect. Photovoltaic power generation employs solar panels composed of a number of cells containing a photovoltaic material. Materials presently used for PV include monocrystalline silicon, polycrystalline silicon, and amorphous silicon, among others.

Photovoltaic systems can be grouped into stand-alone and grid-connected systems. In stand-alone systems, the solar energy yield is matched to the energy demand. Since the solar power often does not coincide in time with the instantaneous power required from the connected loads, storage systems (as batteries) are generally used. If the PV system is supported by an additional power source – for example, a wind or diesel generator – this is known as a hybrid system. In grid-connected systems the public electricity grid functions as energy storage, and all of the energy they generate is fed into the grid.

Wherever it is not possible to install electricity supply from the mains utility grid, or where this is not cost-effective or not desirable, stand-alone photovoltaic systems can be installed. This type of PV systems is generally much smaller than on-grid PV installations.

The three main categories of off-grid PV systems are:

- PV systems providing DC power only,
- PV systems providing AC power through an inverter,
- PV hybrid systems: Diesel, wind or hydro.

The range of applications is constantly growing. There is a great potential for using stand-alone systems in developing countries where vast areas are still frequently not supplied by an electrical grid. Technological innovations and new lower-cost production methods are opening up potential in industrialized countries as well. Such applications include water pumps, telecommunications, solar-home systems, and also larger units for islands.

Developing countries, where many remote villages are often more than five kilometers away from grid power, have begun using photovoltaic energy for rural electrification. With this situation, stand-alone desalination system for water supply projects can be proposed as new applications for stand-alone systems.

5.4.3   *Stand-alone photovoltaic sub-systems*

Stand-alone PV systems consist of different elements; this section presents an essential technical description of each one.

5.4.3.1   *Modules*
The modules in PV arrays are usually configured to give a nominal DC voltage of 12 V, 24 V and up to 48 V, in case of large systems. This means that the modules are usually connected in series. In order to facilitate easier interconnection of the modules, modules with junction boxes should be used rather than modules with plug-in leads (as used to configure the long strings in grid-connected arrays).

5.4.3.2   *Batteries*
Energy storage is required in most stand-alone systems, as energy generation and demand do not generally coincide. The solar power generated during the day is very often not required until the

evening and therefore has to be temporarily stored. Longer periods of overcast weather also have to be catered. Most stand-alone solar systems have batteries except solar water pumping systems.

Rechargeable lead-acid battery is the most common type used in stand-alone solar systems. These are the most cost-effective and can handle large and small charging currents with high efficiency. In PV systems, the storage capacities are generally in the range of 1 to 100 kWh, although a few systems in the MWh range have already been implemented. Other commercially available types of rechargeable batteries are nickel-cadmium, nickel metal hydride and lithium ion batteries.

5.4.3.3 *Charge controller*
In stand-alone systems, the voltage of the PV array should be matched to that of the batteries; this process is regulated by the charge controller. The output charge and nominal voltage from this device must be higher than the battery voltage to guarantee the appropriate charge of batteries. On the other hand, there is a voltage drop through cables and the diodes line, which increases at high temperatures, usually limited to around 1 to 2%. The charge controller therefore measures the battery voltage and protects the battery against overcharging. During low solar radiation periods the PV voltage decreases, then the battery will discharge *via* the array. To prevent this, a reverse current diode is used; this protection element is usually integrated within the charge controller. There are three main kinds of charge controllers: series, shunt (parallel controllers), and maximum power point (MPP).

The tasks of the charge controller are: optimization of the charge of batteries; protection against overcharge; prevention of unwanted discharging; deep discharge limitation; and information on the state of charge of the batteries.

5.4.3.4 *Inverters*
In a PV stand-alone system, the inverter is the electric device that provides the conventional AC loads by converting the direct current into alternating current. The objective of a stand-alone inverter is to enable the operation of a large range of loads: from rugged construction tools through domestic appliances to lighting, or desalination units.

There are three different types of inverters: sine-wave, modified sine-wave and square-wave.

5.4.4 *Stand-alone PV-RO*

The autonomous desalination concept of photovoltaic powered reverse osmosis (PV-RO) consists of a stand-alone PV system that supplies power to a RO desalination plant. Therefore the main loads of this system are the following:

- *Feed water pump*: The raw water, either sea or brackish water, must be collected and pumped to the inlet of the desalination plant in order to supply the required pressure to pass through the pre-treatment of the RO unit. The way of collecting this water is very important for the optimum operation of the unit and the preservation of the RO membranes.
- *Chemical products dosing pumps*: Depending on the quality of the feed water, a chemical pre-treatment, based on dosing different products, is done to avoid scaling (solid salt deposits) and fouling (organic matter deposits). These products are incorporated to the feed flow by a set of small dosing pumps. The most typical products are sodium hypochlorite, to avoid fouling and anti-scaling and acid to prevent scaling.
- *High pressure pump*: Once the feed flow has passed through the physical (filters) and chemical pre-treatments, then the feed water is pressured up to reach the operation pressure. This pressure must be higher than the osmotic pressure, which is proportional to the salinity gradient inside the membrane. So, the higher the salinity, the higher is the required pressure; usual operation values are 60 bar (6 MPa), for seawater, and 10–15 bar (1–1.5 MPa), for brackish water. The high pressure pump is the element with the greatest power demand of the RO plant.
- *Booster pump*: In seawater RO units it is possible to recover the energy from the brine or rejected flow (at about 58 bar or 5.8 MPa) by using an energy recovery device (turbines or

isobaric chambers) that transmits this pressure to the same amount of low-pressure feed flow. As efficiency of these devices is slightly under 100%, the outgoing feed pressure is about 57 bar (5.7 MPa), thus the booster pump provides those an additional pressure of 3 bar (60–57 bar or 6–5.7 MPa) to reach the membranes inlet pressure. The inclusion of the energy recovery device and the associated booster pump implies a significant reduction on the specific energy consumption.

- *Auxiliary loads*: The RO plant includes several complementary elements with low energy consumption but necessary for the operation such as control elements, automatic switches, monitoring transmitters, solenoid valves and others.

5.5 TECHNICAL CHARACTERISTICS OF SELECTED OPERATING SYSTEMS

The Canary Islands Institute of Technology (ITC) has been researching on PV-RO since 1997; as a result of this R&D process an international patent was created (DESSOL®). Once the concept became a mature product, it was possible to transfer the system from the ITC lab facilities to the real world. Two systems are described in this chapter:

- *System in Tunisia*: It is the first installed and is located in the village of Ksar Ghilène. This system has been operating without problems since the commissioning in May 2006.
- *Systems in Morocco*: PV-RO units were installed in four inland rural communities of Morocco. They were commissioned in May 2008 and until now all the units are operative.

5.5.1 *PV-RO system in Tunisia*

The installation to be described was executed within the framework of the Spanish-Tunisian cooperation. It was a project focused on the freshwater supply by a PV driven RO desalination system. The project is located in Ksar Ghilène, an isolated inland village of 300 inhabitants, at the South of Tunisia, in the Sahara desert (Fig. 5.1).

The village is 150 km away from the nearest electrical grid and 60 km away from the nearest freshwater well.

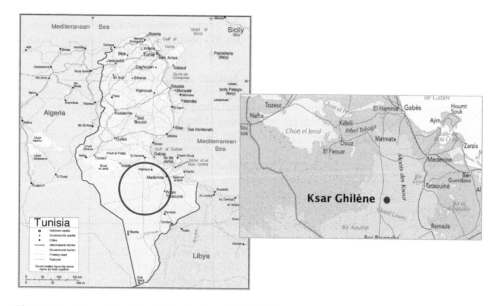

Figure 5.1.   Maps of Tunisia with the location of the PV-RO system.

The partners of this project are the Spanish International Cooperation Agency (AECI), the National Agency for the Control of Energy Consumption (ANME), the Regional Directorate for Agricultural Development of Kèbili (CRDA), and the Government of the Canary Islands (through the General Directorate for Relationships with Africa and the Canary Islands Institute of Technology, ITC).

The nominal capacity for water production is 50 m^3 day^{-1}, but as the power supply depends on the solar radiation, the average freshwater production is 15 m^3 day^{-1}, i.e., the system can operate about 7.5 h day^{-1}. Raw water comes from a brackish well, located in the near oasis. The power supply is provided by a 10.5 kWp PV solar generator with energy accumulation by batteries.

5.5.1.1   *The local data*

The main local information is presented in Table 5.2, indicating that there is low wind potential but excellent solar energy resource.

Concerning the local hydraulic infrastructure, there was already a pumping system in the artesian well and a piping network that distributed the water from the reservoir by gravity.

5.5.1.2   *The building*

One of the peculiarities of the area are the local high temperatures, especially during summer (over 50°C). Therefore, a specific building was designed to take into account this fact; so it was constructed partially buried, with an access ramp to put the equipment in it. On the other hand, the roof and the south wall were used to install the photovoltaic panels, thus the shadow produced by the PV field also contributes to limit the inside temperature. The building is illustrated in Figure 5.2.

Table 5.2.   Basic data of the Ksar Ghilène village (Tunisia).

Concept	Value
Solar radiation (annual average)	5600 kWh m^{-2}
Ambient temperature	0–60°C
Wind speed (annual average)	3.8 m s^{-1}
Brackish water salinity	4280 mg L^{-1}
Brackish water temperature	15–35°C
Daily water consumption (summer)	15 m^3
Capacity of freshwater storage tank	30 m^3

Figure 5.2.   External views of the building for the PV-RO system installed at Ksar Ghilène village, Tunisia.

Table 5.3.   Operation data of the desalination plant (Ksar Ghilène village, Tunisia).

Data	Value	Data	Value
Raw water supply flow	3.00 m³ h⁻¹	Working temperature	19–32°C
Pressure supply	2.35 bar	Feed water salinity	4503 mg L⁻¹
Input flow to membranes	5.20 m³ h⁻¹	Number of membranes	3
Recirculation flow	2.2 m³ h⁻¹	Feed water SDI	<3
Inlet pressure to membranes	11.43 bar	Total power demand	5 kW
Freshwater flow	2.10 m³ h⁻¹	RO power demand	3.57 kW
Total recovery	70%	Total specific energy consumption	2.3 kWh m⁻³

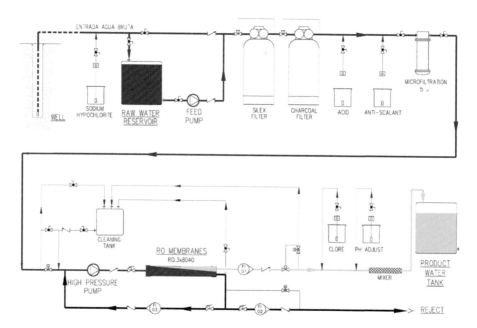

Figure 5.3.   Hydraulic diagram of the RO unit installed at Ksar Ghilène village (Tunisia).

5.5.1.3 *The RO plant*

The RO unit was designed according to the chemical and physical data of the raw water and the local water demand. The raw water temperature is high (more than 35°C) due to the underground conditions of the aquifer. Since this could be slightly over the operation limits of the membranes, it was necessary to include a feed water storage tank inside the building in order to reduce the temperature by a few degrees.

Table 5.3 indicates the main operation parameters of the unit.

The hydraulic diagram of the RO unit is represented in Figure 5.3. The most important circuits of the plant are indicated. Raw water from the well is collected in a buffer tank, from where it is pumped to the pre-treatment system (filters, chemical dosing); then feed water is mixed with part of brine flow (recirculation flow) obtaining the total input flow that is pressurized towards the membranes. The product water is post-treated and then stored; the rejected flow is partially by-passed to the feed line and the rest is conducted to the final disposal point. The recirculation of brine allows increasing the recovery up to 70%. On the other hand, the diagram represents the cleaning circuit as well.

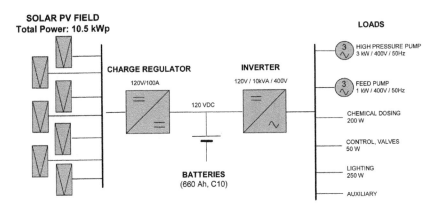

Figure 5.4. Electric diagram of the RO unit installed at Ksar Ghilène village (Tunisia).

Freshwater is stored in an elevated tank to be distributed by gravity to a set of fountains. Brine flow is disposed at a very distant point with no risk of underground water pollution.

5.5.1.4 *The energy generation system*
The installation is composed of seven photovoltaic generators in parallel, with a total nominal (peak) power of 10.5 kWp. Each generator is composed of 10 photovoltaic modules, with an output voltage of 12 VDC and each module has a power of 150 Wp, connected in series. Thus, the total nominal voltage of each generator is 120 VDC and the peak power 1.5 kWp. The modules are assembled on seven metallic structures with a fixed inclination of 40°. Five of the generation lines are located on the roof and the other two at the south wall.

The PV field supplies energy to the stand-alone electric grid, which is composed of a charge controller, a batteries bank, with a 660 Ah (C10) nominal capacity, and a 10 kW inverter. A sketch of this installation is illustrated in Figure 5.4.

The photovoltaic generators are independently wired to the "DC panel", from where a DC power line goes to the charge regulator and from there to the batteries and the inverter. The AC output line of the inverter supplies the 230 V/50 Hz AC panel, to the consumption lines that go to the different loads: lighting (250 W), desalination plant (5000 W), and ventilation (500 W).

5.5.1.5 *The operation of the system*
According to the supplied data and the operating mode, the desalination plant consumed maximal 35 kWh day^{-1} during the summer months, with a water production of 15 m^3 day^{-1} and minimal 16 kWh day^{-1} in the winter months, producing 7.5 m^3 day^{-1} of freshwater.

Figure 5.5 shows the energy flow along four different days. The dark blue line shows the power from the solar photovoltaic field (notice that the first one is for a sunny day, the second one for a day with some clouds at afternoon, the third one for a very low radiation day, and the fourth one for a cloudy day). The red line represents the power load consumed by the RO plant (notice that there is no operation during the third day and partial operation during the last one). The light blue indicates the progress in the capacity of the batteries; during the first two days, the capacity increases when the RO unit is stopped or when the PV power is higher than the demand (noon hours) and decreases when PV power is lower than the demand. During the third day and the first part of the fourth day, all the PV power is used to recharge the batteries, so the capacity rises. During nights there is a minimum consumption due to minor lighting loads and stand-by power converter equipment.

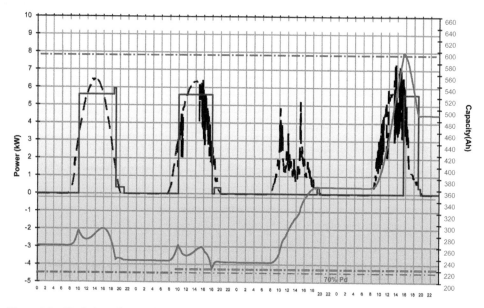

Figure 5.5.   Evolution of power from the PV (dark blue line) RO demanded power (red line) and capacity of batteries (light blue).

5.5.2   *PV-RO systems in Morocco*

The installation of the PV-RO systems of Morocco was carried out within the framework of the ADIRA project (2003–2008)[1] with the collaboration of the local partners (local governments from the provinces of Essaouira and Tiznit, municipal authorities and local associations) and the Moroccan partner of the project (the NGO FM21). The European Commission, the Government of the Canary Islands, through the General Directorate for Relationships with Africa, the Canary Islands Institute of Technology (ITC) and the Governments of Essaouira and Tiznit provinces provided financial support. The geographic location of the four systems is indicated in the map of Figure 5.6.

5.5.2.1   *The local data (Morocco)*
As in the case of Tunisia, most of remote areas with drinking water needs are inland locations where raw water is brackish and taken from a well. The main data for each village are summarized in Table 5.4.

5.5.2.2   *The building*
A common concept was decided for the four cases, consisting of a building hosting the different parts in several rooms: RO unit, batteries, electric converters and related devices, water storage tank, spare parts room and guardian room. The roof of the building was used to locate the PV panels. Figure 5.7 shows a diagram of this construction. Photographs of the four systems can be seen in Figure 5.8.

---

[1]Action co-funded by the European Commission through the MEDA-Water Program, including installation of autonomous desalination units in several locations: Morocco (6 units), Turkey (2 units), Jordan (1 unit), and Cyprus (1 unit); 10 in total, 9 PV-RO systems and one solar thermal driven humidification-dehumidification system.

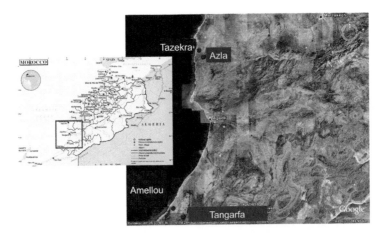

Figure 5.6. Geographic location of the Moroccan villages wherein the four PV-RO systems are installed.

Table 5.4. Characteristics of the Moroccan villages.

	Azla	Tazekra	Amellou	Tangarfa
Province	Essaouira	Essaouira	Ifni	Ifni
Population	300	700	300	300
Electric grid	Weak	Yes	No	Weak
Existing pump in the well	Yes, electric	Yes, electric	No	Yes, diesel
Access	Track 0.5 km	Road	Track 0.5 km	Track 6 km

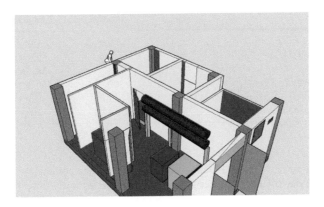

Figure 5.7. Inside view of the building used for the PV-RO systems in Morocco.

5.5.2.3 *The RO plant*

The unitary nominal capacity of the reverse osmosis desalination plant is $1000 \, \mathrm{L \, h^{-1}}$ (for 3 units) and $500 \, \mathrm{L \, h^{-1}}$ (for the 4th unit). The RO plant is able to operate at two different "feed flow – pressure" points, which allows to shift, from partial to total power demand and *vice versa*. This is possible thanks to a frequency converter that modifies the AC frequency supplied to the electric motor coupled to the high pressure pump: the higher the frequency, the more feed flow and pressure available, and consequently more power demand.

Figure 5.8.   External views of the buildings used for the PV-RO units installed in Morocco: (1) Amellou (Ifni Province), (2) Tazekra (Essaouira Province), (3) Tangarfa (Ifni Province) and (4) Azla (Essaouira Province).

Table 5.5.   Operation data of the PV-RO units installed in Morocco.

	Azla	Tazekra	Amellou	Tangarfa
Feed water salinity [g L^{-1}]	2.4	8.2	2.9	2.9
Freshwater flow [m^3 h^{-1}]	1	0.9	0.5–1	0.5
Recovery [%]	76.7	58	38–76.7	76.7
Freshwater conductivity [μS cm^{-1}]	500	300	250–450	140
Operation pressure [bar]	11.3	13.2	9.3–11.2	13.3
Specific Energy demand [kWh m^{-3}]	2.33	3.03	1.86	3.62

In case of 100% power load, the RO unit produces 100% of the nominal water flow; in case of partial power load, the RO unit produces 50% of the desalinated water flow. Table 5.5 presents a set of operation values. As an example of the operation under partial load, the case of Amellou includes a range of values: partial power operation matches the low value of product flow, recovery and pressure, and also, the high value of conductivity. The highest values of energy consumption are for the case of Tazekra (due to the high salinity of feed water) and Tangarfa (due to small capacity).

In all four cases freshwater is stored in an elevated tank to be distributed by gravity to a near fountain, and brine flow is disposed in evaporation ponds covering an area of 100 m^2 per pond.

5.5.2.4   *The energy generation system*

The concept of the photovoltaic system is similar to the case of Tunisia: PV field, charge controller, batteries, inverter and loads. The PV panels are connected in parallel (8 sets) and series

Table 5.6. Electric data of the PV generation systems installed in Morocco.

	Azla	Tazekra	Amellou	Tangarfa
PV installed power [kWp]	4	4	4	2.5
Number of modules [–]	32	32	32	20
PV surface [m²]	33	33	33	19
DC Voltage [V]	48	48	48	48
Batteries capacity [Ah][1]	650	650	650	450
Demanded power [kW]	1.2	1.5	1.7[2]	1.0

[1] Discharge period: 10 h
[2] Feed water pump is PV supplied only for this case

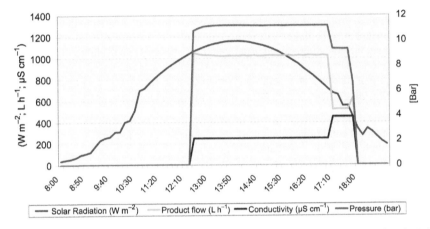

Figure 5.9. Evolution of RO operation variables along one day at Amellou desalination plant for July 15, 2010.

$(4 \times 12$ VDC), so the configuration is 8×4 and nominal output voltage is 48 VDC. Table 5.6 shows of the values of the main technical characteristics.

5.5.2.5 *The operation of the system*

The target of the operation strategy was conceived to maximize the daily flow of water but maintaining the system within safe operation conditions. The control system collects a set of input signals: instantaneous solar radiation, voltage output at the batteries, and state of the RO unit (stopped/operating) and the float switches in the feed and freshwater tanks. From these data, and the specific hour, the control system decides the following possible orders: if the RO unit is stopped, then the control system will start it up; and if the unit is operating, the control system will switch the operation point from maximum to medium power load or *vice versa*.

Figure 5.9 illustrates this variation along one day. The reduction of the solar radiation at the end of the day decreases the available power, so the system reacts by changing the operation point (see the clear reduction in the product water flow from 17:00) and the increment in the product water conductivity (left axis) due to the diminution in feed pressure to the membranes (right axis).

5.6 ECONOMIC AND SOCIAL ISSUES

Leaving aside the arguments in favor of the solar PV-RO, major limitations on the implementation of this kind of stand-alone solution are the site-dependent character and the relative high investment costs of the equipments. One of the main reasons for the current limited use of PV-RO is related to the capital-intensive cost of the project and the low social and economic capacity of the final beneficiary (poor level of training and lack of resources).

This section provides an overview of the capital and running costs of this technology and the interconnection with local, institutional and social issues. Obviously, the answer about how the cost of water affects social aspects is not easy, since it includes a set of multiple elements. Thus, this section tries to identify and describe, in a general manner, this multifaceted answer.

5.6.1   *Water cost analysis*

5.6.1.1   *Generalities*
The cost analysis for a PV-RO project is specific of the site chosen for the installation. The capital cost (investment cost) is the most important cost and requires an adequate work out and planning within the project. Besides, the cost analysis is strongly related to the sustainability of the installation and must be linked to a management plan to warrant its future. Several costs are integrated within the cost of produced water and it is necessary and useful to distinguish them.

As it was described in previous section, a PV-RO installation is composed of several equipments and its design is a function of several key variables. The investment and operational cost are affected by both, but the main parameters to take into account for a cost analysis are the following:

- *Site conditions and isolation*: The site selection requires a deep analysis to know the specific and relevant local conditions which mark the technical and management success of the action and the complexity of the installation. In general, the installation design and the investment requirement are defined by the following items:
  - Quality of feed water,
  - Feed water collection system,
  - Availability of solar resources,
  - Possibility of connection to the electrical grid,
  - Existing basic water infrastructures,
  - Availability of flat area,
  - Climatic conditions and environmental restrictions,
  - Site isolation and location of suppliers to cover the principal maintenance needs.
- *Quality of the feed water and freshwater demand*: The lower feed water salinity is, the lower are the energy and chemical consumptions. The energy required depends on the salinity; the higher the salinity, the higher the pressure required will be. The total quantity of chemicals required (sodium hypochlorite, anti-scaling, hydrochloric acid) is a function of the feed water volume pumped. Product water demand stresses the desalination unit capacity and the total energy required, i.e., the size of PV field and batteries capacity mainly.
- *Social and institutional factors*: Social and political interests should be considered as very influent for the success of the project (ADIRA Partnership, 2008). Additionally, the local training requirement and sustainability of the installation are key factors for the support of the infrastructure after the commissioning. The coordination and response of the stakeholders and beneficiaries are really needed to operate the unit and manage the freshwater supply with minimal disruption and breakdowns.

5.6.1.2   *Capital costs*
From the definition of the previous topics, the main capital investment costs, including logistic and commissioning costs, are:

- *Equipments and accessories*: RO unit, water distribution pipelines, PV solar panels and structures, batteries, electrical connections and power control devices.
- *Civil works*: Building, solar field foundations, freshwater storage, feed water catchment, etc. and land acquisition (if it is required).

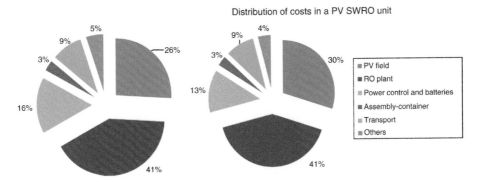

Distribution of costs in a PV SWRO unit

PV field
RO plant
Power control and batteries
Assembly-container
Transport
Others

Figure 5.10.   Pie charts with the general distribution of capital costs: Left: brackish water; right: seawater. Source: Own elaboration considering RO units of 50 m^3 day^{-1} nominal capacity (24 h). (Note that civil works and brine discharge are not included).

- *Other costs*: Those related to transport, training, operation handbooks, taxes, customs clearance, permits and environmental requirements as brine disposal.

Some small-scale solutions (up to 3000 L day^{-1}) are supplied as a compact preinstalled plant with a lower capital costs and minimal ground space requirement in comparison with a building or civil work infrastructure in place.

General distributions of the main investment costs of both seawater and brackish water compact RO-PV units of 50 m^3 day^{-1} nominal capacity (24 h) are shown in the Figure 5.10. Civil works and brine discharge are not included.

During the last years, PV module costs have been reduced considerably (IEA, 2007). This is a clear benefit over the PV-RO investment costs, where the cost of isolated kWp installed has decreased from 10 € Wp^{-1} to less than 6 € Wp^{-1} (including panels, structures, power control, batteries, transport and installation).

Inland installations have the limitation of finding a correct discharge for brine. In some cases a specific sum of the budget goes to build the brine disposal system. There are not many possible solutions: brine is sometimes sent back underground (much deeper than the feed water); in other cases it is disposed in evaporation ponds, and there are examples of mixing brine with feed water to be re-used in irrigation of special crops that are able to grow under high salinity conditions.

5.6.1.3 *Running costs*
In relation to the cost of the water produced for any PV-RO unit, two types of costs can be considered: fixed and variable costs. Fixed costs can be defined as expenses that do not change along the year (amortization, salaries, etc.). Variable costs are paid according to the quantity of operation hours (consumables, spare parts, etc.).

In general, the capital cost is amortized along the installation lifetime, which in the case of desalination unit, is usually considered as 15 years. Currently, if a normal installation of desalination or solar PV field connected to the grid is managed to have at least the fixed and variable costs covered, an off-grid PV-RO project is carried out within specific co-funding and cooperation projects. So, this kind of projects would receive a subsidy for the investment and the final owner is not usually obligated to recover the initial investment. It is a good scheme to support a PV-RO installation as the investor does not have to bear the high investment cost in total. It reduces the final water cost in terms of € m^{-3} as indicated in a specific report on schemes for subsidizing renewable energy driven desalination deliverable (ProDes Project, 2010).

Exploitation costs (Fig. 5.11) include fixed and variable costs. Basically, the O&M costs in this kind of installation are low and consist of labor cost, consumables (chemicals, filters and distillate water for batteries bank replacement) and spare parts (RO membranes, batteries,

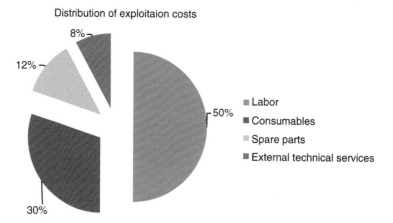

Figure 5.11.   Pie chart with the general distribution of exploitation costs. Source: Own elaboration considering a 50 m³ day⁻¹ nominal capacity (24 hours).

Table 5.7.   Water cost of different PV-RO experiences.

	Case study SWRO-PV[1]	Case study SWRO-PV[1]	Case study BWRO-PV[1] unit	Lampedusa SWRO-PV unit[2]	Azla BWRO-PV unit[3]	Ksar Ghilène BWRO-PV unit[3]
Capacity (L h⁻¹)	417	2083	2083	5000	1000	2083
Total water cost (€ m⁻³)	5.68	3.87	2.09	6.50	4.73	3.00

[1]Cipollina *et al.* (2009); [2]CRES (1996); [3]Own elaboration (SWRO: seawater reverse osmosis; BWRO: brackish water reverse osmosis).

electrical protections, fuses, etc.). O&M costs depend on the annual operation hours and water production. The energy cost for obtaining the freshwater could be included in this classification, but it would be necessary to estimate the cost of producing the energy autonomously, i.e through the solar PV field. The variability of the solar resource and the direct use of this energy in the desalination unit obligate to simplify the calculation and not to consider the electricity cost within the exploitation costs. It should be part of the direct costs of the installation.

Besides, it is necessary to maintain an external technological connection for specific annual revision and complex breakdowns. This is particularly recommendable for sites where the training level of local technicians is low. This kind of service has a considerable impact over the exploitation costs, i.e., over the final cost of the water obtained.

The fixed and variable costs must be covered on long-term to guarantee the sustainability of the unit. Therefore, it is necessary to implement framework conditions in the target sites, which help to support and promote desalination units driven by renewable energies effectively. Several support schemes are developed and discussed within ProDes project (2010). It would be ideal if the final users of water assume part of the exploitation costs according to their economic possibilities; however, this option could not been considered in specific sites where water is given for free.

5.6.1.4   *Concluding remarks*
In conclusion, the cost of water obtained is mainly determined by the real water production, design and investment cost of the installation, site conditions and isolation, cost of the local labor, cost of the land (if it is not negligible) and additional costs (taxes, customs clearance, environmental issues, etc.).

However, each location is unique, so the water cost does not follow the same scenario and items. Table 5.7 shows some examples for costs of produced water in different cases.

The cost of the water produced by an isolated PV-RO unit may be easily competitive in remote areas far from natural water and conventional energy sources compared to desalination using diesel fuel (Bilton *et al.*, 2011) and even more economic than the grid connection for the case of remote areas.

There are several indications inviting to conclude that there will be a low cost scenario on the short to medium term for PV powered RO units. On one hand, PV panels become more and more economical (average cost in EU have decreased from $4.2 \in Wp^{-1}$ in 2001 to $1.2 \in Wp^{-1}$ in 2010; EPIA, 2011) and the performance of last generation RO membranes is very promising and focused on low energy consumption (Li and Wang, 2010), thus future systems will require less PV power; on the other hand, the highly expected increasing demand of PV powered RO systems and the consequent competitiveness among involved companies will hopefully reduce the prices (ProDes Project, 2010). Both PV and RO systems are mature technologies, with a large list of suppliers in many countries. The R&D efforts come together for the development of the components that can be used by all of the manufactures. This will show to the industry that there is large market potential and will help to motivate them to develop products specifically for RE-desalination. More focused R&D efforts are needed to develop the components necessary for the smooth and efficient coupling of the existing desalination and renewable energy technologies. Other R&D priorities include the elements that will make RE-desalination robust for long stand-alone operation in harsh environments (adaptation of pumps and energy recovery systems for efficient operation in small-scale plants, automated and environmental friendly pre- and post-treatment technologies, control systems that optimize the performance and minimize the maintenance requirements).

In relation to the price of water, the subsidized municipal and rural water tariffs are addressed as factors that need to be clarified since they do not consider the possible implementation of this kind of systems by a specific subsidy line. The PV-RO technology is economically feasible in remote locations where the alternatives are limited and expensive. In some remote areas, due to the isolated location, the cost of infrastructure construction to access to conventional water and energy supply could be higher than the costs of a renewable energy powered desalination option. In those cases, there is a current real commercial market for autonomous PV-RO systems. A specific study on this topic concludes that a PV-RO system can be, depending on the local conditions, more cost-effective than the equivalent diesel-based-system (Bilton *et al.*, 2011). For the case of brackish water, the estimated water cost from a PV-RO system is 2.17–2.41 US$ m^{-3}, clearly more economical than using a diesel-powered system (3.34–3.75 US$ m^{-3}). In case of seawater, the best option depends on the location: 4.96–7.01 US$ m^{-3} (estimation for PV supply) and 5.21–5.54 US$ m^{-3} (estimation for diesel supply).

5.6.2 *Influences of the social and institutional local reality*

Since any international cooperation project focusses on local development, the social aspects are one of the key factors for the success of the project. In some situations, social issues are even more important than technical issues, and require special attention.

The social issues are all those elements related to the actors of the project and their specific role and responsibility along the activity. The commitment from the local authorities to support the project is a key factor: their financial contribution, according to their real economic availability, and their involvement in the local activities are completely necessary to get a successful action.

Figure 5.12 is an example for the stakeholders who could be involved in an initiative.

The local collaborator is the most critical actor in these kind of projects. The local partner has the role of contacting directly to the local actors to receive their necessities and visions and also to transmit the points of view of the external partners. The intermediate actor must know the local reality, from the cultural, socio-economic, and political points of view. This is only possible if this entity has a minimum previous experience in the target country (Subiela and Peñate, 2011).

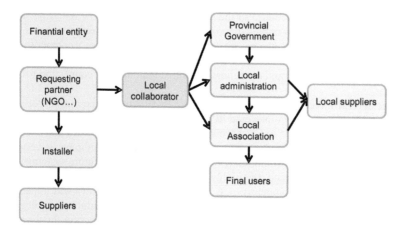

Figure 5.12.   Main actors in an RO-PV cooperation project (Subiela and Peñate, 2010).

Concerning the role of the local actors, it is necessary to include them from the initial steps of the project in order to take into account their opinions and suggestions. This implies also the inclusion of the human resources for the management and basic operations, maintenance and monitoring (Werner and Schafer, 2007).

On the other hand, it is important to mention that the institutional framework of the renewable energies and industrial water production sectors is rather fragmented and complex in several countries. This results in lack of coordination, mistakes and inefficiencies over the PV-RO project, start-up and management. Currently, the institutional structure and the absence of technological culture are a barrier for promoting initiatives which have to be sustainable and be managed directly by the local community.

However, a PV-RO unit facilitates the economic development of the population since the continuous *in situ* provision of drinking water covers a necessity, improving their life conditions considerably:

- *Positive impact on the local economy*: The water availability allows the consolidation of agricultural, cattle and low-level tourist activities (highly dependent on the "water" resource) that represents the sustenance of most of the inhabitants.
- *Positive impact on the social welfare*: Local autonomous drinking water supply assures the human and social development, diminishing the rural exodus towards other zones with more resources. On the other hand, the initiative could improve the social and gender equality. Water is a natural limited resource and a fundamental public good for life and health. The right to have water is indispensable to live with dignity and it is a fundamental condition for the accomplishment of other human rights, reason why measures have to be taken to make sure this right without any discrimination.

The PV-RO concept could be new for the population and administration; so in order to be able to benefit from it, the technology and management and the actors must learn more about it. Thus, training is a key aspect to be provided. More specifically, the involvement of the technical staff of local administration at the outset is very valuable. In relation to the final users, they play an essential role at practical level, since they decide with their behavior and real acceptance how good is the water quality and how useful is the installed system for them.

The ADIRA project handbook recommends a set of steps to be taken into account for the analysis of stakeholders, and their relation within the project (ADIRA, 2008). In conclusion, the definition, in collaboration with the local partners, of a strategy or process for the sustainability of the installation at long-term is strongly recommended. This will guarantee the cost-effectiveness

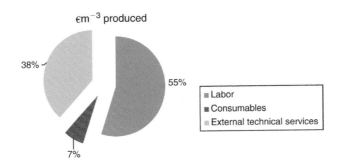

Figure 5.13.   Diagram with the distribution of exploitation costs of Ksar Ghilène PV-RO unit (Tunisia).

along the whole life of the system. The high initial investment requires an external funding, but as in the case of the capital expenses, it is strongly recommended to reach an agreement with the local authorities to obtain their economic contribution to the total running costs. Ideally, the system should be under a complete economic and technical self-management.

5.6.3   *Case of Tunisia*

The total cost of this project was about 264,000 €, of which more than 150,000 € were assigned to equipment and infrastructure investments. From the project global cost, 36,200 € were intended to technical staff, 100,700 € to capital costs – including training – and 50,000 € for civil works and hydraulic installations (Peñate *et al.*, 2007). Some items were designated to administrative expenses and services (transport). The Spanish and Canarian Cooperation (77%) and Tunisian National and Regional Governments (23%) subsidized the project cost.

The real cost of the water produced is estimated to range from 2.5–3.0 € m^{-3}. It is quite interesting if this value is compared with the price that the local population paid for freshwater before the implementation of the RO-PV project (5 € m^{-3}); the local authority has been assuming the costs of water since the beginning, so the population does not pay for their water consumption. The original source of drinking water was a private well, 60 km far away, from where water was transported by tankers to Ksar Ghilène, and then sold to the local population, who went to the delivery point with their containers. The distribution of costs for the water produced by the PV-RO system is shown in Figure 5.13.

Some remarks concerning the exploitation cost: the water cost does not include the depreciation of the equipment and currently, the Regional Agricultural Development Agency (CRDA – Kébili) assumes the total O&M costs of the installation without any cooperation or external financing.

According to the experience, an appropriate tariff system for the users should be applied. Only in this case, it would be necessary to include the capital cost amortization: the water cost would be up to 4.0 € m^{-3} (15 years lifetime). A mixed structure between the subsidy and water tariff should be implemented to make water cost affordable to the local community.

5.6.4   *Case of Morocco*

Within the ADIRA project, six off-grid PV-RO units were commissioned. Four of them were designed and commissioned by ITC. The other two were responsibility of FM21[2], with the technical collaboration of ITC.

In the case of the ITC installations, the total cost of the four units was of 357,000 €. 245,000 € were assigned to equipment, infrastructure, transport and start-up. The local governments

---

[2]FM21: Foundation Marrakech 21, Moroccan NGO, member of the ADIRA partnership.

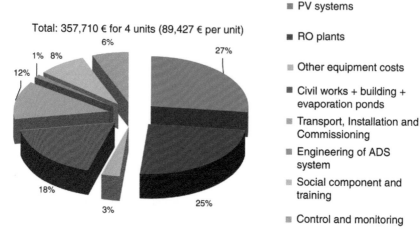

Figure 5.14.　Diagram with the distribution of capital costs of PV-RO units* installed by ITC in Morocco – ADIRA project.

assumed the hydraulic infrastructure and the civil works for the buildings. Almost 30,000 € were assigned to local training and workshops. The project cost was subsidized with about 42,000 € by the European Commission – MEDA-Water Program (55%), Canary Islands Cooperation (19%), local governments (14%) and ITC (12%). The distribution of capital costs is presented in Figure 5.14.

The average total cost of water, including the amortization, is estimated to be $5.45 \, \text{€} \, \text{m}^{-3}$. Currently, the water supply is free. The Canarian Cooperation has subsidized the exploitation costs until December 2011.

5.7　CONCLUSIONS

This chapter presents current technical and economic information about the photovoltaic driven reverse osmosis technology. Despite that there is a long experience in testing this autonomous desalination concept and both photovoltaic and reverse osmosis are mature technologies, not many systems have been implemented under real operation conditions.

The installation of these systems in developing countries implies a set of specific challenges and unexpected problems that do not appear in the units installed for R&D purposes. A planning based on a meditated process of specific and coordinated stages must be carried out in order to avoid as far as possible these difficulties. The implementation must consider not only the technical part, but also the economic, social, and environmental aspects. Therefore this kind of projects is only possible by the incorporation of a wide set of different actors (local and non-local) and experts with specific knowledge and experience in the different technical areas. Moreover, the coordination and good relation among all the involved participants is a key factor for the success of the project.

From the technical point of view, the use of high quality materials, inclusion of energy recovery devices (in seawater cases) and batteries, a tailor-made design and control system, subjected to the local conditions, is strongly recommended in order to develop a system able to operate with minimum maintenance requirements. The experience shows that each target place is different and needs a specific analysis to identify the most suitable option of PV-RO water supply systems. The examples described in this chapter with the theoretical and practical issues for the installations in Morocco and Tunisia show the main principles for the design, implementation, and operation of PV-RO systems. The operation data are also helpful to understand how the control system works.

From the economic point of view, the specific costs of water production by PV-RO are still far away from the water cost of conventional seawater desalination plants. Nevertheless, the future is hopeful. Renewable energies in general and photovoltaic energy, in particular, have been experiencing a great development for the last years, with very positive consequences on the quality and prices of equipment. On the other hand, the oil prices will raise sooner or later, with strong implications in all the economy, but especially in energy intensive industrial processes, as desalination. The growing drinking water demand, due to demographic and climate change reasons, mainly focused on developing countries of Northern Africa and Middle East, will probably lead to the creation of a promising commercial sector of goods and services linked to PV-RO and other stand-alone desalination technologies.

From the social point of view, the cultural and political aspects should be taken into account. The involvement of all the local actors (authorities, associations, beneficiaries, technicians, suppliers) is absolutely necessary since everyone has a specific role. The local training of people for maintenance and operation tasks is particularly critical. At the end, the system must be autonomous, not only from the power supply point of view, but also from the management. The only option for a long-term self-management or sustainability of the project is the commitment of all the involved local actors. Of course, this implies a progressive independence process, in which training of maintenance operators and creation of a local working group is vital.

ACKNOWLEDGMENTS

The authors wish to thank the European Commission (MEDA Water Program), the Spanish Agency for International Cooperation and the Cooperation Area of the Canary Islands Government for their economic contribution to the actions presented in this chapter. Similarly, a special mention is given to the economic and technical support received from the Tunisian partners (National Agency for the Control of Energy Consumption, ANME, and the Regional Directorate for Agricultural Development of Kèbili, CRDA), and Moroccan entities (Governments of the Essaouira, Tiznit and Ifni provinces, the authorities of the rural communities and the local associations). The information provided in this chapter is the result of the effort of many people of the ITC and Las Palmas de Gran Canaria University, involved in the development of photovoltaic energy coupled to reverse osmosis since 1998. Finally, the anonymous participation of women and men, belonging to the target villages, is appreciated; the implementation of the units in Tunisia and Morocco has been possible thanks to them as well.

REFERENCES

ADIRA Partnership: *Handbook (a guide to autonomous desalination system concepts)*. Edited by S. Sözen & S. Teksoy, Istanbul Technical University, 2008 (ADIRA Project was co-funded by EC, MEDA-Water Programme).

Banat, F., Subiela V.J. & Qiblawey, H.: Site selection for the installation of autonomous desalination systems (ADS). *Desalination* 203 (2007), pp. 410–416.

Bellesiotis, V. & Delyannis, E.: The history of renewable energies for water desalination. *Desalination* 128:2 (2000), pp. 149–159.

Bilton, A.M., Wiesman, R., Arif, A.F.M., Zubair, S.M. & Dubowsky, S.: On the feasibility for community-scale photovoltaic-powered reverse osmosis desalination systems for remote locations. *Renew. Energy* 36 (2011), pp. 3246–3256.

CETC (CANMET Energy Technology Centre) – Varennes, RET-Screen International Surface meteorology and Solar Energy (NASA's Earth Science Enterprise Program) http://eosweb.larc.nasa.gov/sse/RETScreen/ (accessed October 2011).

Cipollina, A., Micale, G. & Rizzuti, L.: *Sea water desalination: conventional and renewable energy processes*. Springer, Berlin, Heidelberg, 2009.

CRES (Center of Renewable Energy Sources): *Desalination guide using RE*. EC-THERMIE Program. EC Contract No. SUP-094096-HE, 1996, Athens, Greece.

Delyannis, E.: Historic background of desalination and renewable energies. *Solar Energy* 75:5 (2003), pp. 357–366.

Eltawil, M.A., Zhengming, Z. & Yuan, L.: A review of renewable energy technologies integrated with desalination systems. *Renew. Sustain. Energy Rev.* (2009), pp. 2245–2262.

EPIA (European Photovoltaic Industry Association): Solar photovoltaics competing in the energy sector – on the road to competitiveness. September 2011, www.epia.org/publications (accessed October 2011).

García-Rodriquez, L.: Seawater desalination driven by renewable energies: a review. *Desalination* 143:2 (2002), pp. 103–113.

Ghermandi, A. & Messalem, R.: Solar-driven desalination with reverse osmosis: the state of the art. *Desalin. Water Treatm.* 7 (2009), pp. 285–296.

Gleick, P.H.: Basic water requirements for human activities: meeting basic needs. *Water International* 21 (1996), pp. 83–92

Global Water Intelligence. *The 19th IDA World Wide Desalting Plant Inventory.* Global Information, Inc. Oxford, UK, 2008.

Hanafi, A.: Desalination using renewable energy sources. *Desalination* 97:1–3 (1994), pp. 339–352.

Herold, D., Horstmann V., Neskakis, J., Plettner-Marliani G., Piernavieja, G., Calero, R., *et al.*: Small scale photovoltaic desalination for rural water supply – demonstration plant in Gran Canaria. *Renew. Energy* 14:1–4 (1998), pp. 293–298.

IEA (International Energy Agency): Cost and performance trends in grid-connected photovoltaic systems and case studies report. Report IEA-PVPS T2-06. Photovoltaic power systems programme, 2007, p. 7.

Kalogirou, S.: Seawater desalination using renewable energy sources. *Progr. Energy Combust. Sci.* 31 (2005), pp. 242–281.

Li, D. & Wang, H.: Recent developments in reverse osmosis desalination membranes. *J. Mater. Chem.* 20 (2010), pp. 4551-4566.

Mathioulakis, E., Bellesiotis, V. & Delyannis, E.: 2007, Desalination by using alternative energy: Review and state-of-the-art. *Desalination* 203 (2007), pp. 346–365.

Outzourhit, A.: Application of RE in water desalination projects in rural areas: The ADIRA experience in Morocco. REHYSYS, 2006, *Proceedings of the HY-PA Workshop*, Marrakech, 2006.

Outzourhit, A., Elharrak, N., Aboufirass, M. & Mokhlisse, A.: Autonomous desalination units for fresh water supply in remote rural areas. In: M. El Moujabber, L. Mandi, L., G. Trisorio Luzzi, I. Martin, A. Rabi & R. Rodríguez (eds): Technological perspectives for rational use of water resources in the Mediterranean region. *Options Méditerranéennes* 88, 2009, http://ressources.ciheam.org/om/pdf/a88/00801190.pdf (accessed October 2011)

Peñate, B., Castellano, F. & Ramírez, P.: PV-RO desalination stand-alone system in the village of Ksar Ghilène (Tunisia). *Proceedings of the IDA World Congress-Maspalomas*, 2007.

ProDes project (2008–2010). Deliverables available at www.prodes-project.org (accessed October 2011).

Rogers, P.: Facing the freshwater crisis. *Scientific American*, August 2008.

Sánchez, A.S. & Subiela, V.J.: Analysis of the water, energy, environmental and socioeconomic reality in selected Mediterranean countries (Cyprus, Turkey, Egypt, Jordan and Morocco). *Desalination* 203 (2007), pp. 62–74.

Subiela, V.J., de la Fuente, J., Piernavieja, G. & Peñate, B.: Canary Islands Institute of Technology (ITC) experiences in desalination with renewable energies (1996 – 2008). *Desalination and Water Treatment* 7 (2009), pp. 220–235.

Subiela, V. & Peñate, B.: Proceedings of the NATO Workshop Water Security in the Mediterranean Region (An International Evaluation of Management, Control, and Governance Approaches) Series: *NATO Science for Peace and Security Series C: Environmental Security*, edited by A. Scozzari & B. El Mansouri, 2011, XII.

Tzen, E.: Desalination units powered by RES: opportunities & challenges. *Proceedings of the Seminar Successful desalination RES plants worldwide.* Hammamet, Tunisia, September 2005, available at http://www.adu-res.org/pdf/CRES.pdf (accessed October 2011).

Tzen, E., Theofilloyanakos, D., Sigalas, M. & Karamanis, K.: Design and development of a hybrid autonomous system for seawater desalination. *Desalination* 166 (2004), pp. 267–274.

United Nations: 2005, http://www.un.org/waterforlifedecade/ (accessed October 2011).

United Nations: 2006, http://hdr.undp.org/en/reports/global/hdr2006/ (accessed October 2011).

Werner, M. & Schafer, A.I.: Social aspects of a solar powered desalination unit for remote Australian communities. *Desalination* 203 (2007), pp. 375–393.

WHO: *Water Sanitation and Health.* 2010, http://www.who.int/water_sanitation_health (accessed October 2011).

CHAPTER 6

Wind energy powered technologies for freshwater production: fundamentals and case studies

Eftihia Tzen

> *"Our country is closed in, all mountains that day and night have the low sky as their roof. We have no rivers, we have no wells, we have no springs, only a few cisterns — and these empty — that echo, and that we worship"*
>
> George Seferis, 1900–1971*

6.1 INTRODUCTION

Renewable energy systems convert naturally occurring energy (sunlight, wind, etc.) into usable electrical, mechanical or thermal energy. Most of these systems are well-established and reliable, with a significant number of applications all over the world. The idea of using renewable energy sources (RES) to drive desalination processes is fundamentally attractive, as considerable number of in-depth studies and real-life applications demonstrate (Tzen, 2008).

Wind energy turbines can be coupled with desalination systems in order to provide the necessary energy input, which in itself may become a significant contribution to remote (off-grid) and arid areas. In addition, the desalinated water can be used as temporary energy storage, thus providing a means for the "regulation" of one of the most important inherent characteristics of RES, i.e. their intermittence. Where the system is grid connected, the desalination plant can operate continuously as a conventional plant and the renewable energy source merely acts as a fuel substitute. However, even in such cases, the desalination load could be used for moderating the amount of energy injected to the grid, which tends to become a very useful concept in view of hybrid electricity systems with high penetration of intermittent renewable electricity sources (Tzen, 2009).

The selection of the most suitable desalination technology for a specific site as well as technological combination of RES desalination is an important factor for the success of a project. Three desalination processes require electricity for their operation and can be easily coupled with wind turbine(s). They are reverse osmosis (RO) for the desalination of brackish and seawater, electrodialysis (ED) for the desalination of brackish water and mechanical vapor compression (MVC) for the desalination of seawater.

6.2 WIND ENERGY TECHNOLOGY

Wind energy has the tremendous advantage that it can be harnessed at a very wide range of scales; installations can vary from few watts wind generators up to MW-scale (megawatt scale) wind parks (Fig. 6.1). Wind turbines can be used as stand-alone applications, or they can be connected to a utility power grid (Fig. 6.2). They can also be combined with a photovoltaic (PV) system or a diesel generator and energy storage systems (e.g. batteries), thus forming "hybrid" systems, which are typically used in remote locations where connection to a utility grid is not available (Tzen, 2009).

---

*Source: George Seferis, 1900–1971, Nobel laureate: Mythistorema.

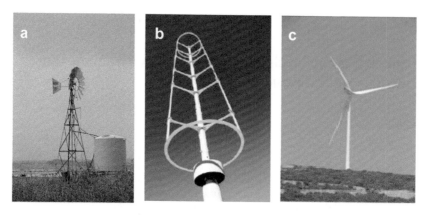

Figure 6.1. Types of wind turbines a) multi-blade turbine, b) vertical axis turbine, c) horizontal axis turbine. Sources: (a) http:/www.wwindea.org, (b) http://news.cnet.com, (c) CRES.

Figure 6.2. Wind parks in Greece: a) Peloponese, b) Thrace (source: CRES, Greece).

Figure 6.3. Two and three blades small wind turbines a) 20 kW wind turbine at Ag. Efstratios Island, Greece, b) 900 W wind turbine at CRES Energy Park, Keratea, Attica (source: CRES, Greece).

The technology of the turbine generators currently in use is only 25 years old and investment in it has so far been rather modest, as compared to other energy sources. Two decades of techno-logical progress have resulted in today's wind turbines being state-of-the-art, modern, modular technology-based and rapid to install. Modern wind turbines have improved dramatically in their power rating, efficiency and reliability.

Wind turbines are basically classified by the position of the spin axis, which can be vertical (VAWT vertical axis wind turbine) or horizontal (HAWT horizontal axis wind turbine). Modern wind turbines are typically of horizontal type (Fig. 6.3).

According to the IEA (International Energy Agency; IEA, 2010), in 2010, worldwide wind power capacity increased, achieving 200,457 MW of installed power. In the EU, wind power installations accounted for 17% of new power installations in 2010, and wind represented 10% of the total EU power generation capacity. That wind power capacity will produce 181 TWh, meeting 5.3% of gross EU final power consumption in an average wind year, avoiding about 115 million tons of CO_2 emissions annually, (IEA, 2010). Table 6.1, presents the wind power installed in Europe at the end of 2010 according to the European Wind Energy Association (EWEA, 2010).

Today, the lessons learnt from the operation of wind power plants, along with continuing R&D, have made wind electricity very close in cost to power from conventional utility generation in some locations. Generation costs have dropped by 50% over the last 15 years, moving closer to the cost of electricity generation from conventional energy sources. By today, wind power generation has come to a historical point, just as installed costs were becoming competitive with other conventional technologies; the investment cost per MW has started increasing for new wind projects. This is believed to be the result of increasing commodity prices (raw material such as steel and copper), the current tightness in the international market for wind turbines and other factors (IEA, 2009).

The economics of a wind scheme depend upon technical, resource and cost parameters. The latter two vary from country to country. In general, the cost of the equipment represents the highest portion of the total investment cost. The costs for electrical and civil engineering infrastructure, transport, management and administration, land, etc. can vary between 25 and 35% of the cost of the installed turbines. The annual operating and maintenance cost accounts for around 1.5% of the initial investment. Also, to the above mentioned cost, a percentage of around 1% per annum of the initial investment should be added, for the cost of insurance (Tzen, 2009). In general, as shown in Table 6.2, average turbine costs vary from a low of 720 € kW^{-1} to a high of 1800 € kW^{-1}. Figure 6.4 presents trends of installed costs for wind projects by country.

The cost of the energy produced by wind farms as have been calculated in each country varies significantly. For the estimation of this cost different assumptions have been made in each country. In Denmark as reported in IEA, the wind electricity onshore in 2009, excluding risk factors and profit, was calculated to be 30 to 50 € MWh^{-1}. The actual cost depends on the wind profile at the location and also on the type of investor. For instance, a private investor will expect a shorter payback period than large investors like utilities. Private investors require 40 to 50 € MWh^{-1} on top of the raw cost (includes profit and risk factor) to invest in wind production.

In Norway, estimates of production cost from sites with good wind conditions (33% capacity factor) suggest a production cost of about 64 € MWh^{-1}, including capital costs (discount rate 8%, 20 years period), operation and maintenance.

Regarding the wind turbine industry, the average rated capacity of new wind turbines installed in 2006 continued the trend towards larger installations. The average rating in 2005 increased to 1.6 MW, while in 2006, the average rating rose to nearly 1.7 MW. In the IEA wind member countries[1] the average size of new wind turbines increased to 1.9 MW in 2009.

The European Wind Industrial Initiative has the objective to make wind one of the cheapest sources of electricity and to enable a smooth and effective integration of massive amounts of wind electricity into the grid. To achieve this, special efforts will be dedicated to greatly increase the power generation capacity of the largest wind turbines (from 5 to 6 MW to 10–20 MW) and to tap into the vast potential of offshore wind (IEA, 2009).

In addition to MW-scale wind turbines, intermediate sizes of 660–850 kW are being manufactured in several countries for single-turbine installations or small wind power plants (Italy, United States).

Although focus has been given to large utility-connected wind parks, there are many other applications, such as autonomous wind energy systems for water pumping, water desalination, as well as local power supply systems in conjunction with diesel generators.

---

[1]IEA member countries are: Australia, Canada, Denmark, Finland, Germany, Greece, Ireland, Italy, Japan, Korea, Mexico, Netherlands, Norway, Portugal, Spain, Sweden, Switzerland, UK and USA.

Table 6.1. Wind power (in MW) installed in Europe by end of 2010 (EWEA, 2010).

	Installed during 2009	Total installed by end 2009	Installed during 2010	Total installed by end 2010
EU capacity [MW]				
Austria	0	995	16	1011
Belgium	149	563	350	911
Bulgaria	57	177	198	375
Cyprus	0	0	82	82
Czech Republic	44	192	23	215
Denmark	334	3465	346	3798
Estonia	64	142	7	149
Finland	4	147	52	197
France	1088	4574	1086	5660
Germany	1917	25777	1493	27214
Greece	102	1087	123	1208
Hungary	74	201	94	295
Ireland	233	1310	118	1428
Italy	1114	4849	948	5797
Latvia	2	28	2	31
Lithuania	37	91	63	154
Luxembourg	0	35	7	42
Malta	0	0	0	0
Netherlands	39	2215	32	2245
Poland	180	725	382	1107
Portugal	673	3535	363	3898
Romania	3	14	448	462
Slovakia	0	3	0	3
Slovenia	0.02	0.03	0	0.03
Spain	2459	19160	1516	20676
Sweden	512	1560	604	2163
United Kingdom	1271	4245	962	5204
Total EU-27	**10499**	**75103**	**9332**	**84324**
Total EU-15	**10038**	**73530**	**8003**	**81452**
Total EU-12	**461**	**1574**	**1298**	**2872**
Of which offshore and nearshore	582	2061	883	2944
Candidate Countries [MW]				
Croatia	10	28	61	89
Fyrom*	0	0	0	0
Turkey	343	801	528	1329
Total	**353**	**829**	**461**	**1290**
EFTA [MW]				
Iceland	0	0	0	0
Liechtenstein	0	0	0	0
Norway	2	431	9	441
Switzerland	4	18	25	42
Total	**6**	**449**	**34**	**483**
Other [MW]				
Faroe Islands	0	4	0	4
Ukraine	4	90	1	87
Russia	0	9	0	9
Total	**4**	**99**	**1**	**101**
Total Europe	**10845**	**76471**	**9918**	**86279**
European Union				84324
EU Candidate Countries				1418
EFTA				478
Total Europe				86279

*Former Yugoslav Republic of Macedonia
Note: Due to previous-year adjustment, 114,77 MW of project decommissioning, re-powering and rounding of figures, the total 2009 end-of-year cumulative capacity is not exactly equivalent to the sum of the 2008 end-of-year total plus the 2009 additions.

Table 6.2.   Estimated average turbine cost and total project cost for 2010 (IEA, 2010); blanks means no data available.

Estimated average turbine cost and total project cost for 2010

Country	Turbine cost [€ kW^{-1}]	Total installed cost [€ kW^{-1}]
Australia	1100 to 1500	1500 to 2500
Austria	1400 to 1800	1700 to 2000
Canada	–	1488 to 1860
China	720	970 to 1020
Denmark		1030 onshore, 2680 offshore
Germany		1336 to 1756 onshore, 3323 to 3561 offshore
Greece		1100 to 1400
Ireland	1100	1800 onshore
Italy	1200	1740
Japan	1500	2250
Mexico	1000 to 1200	1500
Netherlands		1325 onshore, 3200 offshore
Portugal	900 to 1000	1000 to 1400
Spain		1400
Sweden	1400	1600
Switzerland	1450	1885
United States	818 to 1042	1603

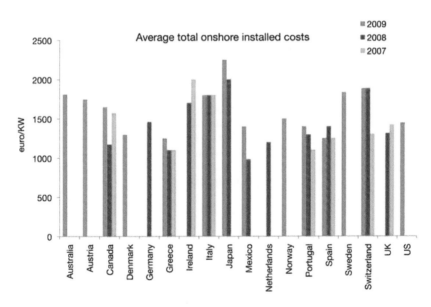

Figure 6.4.   Average total installed costs of wind projects 2007–2009 as reported by IEA wind member countries (IEA, 2009).

With the high priority to develop clean energy sources, many countries are seeing rapid expansion of the small wind industry, especially companies supplying turbines for grid-connected applications. Small-sized turbines, less than 100 kW, are manufactured in Italy, Mexico, Spain, France, Sweden, UK, Denmark and the United States (Tzen, 2009).

6.3 WIND ENERGY FOR FRESHWATER PRODUCTION

Regarding wind and water issues, the first application of the wind energy was the use of wind-mills for water pumping during the 19th century, beginning with the Halladay windmill in 1854 (Fig. 6.1a). Many years later the use of wind turbines for the production of freshwater by desali-nation seems to be an interesting alternative. Today the matching of wind energy and desalination is a well-known aspect and it is market available.

One of the problems of utilizing wind power in desalination applications is the variable nature of the resource. The storage of wind energy in the form of electrical power is only practical when small amounts are involved. Storage batteries increase the total investment cost therefore to run a process of any magnitude on stored electrical energy is not a practical proposition. Variable power input force the desalination plant to operate in non-optimal conditions and may cause operational problems. To avoid the fluctuations inherent in renewable energies, different energy storage systems may be used.

Today, there is a market regarding wind desalination systems, offering stand-alone or grid connected systems of specific sizes. Most of them utilize the reverse osmosis desalination process.

6.3.1 *Wind reverse osmosis systems*

Reverse osmosis (RO) is the most widely used process for seawater desalination. RO process involves the forced passage of water through a membrane against the natural osmotic pressure to accomplish separation of water and ions. The amount of desalinated water that can be obtained ranges between 30 and 75% of the volume of the input water, depending on the initial water quality (brackish or seawater), the quality of the product needed, and the technology and membranes involved. A typical configuration of a RO plant is shown in Figure 6.5.

During the last decades, the energy requirements for the operation of seawater RO units with the use of energy recovery devices have been dramatically reduced to about $3\,kWh\,m^{-3}$.

Other advantages of the RO process are the modular design and compactness of the plants, satisfactory performance in all sizes and easy operation.

As regards to wind energy and reverse osmosis matching, a number of units have been designed and tested. Most of them are stand-alone systems and have been developed within research work. Table 6.3, shows several of the wind-RO installations in Europe.

As early as 1982, a small system was set at Ile du Planier, France where a turbine providing 4 kW is coupled to a $0.5\,m^3\,h^{-1}$ RO desalination unit. The system was designed to operate either *via* a direct coupling or through batteries. Another example where wind energy and reverse osmosis are combined is at the island of Drenec in France, which was installed in 1990. Here a 10 kW wind turbine drives a seawater RO unit producing $0.5\,m^3\,h^{-1}$ of freshwater (ADU RES, 2005).

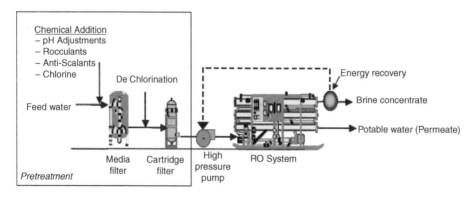

Figure 6.5. Typical configuration and components of an RO plant (Schreck, 2006).

Table 6.3. Wind reverse osmosis applications.

Location	RO capacity [m³ h⁻¹]	Electricity supply	Year of installation
Ile du Planier, France	0.5	4 kW W/T	1982
Island of Suderoog, Germany	0.25–0.37	6 kW W/T	1983
Island of Helgoland, Germany	40	1.2 MW W/T + diesel	1988
Fuerteventura, Spain	2.3	225 kW W/T + 160 kVA diesel, flywheel	1995
Pozo Izquierdo, Gran Canaria, Spain, SDAWES*	8 × 1.0	2 × 230 kW W/T	1995
Therasia Island, Greece, APAS	0.2	15 kW W/T, 440Ah batteries	1995/1996
Syros Island, Greece, JOULE	2.5–37.5	500 kW W/T, stand-alone + grid connected	1998
Keratea, Greece, PAVET	0.13	900 W W/T, 4 kWp PV, batteries	2001/2002
Pozo Izquierdo, Gran Canaria, Spain, AEROGEDESA	0.80	15 kW W/T, 190 Ah batteries	2003/2004
Loughborough Univ., U.K.	0.5	2.5 kW, no battery	2001/2002
Milos Island, Greece, OPC Programme	6 × 600	850 kW W/T, grid connected	2007/2008
Heraklia Island, Greece	3.3	30 kW W/T, floating system, batteries	2007
Delf Univ., The Netherlands	0.2–0.4	Windmill, no battery	2007/2008

W/T: wind turbines; PV: photovoltaic.
*SDAWES: Seawater Desalination with an Autonomous Wind Energy System.

More recently several R&D projects have been carried out, such as the wind desalination system built at Drepanon on a cement plant, near Patras, Greece. The project, including a 35 kW wind turbine, was initiated in 1992, and completed in 1995. The project called for full design and construction of the wind generator turbine (blades, etc.), plus installing two RO units with a production capacity of 5 m³ day⁻¹ and 22 m³ day⁻¹, respectively. Unfortunately since 1995, operational results have been poor due to the low wind regime. A very interesting experience has been carried out at a test facility in Lastours, France, where a 5 kW wind turbine provides energy to a number of batteries (1500 Ah, 24 V). The energy is supplied *via* an inverter to a small RO unit with 1.8 kW nominal power.

The most well-known plants which are larger than the previously mentioned wind powered RO plants are the plants installed in Syros Island, Greece, and in Tenerife, Spain, within a research program of the European Commission. The objective of the project was to develop the concept of a family of modular seawater desalination plants, adaptable to a broad range of regions and making use of the locally available wind energy resources. The family concept was based on a limited number of standardized modules. These modules (advanced technology wind energy converters, hybrid power units, variedly sized seawater desalination units, designed to be capable for off-grid or grid-connected operation, etc.) were equipped with a highly flexible power conditioning and management system. Additionally the concept allows for the parallel production of water and electricity, according to the operator's needs (ADU RES, 2005). The plant on the island of Tenerife was the first to be built and consisted of a 30 kW W/T coupled to two RO modules. The plant on Syros is based on a 500 kW W/T and eight RO modules with an output between 60 and 900 m³ day⁻¹ of freshwater (Fig. 6.6). The smaller unit on Tenerife was used to test some of the concepts, prior to the construction of the larger plant in Syros (see Table 6.4).

At the Syros project, the 500 kW W/T is coupled to a grid management system which allows for electricity frequency and voltage control, and self-adaptation of the wind energy converter to the weak electricity grid of the island (Fig. 6.7). The electricity from the wind turbine is buffered in the energy storage system, which includes a diesel generator, batteries and a flywheel. The output is fed into the RO unit and the electricity grid (Tzen, 2009).

Figure 6.6.   Wind powered RO unit at Syros Island: a) 500 kW W/T, b) RO plant.

Table 6.4.   Technical characteristics of the two plants in Tenerife and Syros islands (ADU RES, 2005).

Plant	Tenerife, Spain	Syros, Greece
Installation place	Wind park of Granadilla	Syros
WEC characteristics	ENERCON E-12 with magnet synchronous generator, 30 kW nominal power and a passive yaw system	ENERCON-E40 with synchronous ring generator, 500 kW nominal power and a grid management system
Storage system	Does not include the optional energy storage system	Energy storage system
Number of RO blocks	1	8
Max. power of RO unit	200 kW	200 kW
Desalination capacity	60–110 m³ day⁻¹	60–900 m³ day⁻¹
Water storage	1 × 14 m³ tank of pre-filtered water 1 × 14 m³ tank for produced water	

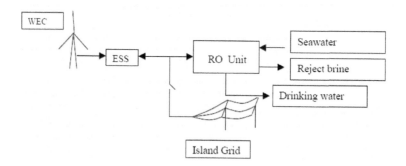

Figure 6.7.   Wind-RO plant design at Syros Island.

One more wind-RO project was carried out in Fuerteventura Island in the period of 1995–2002. The project was proposed by the University of Las Palmas de Gran Canaria (ULPGC) with the further collaboration of the Canary Islands Institute of Technology (ITC), and the financial contribution of the European Commission, the municipal government, the Fuerteventura Island Water Council, and the Regional Industry Ministry (Subiela *et al.*, 2009).

The project studied the operational strategies of a wind-diesel system which had been installed in an isolated fishing village community with a permanent population of 60 inhabitants, on the

Figure 6.8.   View of the project installations in the Punta Jandía fishing village in Fuerteventura (source: ITC, Gran Canaria).

island of Fuerteventura (Canary Islands). The system consisted of a wind power-diesel system to supply power, water, cooling and ice to the Punta Jandía fishing village (Fig. 6.8).

The power supply system in Fuerteventura consists of a 225 kW wind energy converter and two 160 kVA diesel engines with flywheel and synchronous generator of 75 kVA each one, to produce electricity for fishing conservation, a 25 m^3 day^{-1} sewage water treatment plant, a 56 m^3 day^{-1} seawater desalination plant, as well as public and private lighting, etc. for an average of 300 people (CRES, 1998).

A stand-alone wind-RO unit was installed in Therasia Island in Greece in 1997. The seawater RO unit had a capacity of 0.2 m$^{3-1}$ and the nominal power of the wind turbine was of 15 kW. The project was of significant importance since it was the second water resource of the island. The first water resource was the transportation of water by tankers. The cost of the transported water was too high (15 ECU m^{-3} in 1997) and the water quality unknown. The system is abandoned (CRES, 1998).

An autonomous hybrid solar/wind system has been developed in around 1999 by the Planning, Development and Technology Division of Israel Electric Corporation. The purpose of this system was to develop and promote an autonomous desalination system for remote and isolated areas, which are devoid of water resources (ADU RES, 2005). The project was partially financed within the 4th Framework Programme of the European Commission. The system consists of a reverse osmosis unit, designed to produce 3 m^3 day^{-1} freshwater from brackish water which has a salinity of 3500 to 5000 mg/L (TDS, total dissolved solids). The feed water is pumped from a nearby brackish water well. A water storage tank of 5 m^3 stores the produced water. The system has been designed based on the premise that the average on-site wind velocity is about 4–5 m s^{-1} and an isolation level[2] of about 5–5.5 kWh m^{-2} day^{-1}. The power supply system consists of a 3.5 kWp PV system and a 600 W wind generator, which is used only as backup. A DC-AC power inverter is used to invert the DC power from the battery back. The cost of the freshwater produced has been estimated to about 7.5 ECU m^{-3} (1999).

In 1999, within SDAWES project ("Seawater Desalination by an Autonomous Wind Energy System") another wind-RO unit was installed and tested in Pozo Izquierdo, Gran Canaria Island. The wind turbine powered RO plant was part of a much larger project involving three desalination processes funded by the EC. The SDAWES project was developed within a program of the EC (JOR3-CT95-0077) and was coordinated by Canary Islands Institute of Technology, ITC, in Spain. Commissioning and testing was carried out from 1999 to 2002. The plant combines

---

[2]Insolation (incoming solar radiation) is the amount of solar radiation incident on any surface – for our purposes, we will be comparing insolation levels on the surface of the earth. The amount of insolation received at the surface of the earth is controlled by the angle of the sun, the state of the atmosphere, altitude, and geographic location.

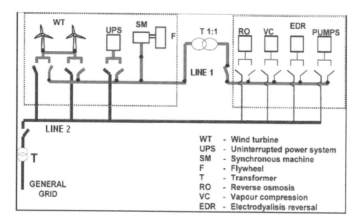

Figure 6.9.   Configuration of the wind desalination system, SDAWES project, in Pozo Izquierdo, Gran Canaria Island, Spain.

three different desalination systems, a reverse osmosis (8×25 m^3 day^{-1}), a mechanical vapor compression (50 m^3 day^{-1}) and an electrodialysis reversal (190 m^3 day^{-1}) to an off-grid wind farm (first in the world connected to a desalination plant) in order to produce freshwater; total nominal water production capacity of 440 m^3 day^{-1} (Subiela et al., 2009).

A shown in Figure 6.9, the system is based on two electric circuits, the blue line (Line 1), which shows the supply from the off-grid wind farm and the red line (Line 2), shows the supply from the general grid. Concerning the RO unit, product water flow and conductivity are slightly affected by the variations in frequency of the electrical grid.

The daily average conductivities are practically unaffected as the acceptable decreases in electrical frequency (52–48 Hz) do not produce large increases in the permeate electrical conductivity (approximately 28 μS cm^{-1}). However, punctual fluctuations were detected in the short periods with unstable frequency (operation under minimum wind speed). According to ITC simulation results, flow fluctuations created by the variable power supply could be favorable in terms of reducing the concentration polarization in RO, depending on the frequency and amplitude of fluctuations.

According to the researchers, one of the main outcomes of the tests results is that RO technology is the most suitable one for coupling to an off-grid wind farm. Nevertheless, each desalination technology has possibilities to improve the operation with off-grid wind farms by developing specific designs.

In 2001, within a national program, Operational Program for Competitiveness, Measure 4.3, of the Greek Ministry of Development, the Centre for Renewable Energy Sources & Saving, CRES developed a stand-alone hybrid RO plant. This project deals with the design and development of a small autonomous hybrid seawater-RO unit which was able to provide freshwater to a remote area with about 10 inhabitants having in mind that Greece has a great number of small islands with few inhabitants which in most cases are remote and isolated with electricity and water supply problems.

The hybrid RO system mainly consists of 3.96 kWp photovoltaic generators, a 900 W wind generator, a 130 L h^{-1} seawater-RO unit, a battery bank of 1800 Ah/100 h and two inverters of 1.5 kW and 4 kW nominal power (see Figs. 6.10 and 6.11) (Tzen, et al., 2008). The system is still in operation for demonstration purposes.

A small-scale wind-powered seawater-RO system without batteries was demonstrated in 2003 at CREST, Loughborough University, UK. The system operated at variable flow and variable pressure in accordance with the variable power available in the wind. The system pressure ranged between 38 and 51 bar (3.8–5.1 MPa) during this test (ADU RES, 2005). Energy recovery was provided by a Clark pump. The feed water used for testing was pure NaCl solution with a TDS

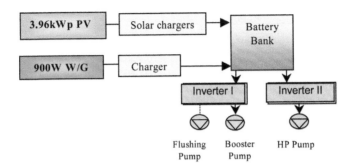

Figure 6.10.   Configuration of CRES hybrid RO unit.

Figure 6.11.   Hybrid seawater-RO unit at CRES: a) 130 L h^{-1} SWRO unit, b) the battery storage system, inverters.

of around 32,800 mg L^{-1} to emulate the osmotic pressure of seawater. The concentration of the product water varied between 470 and 800 mg L^{-1}. The wind turbine is rated at 2.5 kW, gearless with variable-speed permanent-magnet generator. It has 3 blades, a rotor diameter of 3.5 m and a rated rotor speed of 300 rpm. It is mounted on a 13 m un-guyed tapered steel tower. The AC from the generator is rectified and limited (voltage regulator), then fed directly to the inverters, (see Fig. 6.12). No batteries or other energy storage or backup are included. This drives the plant to a frequent replacement of the membrane.

Furthermore, a stand-alone wind-RO was installed in Pozo Izquierdo, Gran Canaria Island in 2003. The plant started its operation at the end of 2004. The project has been financed by the Government of the Canary Islands (FEDER) and has been carried out by Instituto Tecnologico de Canarias, ITC. The scope of the project was the electrical coupling of a commercial wind turbine to a RO unit for seawater desalination, operating under a constant regime and managing the storage and available wind energy use through minimum battery bank.

The system consists of a 15 kW Wind turbine, a 190 Ah lead acid battery bank, an inverter and a 0.80 m^3 h^{-1} RO unit (Fig. 6.13). The RO unit consists of 2 pressure vessels in series with two membranes each (SW30-4040). The salinity of the feed water is 35.500 mg L^{-1} TDS. The produced water has a salinity of less than 500 mg L^{-1} TDS. The nominal pressure of operation of the RO unit is 55 bar (5.5 MPa) and the recovery ratio is of 37%. The high-pressure pump, which is the main load of the RO unit, is 7.5 kW. The pump motor is a three-phase motor. The specific energy consumption of the RO unit is 7 kWh m^{-3}.

In 2008, a wind desalination plant was installed and began its operation in the island of Milos in Greece (see Fig. 6.14). The project is financed by the Operational Program for Competitiveness, Measure 6.3 of the Greek Ministry of Development. The wind-RO plant consists of an 850 kW grid-connected wind turbine to cover the energy consumed by a RO unit for seawater desalination

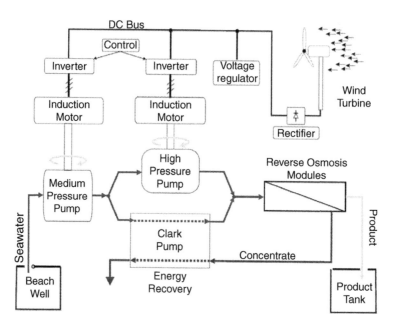

Figure 6.12.   Block diagram of the wind-RO pilot unit at CRES (source: CREST, UK).

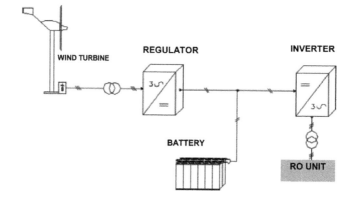

Figure 6.13.   Typical configuration of an autonomous wind-RO system.

(Bouzas, 2010). In 2009, the unit increased its capacity from 2000 m^3 day^{-1} to 3600 m^3 day^{-1} of potable water (Ifantis and Ifantis, 2010).

This production is sufficient to cover the water needs of Milos Island. Until 2008 the island covered its water needs with the transportation of water with tankers[3], like most of the Cyclades islands. The water transportation cost up to 2009 for Cyclades and Dodecanese Aegean islands ranged from 5–8 € m^{-3}.[4] During 2010 the water transport cost in several Aegean islands, increased to around 12.5 €/m^3. The project is designed to cover the continual increasing needs for potable water on the island, supplying high quality water to comply with the highest standards on a 24 h basis, at a remarkably lower cost than the present. As it is mentioned in the literature (Ifantis and Ifantis, 2010) the cost of the produced water from the wind-RO plant is of 1.8 € m^{-3}.

---

[3] Water is not used for drinking.
[4] Mainly depending on the distance from the mainland.

Figure 6.14. View of the wind-RO unit in Milos Island, Greece: a) the wind park, b) the RO unit, c) the control room of the RO unit, and d) the RO modules in a container (source: ITA S.A., Greece).

The RO plant is located at the old betonies' mine at "Vouno Triovassalou" on the island of Milos. The project consists of the following units:

- 3600 m^3 day^{-1} (6 × 600 m^3 day^{-1}) RO unit for seawater desalination
- Potable water storage tanks of 3000 m^3 capacity
- Transfer pipeline for potable water from RO unit to the storage tanks
- 850 kW wind turbine (the W/T is installed in an already existing wind park, of 3 W/T); the average yearly electricity generation of the wind turbine is estimated to 1,800,000 kWh
- Supervisory Control and Data Acquisition Systems (SCADA) to balance the operation of the units according to the electrical load of the island of Milos and the operating status of the thermal plant, (Public Power Corporation, PPC) (Fig. 6.15).

With the use of the earliest technology energy recovery devices, and the use of high efficiency pumps, the specific energy consumption of the RO unit is around 2.5 kWh m^{-3} (Bouzas, 2010).

6.3.2 *Wind mechanical vapor compression systems*

Vapor compression (VC) distillation is a thermal process that has typically been used for small and medium-scale seawater desalination units. VC distillation takes advantage of the principle of reducing the boiling point temperature by reducing the ambient pressure, but the heat for evaporating the water comes from the compression of vapor, rather than from the direct exchange of heat from steam produced in a boiler. The vapor can be compressed by either a mechanical compressor

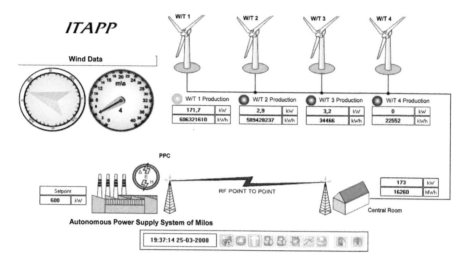

Figure 6.15.   The SCADA system provides the data for the proper operation and collaboration of the wind
turbines with the electricity grid (source: ITA S.A., Greece).

Figure 6.16.   Conventional vapor compression plants a) MVC plant in Sicily (source: SIDEM), b) 25 TVC
plant in Suez (source: Weir Westgarth Ltd).

or by the use of a steam jet thermo-compressor. In most cases, a mechanical compressor is used.
Two methods of compression are employed (Fig. 6.16):

- Mechanical vapor compression (MVC)
- Thermal vapor compression (TVC)

In MVC, the compressor is operated by an electric motor or diesel engine. High-pressure steam
is used to compress the vapor generated in the vessel. The compressed steam is then used as the
heat source for further vaporization of the feed water (Figs. 6.17 and 6.18). The unit consists of
an evaporator, a vapor compressor and the heat recovery exchangers. In the MVC process the
vapor produced inside the evaporator is directly withdrawn by the compressor which increases its
enthalpy, thus allowing its use as heating steam.

The distillate condensed in the tube side of the evaporator is cooled by a plate heat exchanger
(ideal to work with high salinity water and with low temperature differences) where the feed
seawater is preheated. The same is done for the exiting brine, thus recovering significant amount

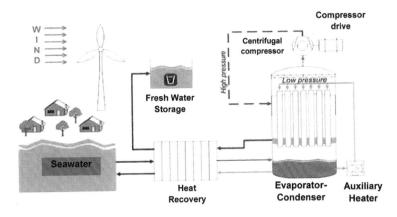

Figure 6.17.   Configuration of a wind-MVC plant (source: WME, Germany).

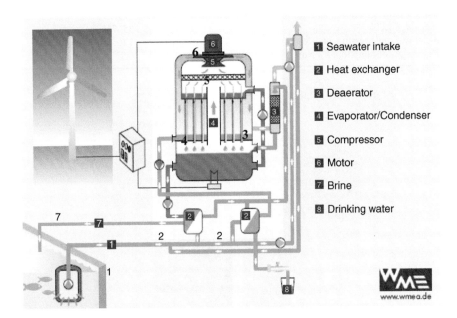

Figure 6.18.   Technical specifications of a wind-MVC plant (source: WME, Germany).

of process heat. MVC units are usually built in the range of 250 to 2500 m³ day⁻¹ and used for tourist resorts, small industries, and remote sites.

Regarding wind-MVC coupling only a small number of applications are known. A pilot plant was installed at the German island of Borkum in 1991 where a wind turbine with a nominal power of 45 kW was coupled to a mechanical vapor compression (MVC) evaporator in a system capable of desalinating seawater and producing up to 48 m³ day⁻¹ of freshwater. The compressor required 36 kW power, the system was controlled by varying the compressor speed, and assisted by a resistance heating where the compressor runs at its speed limit.

The experience was followed by another larger plant at the island of Rugen in 1995. The wind turbine was rated at 300 kW and the MVC unit at a maximum water capacity of 12.5 m³ h⁻¹. Again a resistance heating was used for auxiliary power when required. According to the available data, the specific energy consumption ranged between 9 and 20 kWh m⁻³.

Table 6.5.   General data of the SDAWES project (Subiela *et al.*, 2009).

Energy source	2 wind turbines of 230 kW each
Flywheel, rpm	1500
Synchronous machine, kVA	100
Uninterrupted power system unit (UPS), kW	7.5
8 RO desalination units (seawater)	
Nominal production capacity, $m^3 h^{-1}$	1
Nominal working pressure, bar (MPa)	60–70 (6.0–7.0)
Conversion rate, %	30
Specific energy consumption, $kWh\,m^{-3}$	7.2
Vapor compression plant (seawater)	
Nominal production capacity, $m^3 h^{-1}$	2
Conversion, %	50
Nominal working pressure at 62°C, bar (MPa)	0.2 (0.02)
Specific energy consumption, $kWh\,m^{-3}$	16
Electrodialysis reversal (EDR) (brackish water)	
Nominal production capacity, $m^3 h^{-1}$	3–7.9
Conversion rate, %	35–75
Nominal working pressure, bar (MPa)	1 (0.1)
Specific energy consumption, $kWh\,m^{-3}$	3.3

Figure 6.19.   Wind-MVC system in Gran Canaria, Spain (source: R. Morris).
Technical data of the wind-MVC system:
MVC capacity: $50\,m^3\,day^{-1}$
Feed water: seawater
Nominal power: $2 \times 230\,kW$ W/T
Pressure op.:   0.2 bar (0.02 MPa)
Temperature op.:   62°C
Compressor nom. power: 30 kW

In 1999, within SDAWES project (see Table 6.5) another wind-MVC unit was installed and tested in Pozo Izquierdo, Gran Canaria Island (Fig. 6.19). The MVC plant for seawater desalination has a capacity of $50\,m^3\,day^{-1}$. The nominal power of the wind turbines is $2 \times 230\,kW$.

Operating experience of the wind-MVC plant in Gran Canaria till date is limited as the unit was only commissioned in 1999. There were initial problems with scaling of the heat

Figure 6.20.   Autonomous wind-MVC plant in Symi Island, Greece: a) the 800 kW wind turbine, b) the MVC desalination unit.

transfer surfaces but these appear to have been overcome by using the correct chemical additive (Morris & Associates, 1999). The specific energy consumption of the unit was high at around $15–16 \, kWh \, m^{-3}$. This reflects the small-scale size of the unit and the heat transfer surface area.

According to the institute results, one of the main outcomes is that RO technology is the most suitable one for coupling to an off-grid wind farm. Nevertheless, each desalination technology has possibilities to improve the operation with off-grid wind farms by developing specific designs.

In 2008, a stand-alone wind-MVC system was installed in Symi Island in Greece (Fig. 6.20). The project was financed by the national program Operational Program for Competitiveness, Measure 6.3 of the Greek Ministry for Development. The desalination unit has a yearly production capacity of 90,000 m^3 of high purity water and the wind turbine has a nominal power of 800 kW. The specific energy consumption of the MVC unit is about $14.5 \, kWh \, m^{-3}$.

6.3.3   *Wind electrodialysis systems*

Electrodialysis is unique among all of the desalination processes, its main power requirements are of direct power (DC-direct current requirement). Alternating power (AC-alternating current) is only required for the operation of AC pumps or reversing controls. It is a membrane process like reverse osmosis, however, it differs in principle. In RO a pump pushes water through membranes leaving the salts behind, while in electrodialysis (ED), salts are drawn through membranes leaving desalted water behind.

A modification to the basic electrodialysis process is the electrodialysis reversal (EDR). An EDR unit operates on the same general principle as a standard ED plant, except that both, the product and the brine channels, are identical in construction. In this process the polarity of the electrodes changes periodically in time, reversing the flow through the membranes. Immediately following the reversal of polarity and flow, the product water is dumped until the stack and lines are flushed out and the desired water quality is restored.

One of the barriers of electrodialysis, is that the energy requirements are directly proportional to the quantity of salts removed, thus technology is attractive only for the desalination of brackish water (less than $3500 \, mg \, L^{-1}$ TDS). In general, the total energy consumption, under ambient temperature conditions and assuming product water of $500 \, mg \, L^{-1}$ TDS, would be around 1.5 and $4 \, kWh \, m^{-3}$ for a feed water of 1500 to $3500 \, mg \, L^{-1}$ TDS, respectively.

So far, most research on electrodialysis with renewable energies sources has been carried out on ED with photovoltaics. Little data have been reported for the wind-electrodialysis matching.

As mentioned before, within SDAWES project a wind-EDR system was examined. The EDR unit has a capacity ranged from 3 to around $8 \, m^3 \, h^{-1}$ and an average specific energy consumption of $3.3 \, kWh \, m^{-3}$. The electrical conductivity of the brackish feed water ranged from 2500 to

$7500 \, \mu S \, cm^{-1}$, while the product water conductivity ranged between 200 and $500 \, \mu S \, cm^{-1}$ (Veza et al., 2004).

The EDR plant was tested in on-grid and off-grid operation. The two operations modes were examined and compared. A control system has been developed for automatic management and optimization of plant operation, both for on-grid and off-grid modes. The control system was capable of setting the optimum operating condition, based on the available power and the criteria predefined by the plant operator (Carta et al., 2004).

After having previously analyzed the behavior of the system on-grid, the following stage was to develop an operational envelope for the electrodialysis reversal unit while operating off-grid, i.e., only coupled to the wind farm. The unit included power converters for the membrane stacks (DC-drivers) and variable frequency drivers (VFD) for the feed pumps. The tests were carried out to establish the power intervals for the EDR unit depending on the product flow rate specified as well as water quality.

According to the researchers of ITC and of University of Las Palmas, the desalination unit showed good flexibility, adapting smoothly to variations in wind power, even when sudden drops or rises occurred. The control system, slightly modified from a standard design, can cope with such sudden variations.

Good agreement between performance simulated with software and the actual operating performance was observed. The presence of harmonics in the electric system due to DC drivers and due to the variable frequency drivers (VFD) may become harmful for the control and electric system, and care must be taken through appropriate mitigating measures (Veza et al., 2004).

6.4   WIND DESALINATION MARKET

In recent years several suppliers provide in the market modular RES desalination solutions, most of them of standard and large sizes. Regarding wind-RO matching there are small number of companies which provide autonomous and/or grid connected systems in a range of around 175 to $2000 \, m^3 \, day^{-1}$ for brackish and seawater desalination. Concerning wind-MVC seawater desalination, there is only one company that provides compact systems for autonomous or grid operation (Tzen, 2008).

In general, commercialization of RES desalination is a relatively new area and little is known about the size of the potential market and the types of RES desalination plants that are most suitable for different parts of the market. Without a comprehensive market analysis it is difficult to determine (i) where and how to enter the market, (ii) how long it may take to receive a return on investments and (iii) how large the return on investment and therefore the magnitude of risk associated with investment in the technology may be. Although that there are (i) several authoritative studies that show that the need for desalination technologies is growing and (ii) an increasing general and governmental support for renewable energy technologies, this is inadequate to define the market for RES desalination. The need of RES desalination is not congruent with the demand. The range of RES desalination technologies is large, with each technology having particular characteristics which would need to be matched to a market analysis to enable investment decisions to be made; this level of detailed analysis is currently missing (Papapetrou et al., 2010).

6.5   CONCLUSIONS

The matching of the desalination process to a renewable energy source is technically feasible; however, special care is required on parameters regarding the selection of the site for the installation of the system, the RES potential, selection of the proper desalination technology, energy requirements and availability, simplicity and automation of the system, optimum design and staff availability (Tzen and Morris, 2003).

Desalination processes are best suited to continuous operation. In contrary, the majority of the renewable energy sources is distinctly non-continuous and is in fact intermittent often on a diurnal basis. A renewable energy driven desalination plant can be designed to operate coupled to the grid and off-grid (stand-alone or autonomous). The matching of the desalination process to a renewable energy source is fairly complex, especially regarding autonomous operation.

Each desalination system has specific problems when it is connected to a variable power system. Reverse osmosis has to deal with the sensitivity of the membranes regarding fouling, scaling, as well as unpredictable phenomena due to start-stop cycles and partial load operation during periods of oscillating power supply (Morris and Associates, 1999). On the other hand, units, which include storage back up systems like battery banks, increase system's initial cost and in hard climate conditions the maintenance requirements. Though, data quantifying battery failure and frequent replacement are not considered. Another example on the intermittent operation of the desalination systems concerns the vapor compression system. VC has considerable thermal inertia and consumes much energy to get to the nominal working point. Also, scaling problems have to be considered due to discontinuous power supply.

The use of battery storage seems in small-scale stand-alone RES desalination systems the best solution. Larger units are probably better connected to the grid. The consumed energy of the desalination unit, as in the case of Milos Island, could be covered by an on-grid RE power source. The environmental gain, especially for the avoidance of extra energy load to the local conventional power supply plants of the remote or isolated areas, is of significant importance.

The success of an installation depends on the quantity and the quality of the produced water, the energy consumption and the final unit water cost. As mentioned in the literature the cost of water produced by wind powered RO systems is in the range of 3–$7\,€\,m^{-3}$ for small RO plants (less than $100\,m^3\,day^{-1}$), and estimated in 1.5–$4\,€\,m^{-3}$ for medium capacity RO units (1000–$2500\,m^3\,day^{-1}$) (Papapetrou *et al.*, 2010). The final unit water cost of such systems depends on several parameters such as the site, the wind potential, the feed water quality, etc. This cost can be easily compared with the cost from other water sources, such as bottled water or with the water which is transported by tankers or trucks.

REFERENCES

ADU RES: Report on the status of autonomous desalination units based on RE Systems. INCO-CT-2004-509093, ADU RES Project, 2005.

Bouzas, I.: Desalination with the use of wind energy in Milos island. International Technological Applications (ITA) S.A., PRODES Workshop, Renewable energy Technologies with Desalination: Technologies Progress-Legislation-Funding Schemes, CRES, Athens, September 2010.

Carta, J.A, Gonzalez, J. & Subiela, V.: The SDAWES project: an ambitious R&D prototype for wind powered desalination. *Desalination* 161 (2004), pp. 33–48.

CRES: *Desalination guide using renewable energies*. Centre for Renewable Energy Sources (CRES), Pikermi, Greece, 1998.

EWEA: *Powering the energy debate*. The European Wind Energy Association Annual Report 2009, ISSN: 2032-9024, 2010.

IEA: *Wind Energy 2009*. IEA Annual Report, July 2010, ISBN 0-9786383-4-4.

IEA: *Wind Energy 2010*. IEA Annual Report, July 2011, ISBN 0-9786383-5-2.

Ifantis, N. & Ifantis, A.: Coupling of wind energy with reverse osmosis: the case of Milos *PRODES Workshop*, Promotion of Desalination Technologies with RES for a Sustainable Development, AUA, Athens, Greece, March 2010.

Morris, R. & Associates: *Renewable energy powered desalination systems in the Mediterranean Region*. UNESCO, Scotland, July 1999.

Papapetrou, M., Wieghaus, M. & Biercamp, Ch.: Roadmap for development of desalination powered by renewable energy, promotion of renewable energy for water production through desalination, PRODES Project, ISBN 978-3-8396-0147-1, 2010.

Schreck, Sc.: *Integrated wind energy/desalination system*. National Renewable Energy Laboratory, NREL/SR-500-39485, 2006.

Subiela, V.J., de la Fuente, J.A., Piernavieja, G. & Penate, B.: Experiences in desalination with renewable energy sources (1996–2008). *Desalin. Water Treat.* 7 (2009), pp. 220–235.

Tzen, E. & Morris, R.: Renewable energy sources for desalination. *Solar Energy* 75 (2003), pp. 375–379.

Tzen, E., Theofiloyianakos, D. & Kologgios, Z.: Autonomous reverse osmosis units driven by RES, experiences and lessons learned. *Desalination* 221 (2008), pp. 29–36.

Tzen, E.: Renewable energy sources for seawater desalination—Present status and future prospects. In: D.J. Delgado & P. Moreno (eds.): *Desalination research progress*. NOVA Science Publishers Inc., NY, 2008.

Tzen, E.: Wind and wave energy for reverse osmosis. In: A. Cipollina, G. Micale & L. Rizzuti (eds): *Seawater desalination*. Springer Publications, 2009, pp. 213–245.

Veza, J.M., Pefiate, B. & Castellano, F.: Electrodialysis desalination designed for off-grid wind energy, *Desalination* 160 (2004), pp. 211–221.

CHAPTER 7

Geothermal water treatment – preliminary experiences from Poland with a global overview of membrane and hybrid desalination technologies

Wiesław Bujakowski, Barbara Tomaszewska & Michał Bodzek

"We live in the hope and faith that, by the advance of molecular physics, we shall by-and-by be able to see our way as clearly from the constituents of water to the properties of water, as we are now able to deduce the operations of a watch from the form of its parts and the manner in which they are put together."

T.H. Huxley (1869)*

7.1 INTRODUCTION

Drinking water shortages in many regions of the world have contributed to the development of water treatment technologies. Simple treatment methods such as filtration, coagulation, sedimentation and high performance membrane and thermal desalination technologies have been gradually modified and upgraded since the mid-20th century. Hybrid systems that combine the advantages of multiple desalination technologies are also increasingly coming into focus. They have become a widely used method of producing water for drinking and household purposes.

One-half to two-thirds (depending on the source) of the total number of water desalination plants in the world are operated in the dry regions of the Middle East (Tsiourtis, 2001; Sadhukhan *et al.*, 1999; Bodzek and Konieczny, 2005; Morin, 1995; Eltawil *et al.*, 2009; Mezher *et al.*, 2011).

Compared to other European countries, Poland has scarce drinking water resources and exhibits significant variation in annual runoff. Shortages are reflected by the absence of groundwater reservoirs in some regions and the significant quantitative and qualitative anthropogenic pressure to which major aquifers are often subjected. On the other hand, the geothermal water resources present in sedimentary/structural basins, mostly in the Polish Lowlands and the Podhale geothermal system (Podhale basin) (Fig. 7.1) not only provide a valuable source of renewable energy (which is utilized although only to a limited extent) but can also be used for many other purposes, for example to drinking water production or for production of salt with balneological and economic importance.

The geothermal water reservoirs that are currently being exploited in Poland (Polish Lowlands and Podhale basin) exhibit varied physical properties (temperature, pH, electrical conductivity) and chemical composition. Both freshwaters with low dissolved mineral content (Mszczonów, Podhale – Zakopane-Antałówka, Zakopane-Szymoszkowa, Bukowina Tatrzańska, Cieplice Zdrój, Lądek Zdrój) and brines with mineralization exceeding $100\,g\,L^{-1}$ (Pyrzyce, Stargard Szczeciński, Ustroń) are exploited in the country (Fig. 7.1) (Bujakowski and Barbacki, 2004; Bujakowski and Tomaszewska, 2007; Kępińska, 2006). In the first case, open drain system operation (without injecting cooled water into the formation) has significantly improved the economic performance of the enterprise. Where water from geothermal sources is used as drinking water this also contributes to improving freshwater management. The extraction of brackish and saline water

---

*Source: T.H. Huxley: *On the Physical Basis of Life*, 1869.

is associated with serious problems, including the corrosion and clogging of geothermal water installations that significantly affect the cost of exploiting geothermal energy. The precipitation of secondary minerals from water reduces the productivity and absorption capacity of the wells used to inject cooled water, limiting the flow rate of geothermal fluids within the installation and ultimately shortening its lifespan.

These considerations gave rise to a research program designed to investigate the opportunities for the comprehensive utilization of cooled geothermal water resources and optimizing the operation of existing systems. The program involves, *inter alia*, the assessment of the possibility of desalinating geothermal waters to improve the overall drinking water balance in selected regions of the country. The use of water desalination technology to reduce corrosion and mineral precipitation in geothermal plants is also being assessed. Mixing raw geothermal water cooled in heat exchangers with "desalinated water" in suitable proportions may improve plant operation by eliminating the phenomena that lead to the clogging of geothermal systems. The salt obtained during the geothermal water purification process may prove a valuable product with balneological and economic importance (Bujakowski and Tomaszewska, 2007; Bujakowski *et al.*, 2010). The study presented is utilitarian in character and concerns two major research areas: geothermics (methods for using cooled water) and hydrogeology (water management).

Geothermal energy will be used for production of drinking water in two stages (i) the use of geothermal waters in a membrane system (optimum temperature of geothermal water for

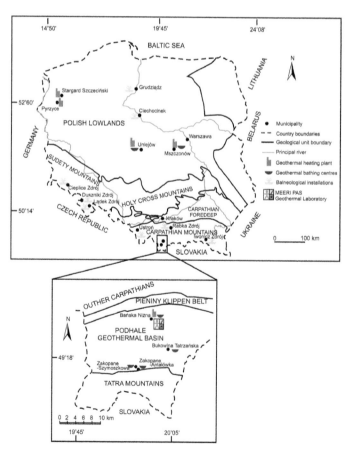

Figure 7.1.   Location of geothermal facilities in Poland against the background of geothermal units (modified from Bujakowski, 2010).

desalination processes) and (ii) concentrate of retentate (using water with a temperature of 80°C). In the second case geothermal energy (heat) will be supported by solar energy. The geothermal water treatment project is strictly of research nature.

The highest quality geothermal resources in Poland are present within the Podhale geothermal system in the southern part of the country (Fig. 7.1). This is due to the high artesian flow rates from wells which range from 90–550 m^3 h^{-1} and the renewable nature of the reservoir present in the region (the Tatra area provides recharge of the aquifers). At the same time, this is an area with a deficit of freshwater. This was the reason why pilot studies concerning the desalination of geothermal water were launched in that region. The selection of the most suitable water treatment technology was preceded by a detailed analysis of seawater desalination methods used globally. The technology was selected on the basis of several factors, the most important of which were the physical properties and chemical composition of water.

7.2   GLOBAL OVERVIEW OF MEMBRANE TECHNOLOGIES

The choice of desalination methods is affected by many factors, the principal being water salinity. The most popular technology in the United States is membrane-based desalination (Younos and Tulou, 2005). In terms of capacity, most desalination plants are installed in the Near and Middle East, where evaporative technologies continue to prevail (Fritzmann *et al.*, 2007). In Europe, all recently constructed plants are based on reverse osmosis (Table 7.1) (Fritzmann *et al.*, 2007).

The total global capacity of all plants currently online is 66.5 million m^3 d^{-1}, an increase of 8.8% over the 2010 inventory. The total worldwide capacity of all desalination plants, including those online, under construction and/or contracted, stands at 77.4 million m^3 d^{-1} (H$_2$O Applying, 2011). In 2001, the total worldwide desalination plant capacity amounted to 32.4 × 10^6 m^3 d^{-1} which was almost three times higher than 10 years earlier (Bodzek *et al.*, 2009). About 80% of the world's desalination capacity is provided by two technologies: Multi-stage flash evaporation (MSF), and reverse osmosis (RO). Figure 7.2 shows the global desalination capacity by process, highlighting the high capacity shares of RO and MSF (Mezher *et al.*, 2011). In practice, although these plants operate in 120 countries, 65% of their output is produced in the Persian Gulf region (Bodzek *et al.*, 2009). There are 45 MSF plants, 32 MED (multi-effect desalination) plants and 41 RO plants in the Gulf region (Landais, 2009). The use of electrodialysis (ED) technology is limited primarily to treat saline water in relatively small-scale desalination plants. There is an increasing trend to replace distillation by membrane methods, primarily reverse osmosis (RO).

Table 7.1.   Recently constructed RO-based desalination plants (based on Fritzmann *et al.*, 2007).

Country/city	Capacity (thousand m^3 d^{-1})	Construction year
Cyprus		
Larnaca	56	2001
Dhekelia	40	1996
Spain		
Carboneras	123	2002
Marbella	55	1996
Canal de Alicante	63	2006
Cartagena	140	2006
Spain – Canary Islands		
Las Palmas	78	1969–2004
Telle	35	2001
Lanzarote	20	2000
Spain – Mallorca		
Bahia de Palma	67.5	2001

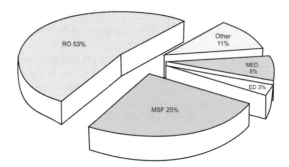

Figure 7.2.   Worldwide capacities (2009) of various processes used in desalination plants (modified from Mezher *et al.*, 2011).

7.2.1   *Types of membrane processes*

The history of water desalination using membrane-based methods dates back almost 100 years (Younos and Tulou, 2005). Membrane-based desalination primarily employs reverse osmosis and electrodialysis, which is based on the use of ion-exchange membranes to separate ions (Tsiourtis, 2001).

Membrane processes, which are driven by the pressure difference (Δp) between both sides of a membrane (microfiltration, ultrafiltration, nanofiltration and reverse osmosis) (Bodzek *et al.*, 1997; Narębska, 1997; Gawroński, 2004), are mostly used for concentrating and/or purifying dilute aqueous solutions. Under the pressure applied, the solvent and low molecular weight solutes pass through the membrane, while other molecules with higher molecular weight, colloids and fine suspensions cannot cross it. Depending on whether microfiltration, ultrafiltration, nanofiltration or reverse osmosis is applied, particles with increasingly smaller molecular weights are blocked (Fig. 7.3). The area of application determines the size of the particles that the membrane stops. Technologies based on the solution-diffusion mechanism range from reverse osmosis with the densest membranes that are only permeable to water, through nanofiltration, with membranes that make it possible to separate ions with different valences and ultrafiltration with membranes that stop fine suspensions, colloids, bacteria and viruses, to microfiltration membranes with the largest pores that are able to stop macrosuspensions.

Within the present technological and theoretical framework, membrane technologies are utilized in the following areas (Bodzek and Konieczny, 2005; Bodzek and Konieczny, 2006a,b; Bodzek *et al.*, 1997):

- reverse osmosis (RO) is used to stop ions and most low molecular weight organic compounds; it is primarily employed in the desalination of water and wastewater as well as to remove metal ions, inorganic anions and other low molecular weight organic compounds;
- a nanofiltration membrane (NF) stops colloids, many low and medium molecular weight organic compounds and divalent ions; it can be used to soften water and to remove organic and inorganic micropollutants from water and wastewater;
- ultrafiltration (UF) and microfiltration (MF) membranes serve as a barrier to dispersed substances, including colloids and micro-organisms. The membranes can be used to remove these from water and wastewater, as well as within the framework of membrane-based and thermal desalination and demineralization processes. Moreover, they are utilized in integrated/hybrid arrangements in processes such as coagulation – UF/MF, powdered activated carbon adsorption – UF/MF, biological filtration – UF/MF, oxygenation (ozonation and others) – UF/MF and in membrane bioreactors.

A promising technology for the desalination of seawater or brackish water is electrodialysis (ED) and electrodialysis reversal (EDR), although the production of potable water from brackish

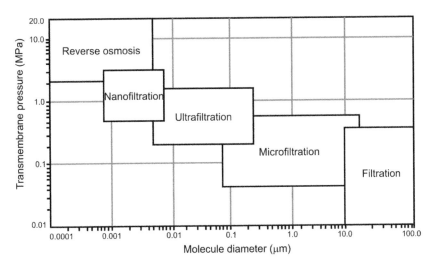

Figure 7.3.   Sizes of particles blocked in microfiltration, ultrafiltration, nanofiltration and reverse osmosis processes (modified from Bodzek and Konieczny, 2005).

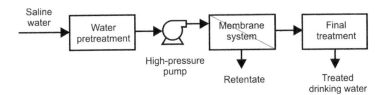

Figure 7.4.   RO water desalination plant – simplified diagram.

water sources is presently the largest single application of electrodialysis (Van der Bruggen and Vandecasteele, 2002; Strathmann, 2010; Eltawil *et al.*, 2009; Al-Karaghouli *et al.*, 2010). The membrane distillation (MD) process which was described in literature in the late 1960s has also not yet been commercialized as a water desalination process so far, mainly due to the absence of suitable membranes and unfavorable opinions concerning the economic viability of the process (Ryabtsev, 2001; El-Bourawi *et al.*, 2006; Susanto, 2011).

7.2.1.1   *Reverse osmosis*

The desalination of brackish water was the first successful large-scale application of reverse osmosis (Van der Bruggen and Vandecasteele, 2002), and was first used in the industry in the late 1960s. During the next decade, new RO membranes with relatively high permeability appeared that were suitable for seawater desalination. In the 1980s, reverse osmosis became a competitor for distillation-based methods. Currently, more than 90% of the RO facilities in operation produce drinking and household water as well as extra pure water for the power industry, manufacture of semiconductors, etc. (Baker, 2001; Mulder, 1991).

   A typical reverse osmosis water desalination plant consists of a raw water pretreatment system, a membrane desalination system including a high pressure pump and a final treatment system whose purpose is to ensure that the legal requirements applicable to drinking and household water in the jurisdiction in question are met. Figure 7.4 presents the most important components of RO desalination plant (Bodzek and Konieczny, 2005; Aim and Vladan, 1989).

   The advantage of reverse osmosis is the low cost of desalinated water, which is around US$ 0.5–0.7 per m^3 as compared to US$ 1.0–1.4 per m^3 for multi-stage flash evaporation

(MSF) and multi-effect distillation (MED) (Malek *et al.*, 1996; Blank *et al.*, 2007; Reddy and Ghaffour, 2007; Mathioulakis *et al.*, 2007; Gerstandt *et al.*, 2008; Alishiri, 2008; Karagiannis and Soldatos, 2008). Currently, owing to the reduction in cost that has been achieved during the last 20 years, it is a generally accepted method.

The RO process is affected by a particular problem associated with its performance – fouling, i.e. the permanent, often irreversible, change in membrane permeability caused by many different factors. Fouling is the deposition of organic matter (suspended matter, colloids, soluble macromolecular compounds, salts) on the membrane surface and/or in the pores, which reduces membrane permeability. It is caused by organic and inorganic substances as well as suspended matter (Bodzek and Konieczny, 2005; Bodzek *et al.*, 1997; Fritzmann *et al.*, 2007; Mulder, 1991; Kołtuniewicz, 1996). Another operational problem is membrane scaling (solid salt deposits) caused by substances such as $CaCO_3$, $CaSO_4$ and $BaSO_4$; the intensity of this phenomenon depends on the permeate/raw water volume ratio. At a desalinated water recovery ratio of 50%, this phenomenon can be effectively reduced by adding divalent ion complexing agents (so-called antiscalants) to water.

As illustrated by the considerations above, the pretreatment of raw water is the main factor determining the success or failure of a desalination plant. During the last ten years, low-pressure membrane processes, namely ultrafiltration (UF) and microfiltration (MF), have proven to be suitable means of removing dispersed substances, some organic compounds and microbiological contaminants, including pathogenic substances (Van der Bruggen and Vandecasteele, 2002; Truby, 2000; Van Hoof *et al.*, 2001; Brehand *et al.*, 2002a,b). The use of UF/MF as a water pretreatment method before RO desalination makes it possible to reduce investment expenditure by 20 and 12.5%, respectively, for the desalination of brackish water and seawater compared to traditional solutions. Significant savings can also be achieved with respect to operating costs (Van der Bruggen and Vandecasteele, 2002).

An example of effective use of UF in the pretreatment of water preceding desalination has been provided by the pilot test results obtained in England during the filtration of seawater from the North Sea (Murrer and Rosberg, 1998). Microbiological tests have shown that UF removes microorganisms effectively, thus significantly decreasing the biological fouling of RO membranes.

Around 80% of membranes currently used for RO purposes are composite membranes (Alley, 2003). Two trends can be observed in studies conducted during the last ten years: the introduction of low-pressure RO membranes used for desalination and the use of high-pressure membranes that offer higher efficiency compared to conventional ones (Van der Bruggen and Vandecasteele, 2002; Matsuura, 2001). As far as modifications of RO desalination membranes are concerned, work is underway on commercial development of chlorine-resistant membranes that would eliminate the need for dechlorination of the RO feed and rechlorination after the membrane system, reducing the overall cost of RO. Improving the service life of membranes, reducing the scope of water pretreatment and minimizing fouling and scaling particularly for surface water sources, also further optimization and development in RO membrane technology (membrane manufacturers are developing new RO membranes with higher boron rejections) are still under way (Greenlee *et al.*, 2009).

In May 17, 2010 the Hadera seawater desalination plant, the largest RO facility that can produce 127 million cubic meters of water per year at a price of $0.57 per cubic meter opened in Kadima, Israel (IDE Technologies Ltd., 2011). The second largest RO installation in the world, Ashkelon plant, located at the Mediterranean Sea, with capacity of $330,000\,m^3\,day^{-1}$ (120 million m^3 per year) has been in operation in Israel since 2005 (Sauvet-Goichon, 2007). The operation costs of RO reduced over the years due to development of low-cost efficient membranes and usage of pressure recovery devices. Seawater RO cost has gone down to about $0.53\,US\$\,m^{-3}$ in Ashkelon (Sauvet-Goichon, 2007; Mezher *et al.*, 2011) (Table 7.2).

7.2.1.2   *Electrodialysis*
Electrodialysis, which was developed in the 1950s, has primarily been used for the desalination of brackish water. In the case of seawater, the cost of ED desalination is as high as energy

Table 7.2.   Electrical and thermal energy consumption of major desalination processes and cost of water (based on Mezhre *et al.*, 2011).

Parameter	MSF	MED	RO	ED	VC
Energy requirement	Elec.: 3.5–5.0 kWh m^{-3} Th: 69–83 kWh m^{-3}	Elec: 1.5–0.5 kWh m^{-3} Th: 41–61 kWh m^{-3}	SW: 4–8 kWh m^{-3} BW: 2–3 kWh m^{-3}	SW: 17 kWh m^{-3} BW: 3–7 kWh m^{-3}	7.5–13 kWh m^{-3}
Cost of water	0.9–1.5 US$ m^{-3}, the cost reduces with cogeneration and unit capacity	Around 1 US$ m^{-3}, 0.83 US$ m^{-3} for Jubail II plant	0.99 US$ m^{-3} for SW., 0.53 US$ m^{-3} Ashkelon 0.2–0.7 US$ m^{-3} for BW	About 0.6 US$ m^{-3} for BW	Unit cost is larger than MSF and MED

SW: seawater; BW: brackish water; Elec: electrical energy; Th: thermal energy.

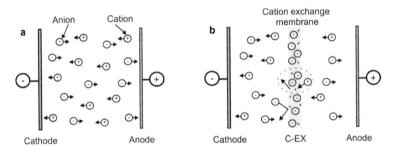

Figure 7.5.   (a) Ions transported in the electric field and (b) through a cation exchange membrane (modified from El-Dessouky and Ettouney, 2000; Bodzek, 1999; Strathmann, 2004).

consumption which is proportional to the amount of salt that is carried through the membrane (Table 7.2). Where concentrated solutions are involved, the contribution of secondary processes also rises, significantly reducing the energy efficiency of the process. Electrodialysis is mainly used in small to medium size plants with capacities of less than a few 100 m^3 day^{-1} to more than 20,000 m^3 day^{-1} with a brackish water salinity of 1000 to 5000 mg L^{-1} total dissolved solids. Reverse osmosis is considered to have an economic advantage for the desalination of water with total dissolved salts in excess of 10,000 mg L^{-1} (Strathmann, 2010).

In the ED process, ion exchange membranes are used in such a way that when placed in an electric field, they allow the transport of cations and anions in the raw saline water. (Fig. 7.5) (El-Dessouky and Ettouney, 2000; Bodzek, 1999; Strathmann, 2004). Ion exchange membranes are semipermeable structures – spatial polymeric networks with integrated ion exchange groups.

Cation exchange membranes include highly concentrated negatively charged groups (e.g. SO_3^-) that are strongly bound to the polymeric network, while anion exchange membranes include positively charged groups (e.g. NH_3^+).

The position of electrodialysis was strengthened after the Ionics Company had developed polarity reversal electrodialysis, so-called electrodialysis reversal (EDR) (Meller, 1984). This solution brings a number of advantages, preventing membrane fouling and scaling.

The main area of application of electrodialysis with monopolar membranes is desalinating brackish water in order to obtain drinking water (Wiśniewski, 2001; Strathman *et al.*, 2006). If salt concentrations in feed water range from 1000 to 2000 mg L^{-1}, then it is possible to obtain usable water in a single-stage process with a water recovery ratio of up to 85% (Wiśniewski, 2001; Heshka, 1977; Thompson *et al.*, 1977). Generally, ED is used to desalinate brackish water containing 1000–5000 mg L^{-1} of dissolved solids (Sadhunkhan *et al.*, 1999).

The largest number of brackish water desalination plants employing electrodialysis operate in the U.S., Japan and China (Wiśniewski, 2001; Strathmann, 2010). At these facilities, desalination is based on the electrodialysis reversal (EDR) process which makes it possible to avoid the precipitation of unwanted salts (CaCO$_3$ and CaSO$_4$) in electrodialyzer chambers. In the U.S., there were more than 30 plants that were producing drinking water from brackish water by EDR, with a capacity of about 150,000 m^3 day^{-1} in the 1970s (Leitner and Leitner-Murnay, 1977). Between 1992 and 1996, the capacity of such plants grew by 100%, but this brackish water desalination technology was still much less widely used than reverse osmosis (Wiśniewski, 2001). Electrodialysis was one of the most important processes used in Japan. One of the largest ED plants was opened in Oshima, Tokyo in 1990 (Hamada, 1995). In this plant raw water with a salinity of 1600 mg L^{-1} was purified using two-stage electrodialysis. Plant capacity (1850 m^3 day^{-1}) producing usable water with a salt content of approximately 420 mg L^{-1}, and the water recovery ratio in the process employed exceeds 86% (Hamada, 1995). In China, the EDR process has been successfully used for thirty years in the desalination of brackish water with high sulfate hardness (Song *et al.*, 1995).

The electrodialysis process can be used to desalinate seawater, but multi-stage electrodialysis is required in order to obtain drinking water. This makes it possible to obtain usable water with a salt content of 700 mg L^{-1} from seawater with a saline concentration of 35,000 mg L^{-1}.

ED is primarily applied on brackish water with low TDS hence the cost is low (about 0.6 US$ m^{-3}). ED has a high water recovery of 85–94% and a concentrate of 140–600 mg L^{-1} TDS. This means that the concentrate has a great impact on the environment, and also there is a possibility of leakage in membrane stacks. ED has been on a steady low increasing trend since year 2000 (Mezher *et al.*, 2011). Large plants with a capacity of 20,000 to more than 200,000 tons of salt per year are in operation in Japan (Strathmann, 2010).

7.2.1.3   *Membrane distillation*

Favorable results from studies of the membrane distillation (MD) process, revealing the broad scope of its applications and the possibility of achieving separation in cases where other methods fail, have contributed to the growing interest in the implementation of this process. Several pilot tests are being conducted on new ones (Gryta, 2003; Zakrzewska-Trznadel *et al.*, 1999; Wirth and Cabassud, 2002; Banat and Simandl, 1998).

Membrane distillation (MD) is a process of water evaporation through a porous lyophobic membrane which forms a non-selective physical barrier (Tomaszewska *et al.*, 1995; Tomaszewska, 1996). The hydrophobic membrane usually separates aqueous solutions with different temperatures and compositions. The process is driven by the vapor pressure difference resulting from the temperature difference across the membrane, mass transport taking place towards the lower-temperature flow. The principle behind the MD process is presented in Figure 7.6.

Depending on the way how vapor pressure difference as driving force and vapor condensation are provided, four different configurations of MD are currently known (Susanto, 2011):

- Direct contact membrane distillation (DCMD). In this type, water or an aqueous solution having lower temperature than liquid in feed side is used as condensing fluid in permeate side. The trans-membrane temperature difference results in vapor pressure difference as a driving force. Even though its heat loss is the greatest, this configuration is the simplest among all configurations of MD.
- Air gap membrane distillation (AGMD). In this configuration, to reduce heat loss, a stagnant air gap is inserted between the membrane and a condensing surface. However, this stagnant air adds a mass transfer resistant.

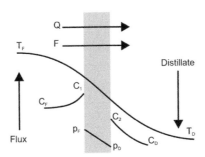

Figure 7.6. Direct contact membrane distillation (C: concentration, T: temperature, p: vapor pressure, Q: heat, F: flux) (modified from El-Dessouky and Ettouney, 2000; Bodzek, 1999; Strathmann, 2004).

- Sweeping gas membrane distillation (SGMD). To minimize heat loss in DCMD and mass transfer resistant in SGMD, a cold gas (inert) is used in permeate side to sweep the vapor molecules and carry to outside the membrane module for condensation. Nevertheless, operational cost will definetly increase due to external condensation system.
- Vacuum membrane distillation (VMD). In this type, permeate side is vacuumed yielding lower vapor pressure than in the feed side. Consequently, heat loss can be reduced and permeate flux can be increased (it should be noted that the applied vacuum pressure should not exceed the saturation pressure of volatile molecules).

Water desalination by MD is less economic than other methods, e.g. state-of-the-art RO membranes (Tomaszewska, 2001; Scott 1997). The situation changes where water with a high salt content is to be desalinated. In this case, a combination of RO with MD proves advantageous, allowing for high water recovery (Gryta, 2003). In order for 1 m^3 of water to evaporate in an MD installation, 600–690 kWh of energy is required (Gryta, 2003; Zakrzewska-Trznadel et al., 1999; Godino et al., 1996). Using heat recovery modules, this amount can be reduced to 150–180 kWh (Zakrzewska-Trznadel et al., 1999). Due to the high energy consumption of the MD process, the cost of water produced strongly depends on the cost of the energy supplied and the temperature at which the process takes place.

Solar energy is the most interesting alternative energy for MD. Recently, Band and Jwaied (2008) reported an economic assessment of solar powered MD for potable water production in arid area. Based on the calculations, the estimated cost of potable water produced by the compact unit is 15 US\$ m^{-3}, and 18 US\$ m^{-3} for the large unit (Susanto, 2011).

7.3 HYBRID DESALINATION PROCESSES

The possibility of combining different desalination processes in order to achieve a synergetic effect has been suggested for several years (Tsiourtis, 2001; Van der Bruggen and Vandecasteele, 2002; Fritzmann et al., 2007; Matsuura, 2001; El-Sayed et al., 1998; Hamed, 2005); it is usually recommended that hybrid systems be developed combining membrane technologies with conventional separation systems as well as with other membrane-based systems. Solutions of this type often cost less than systems based on a single process. Three types of such arrangement can be distinguished (Fritzmann et al., 2007; Wilf, 2007; Awerbuch, 2005; Hamed, 2005; Fazle Mahbub et al., 2009; Helal, 2009):

- simple hybrid systems;
- integrated hybrid systems;
- hybrid systems that combine electricity generation with desalinated water production.

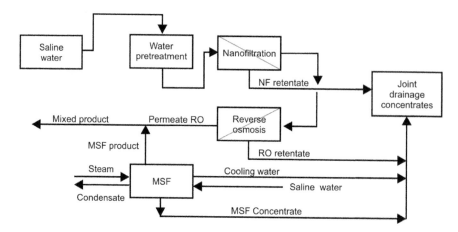

Figure 7.7.   Simplified diagram of a hybrid NF/RO/MSF desalination plant (modified from Helal, 2009).

According to General des Eaux and USFilter (Matsuura, 2001), there are several possible methods for combining membrane processes in hybrid systems; the most important among these being:

- combining reverse osmosis with thermal methods and membrane distillation;
- replacing the conventional water pretreatment system with ultrafiltration or microfiltration;
- the application of nanofiltration in order to soften raw water before distillation or RO.

The use of ultrafiltration or microfiltration processes for the pretreatment of raw water is also widespread in the case of membrane desalination methods. The use of nanofiltration membranes is also considered advantageous, since they exhibit very high divalent ion retention coefficients. Monovalent ion retention is limited. The application of NF as pretreatment before introducing water to an RO and/or thermal facility makes it possible to achieve the following (Bodzek and Konieczny, 2005):

- the elimination of turbidity, bacteria and sediment-forming ions and a reduction in the mineralization of the solutions fed into the MSF plant;
- the operation of thermal installations without the need to use any additional chemicals;
- an increase in permeate recovery by 70% and distillate recovery by 80%;
- reduced electricity consumption during the production of drinking water;
- the possibility of operating thermal facilities at higher temperatures ranging from 120 to 160°C with no or minimum need for additional chemicals.

In pilot studies conducted at the Saline Water Conversion Corporation (SWCC) laboratories in Saudi Arabia, the performance of a triple hybrid NF/RO/MSF system with a capacity of 20 m³ day⁻¹ was tested (Awerbuch, 2005; Helal, 2009). The MSF section operated effectively at a brine temperature of 130°C for 50 days without adding antiscalants. During the operation, the product recovery ratio increased from 35% (a value typical for MSF plants) to 70%. A simplified diagram of the pilot triple hybrid NF/RO/MSF plant is shown in Figure 7.7 (Helal, 2009).

A large hybrid SWRO/MSF desalination plant with a capacity of 450,000 m³ day⁻¹ has recently been constructed in Fujairah, United Arab Emirates (Hamed, 2005). Its specifications are provided in Table 7.3 (Helal, 2009).

Table 7.3.   Selected design parameters of the Al Fujairah
plant (modified from Helal, 2009).

Parameter	Unit	Value
Electricity production		
gross	MW	662
net	MW	500
Water production	m^3 day^{-1}	454600
Desalination setup		
MSF	m^3 day^{-1}	284125
RO	m^3 day^{-1}	170475

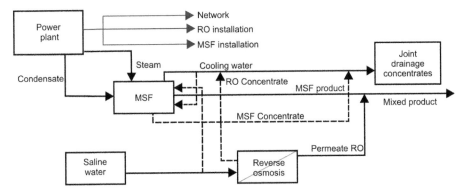

Figure 7.8.   Simplified diagram of a triple hybrid power plant/RO/MSF desalination system (modified from
Fritzmann *et al.*, 2007; Wilf, 2007; Awerbuch, 2005; Hamed, 2005; Fazle Mahbub *et al.*, 2009;
Helal, 2009).

With respect to desalination and electricity generation, hybrid desalination systems that combine both thermal and membrane desalination processes with electricity generation are also
considered a technologically and economically viable alternative (Awerbuch, 2005). Hybrid configurations (RO/thermal/power plant) are flexible in operation, consume less energy, involve
lower investment expenditure, exhibit better availability and enable improved management of
water and electricity. These systems are currently considered a real alternative to dual systems
using thermal methods. Advantages of triple hybrid electricity-MSF-SWRO systems (Fig. 7.8)
over dual electricity-MSF and simple MSF or RO systems have been described in many publications (Fritzmann *et al.*, 2007; Wilf, 2007; Awerbuch, 2005; Hamed, 2005; Fazle Mahbub *et al.*,
2009; Helal, 2009).

Hybrid systems that combine MSF and reverse osmosis technologies with energy generation
make it possible to achieve several benefits; the most important among them being:

- the integration of some elements common to RO and MSF installations;
- the mixing of desalinated water from the RO and MSF processes, making it easier to obtain
  drinking water or water for industrial purposes that meets the standards required by regulations;
- the use of waste heat from the MSF plant and/or power plant for power supply of the RO plant;
- the possibility of using low pressure water vapor generated by the MSF plant and/or the power
  plant to degas the water before it is fed to the membranes;
- the efficient use of electricity and desalinated water.

7.4   FRAMEWORK FOR DESALINATING GEOTHERMAL WATER IN POLAND

Geothermal water and energy are used for various purposes: heating and cooling buildings, drying building materials and agricultural products, rearing livestock, thermophilic fish farming, soil heating, recreation and balneology (Ney, 1997; Lund, 2010; Lund et al., 2005, 2010; Bujakowski, 2005, 2010; Kępińska, 2005). In some cases, after heat recovery, geothermal water is introduced into the water supply network if it satisfies applicable physico-chemical and health requirements (Bujakowski and Tomaszewska, 2007; Tomaszewska, 2009; Bujakowski et al., 2010; Tomaszewska and Pająk, 2010).

A model procedure for the management of geothermal water and energy should involve their comprehensive utilization in order to optimize the operation of geothermal systems. The studies currently underway focus on improving the economic performance of existing facilities and streamlining their operation.

Theoretical research concerning the opportunities for geothermal water desalination, including consideration of variability in the physical properties and chemical composition of the geothermal waters extracted in Poland, started in 2008. A project was developed at that time which included technology that has now gone into the pilot stage. One of the main research topics is the evaluation of the feasibility of using geothermal water resources in order to enhance drinking water management through the use of desalinated water (permeate). Another research field is related to the assessment of whether the water desalination technology can be used in order to reduce corrosion and mineral precipitation in geothermal installations. It appears that the mixing of raw geothermal water cooled in heat exchangers with the permeate in suitable proportions may improve the performance of plant. At the same time the recovery of mineral substances from the concentrate may be of balneological and economic importance (Bujakowski and Tomaszewska, 2007; Tomaszewska, 2009; Bujakowski et al., 2010).

Pilot water desalination tests have been conducted on a semi-industrial scale at two geothermal facilities (first in Podhale Basin and second in Polish Lowlands). In pilot tests the renewable energy (heat) is used for drinking water production:

- the use of geothermal water heat for optimal operation of RO at 30°C, since geothermal water can secure this level of temperature;
- geothermal and solar energy (heat) planned to be used to concentrate retentate obtained from RO, which is a second level recovery of distillate. This does not exclude to gain in this way concentrate for commercial purposes such as medical, therapeutics and cosmetics.

Geothermal energy will be therefore used for production of drinking water in two stages: first – the use of geothermal waters in a membrane system (optimum temperature of geothermal water for desalination processes) and second – concentrate of retentate (using water temperature of 80°C). In the second case geothermal energy (heat) will be supported by solar energy.

7.4.1   *Presence and quality of geothermal waters in Poland*

About two-thirds of the territory of Poland is considered promising in terms of the technological feasibility of developing its geothermal energy potential (Ney, 1997). Geothermal energy resources are associated with groundwater present at depths of up to 3000 m b.g.s. within certain regional geological units: the Polish Lowlands, the Carpathians, the Carpathian foredeep and the Sudetes (Fig. 7.1). The studies so far conducted in Poland and confirmed by a number of research and investment projects, indicate that the comprehensive utilization of geothermal waters should be based primarily on the potential accumulated in the Lower Cretaceous and Lower Jurassic reservoirs of the Polish Lowlands and the Middle Triassic, Jurassic and Middle Eocene reservoirs of the Podhale geothermal basin within the Inner Carpathians (Chowaniec, 2003; Bujakowski and Barbacki, 2004; Bujakowski, 2005, 2010; Bujakowski et al., 2006; Górecki et al., 2010; Kępińska, 2006; Tomaszewska, 2009).

The Lower Jurassic geothermal reservoir consists of a fine and mixed grain-size sand and sandstone layer of 10 to 650 m thickness; depending on the depth, the water within the reservoir exhibits a mineralization ranging from 2 to over $100 \, g \, L^{-1}$ and its temperature ranges from 40 to 80°C (depth from -1500 to -2000 m a.s.l.). The reservoir area is $158,600 \, km^2$, i.e. slightly less than 51% of the area of Poland. In the vast areas covered by Middle Jurassic aquifer altitudes are -1000 m a.s.l. and the deepest occurrences were reported from central part country to -3900 m a.s.l. The Lower Cretaceous reservoir has an area of $115,521 \, km^2$, i.e. 40% of the area of Poland. It forms a complex of discontinuous, interspersed sandy, sandy-marly and sandy-mudstone layers with thicknesses ranging from a few to 300 m. The TDS values do not exceed $40 \, g \, L^{-3}$, and the temperature is varying from 20 to about 80°C. The top part of Lower Cretaceous formation occurs at various altitudes from about $+250$ m a.s.l. to below -2500 m a.s.l. Within the dominant part of the reservoir, water temperature ranges from 20 to 40°C, only in some areas temperatures rise to over 50°C being related to the deepest structural depressions (Górecki, 2006; Górecki, 2010; Szczepański, 1990, 1995; Strzetelski, 1990). Aquifers within the Lower Jurassic and Lower Cretaceous formations of the Polish Lowlands exhibit high hydraulic permeability and porosity of the reservoir, which means that the capacity of hydrogeothermal wells discharges to be high (from 25 to more than $200 \, m^3 \, h^{-1}$).

The most favorable hydrogeological conditions for the presence of geothermal waters in Poland exist within the Podhale geothermal system (Podhale basin). This geothermal reservoir consists of several aquifers present within Triassic limestone and dolomite, Jurassic sandstone and carbonate rocks and within Eocene carbonate formations (Sokołowski, 1973; Małecka, 1981; Kępińska, 2006; Chowaniec, 2003). The aquifers lie directly beneath the insulating cover of Podhale flysch (Upper Eocene–Oligocene). The thickness of the reservoir rocks ranges from 100 to 700 m. Maximum artesian flow rates from wells range from 90 to $550 \, m^3 \, h^{-1}$, and water temperatures range from 20 to 90°C depending on the depth from 660 m b.g.s. to 3500 m b.g.s. The mineralization of geothermal waters ranges from less than $200 \, mg \, L^{-1}$ within the Tatra massif to $3000 \, mg \, L^{-1}$ (TDS) in the northern part of the reservoir. The Tatra area provides the recharge for the aquifers, so the resources found there are of a renewable nature (Kępińska, 2006; Chowaniec, 2003; Bujakowski and Barbacki, 2004; Bujakowski *et al.*, 2006; Bujakowski, 2010; Tomaszewska, 2009). The geothermal waters sourced in the area are mainly used for heating (PEC Geotermia Podhalańska S.A.) and recreational purposes (Zakopane-Antałówka, Zakopane-Szymoszkowa, Bukowina Tatrzańska, Szaflary).

The research and implementation work carried out in Poland since the mid-1980s has so far resulted in the commissioning of 16 geothermal facilities extracting geothermal waters with temperatures exceeding 20°C which are used for heating, medicinal and recreational purposes (Bujakowski, 2010). In Table 7.4, the disposable energy resources of major geothermal reservoirs in Poland are listed, while Table 7.5 presents the main parameters of the facilities currently in operation.

In terms of salinity, the geothermal waters extracted in Poland can be divided into four groups:

- Group I: very high mineralized waters (Pyrzyce, Stargard Szczeciński, Ustroń, Grudziądz) with a concentration of dissolved minerals ranging from 70 to $135 \, g \, L^{-1}$, the main components being sodium and chloride. These geothermal waters are sourced from Lower Jurassic sandstone aquifer formations (in the northwest of Poland – Pyrzyce, Stargard Szczeciński, depth from 1500 to 2600 m b.g.s., temperature ranges 60–87°C) and from Devonian limestone and dolomite (Outer Carpathians – Ustroń, depth 1760 m b.g.s., temperature 28°C). They exhibit an increased content of dissolved sulfate, calcium and magnesium, iron, iodine, bromide, boron, strontium and fluoride.
- Group II: high mineralized waters with high concentrations of dissolved minerals (Ciechocinek, Rabka Zdrój), which are also categorized as chloride-sodium waters with a concentration of solutes ranging from 22 to $58 \, g \, L^{-1}$. These are waters associated with Lower Jurassic aquifers of the Polish Lowlands (Ciechocinek, depth 1450 m b.g.s., temperature 27–32°C) and Paleogene aquifers of the Carpathian overthrust (Rabka Zdrój, depth of aquifer 1200 m b.g.s., temperature

Table 7.4.   Disposable energy resources within major groundwater basins in Poland (based on Bujakowski, 2005, 2010; Górecki, 2006).

Reservoir age	Reservoir area [km²]	Percentage of country area [%]	Temperature [°C]	Disposable energy resources [TJ year⁻¹]
Main hydrogeothermal horizons, Polish Lowlands				
Lower Cretaceous	115 521	36.9	40–100	382 000
Upper Jurassic	198 975	63.6	40–100	224 000
Middle Jurassic	202 225	64.7	40–100	999 000
Lower Jurassic	158 600	50.7	40–100	1 731 000
Upper Triassic	175 900	56.3	40–100	761 000
Lower Triassic	229 525	73.4	40–160	2 585 000
Lower Permian	101 013	32.5	50–220	2 030 000
Carboniferous	46 709	14.9	30–240	526 000
Devonian	48 424	15.5	40–170	374 000
Total	–	–	–	9 219 000
Podhale geothermal basin				
Triassic/Jurassic/Paleogene	475	0.15	293–373	1490

Table 7.5.   Main energy parameters and water mineralization at geothermal and balneological facilities in Poland; for localities see Figure 7.1 (based on Kępińska, 2005; Bujakowski, 2005, 2010; Tomaszewska, 2009).

Location	Admissible volume of extracted water [m³ h⁻¹]	Temperature [°C]	Mineralization [g L⁻¹]	Capacity total/ geothermal (MWₜ)	Groundwater extraction [m³ year⁻¹]	[m³ h⁻¹]
Heating installations (temp. >298 K)						
Mszczonów	60	42	0.5	7.4/1.1	283 509	32.4
Podhale						
Bańska Niżna	670	86	2.9	80.5/15.5	2 977 218	339.9
Zakopane-Antałówka	130	33.5	0.4	2.6/2.6	292 709	33.4
Zakopane-Szymoszkowa	80	27	0.5	1.2/1.2	no data	no data
Bukowina Tatrzańska	40	64.5	0.15	2.41/2.26	88 298	10.1
Pyrzyce	340	61	135	48.0/15.0	621 879	71.0
Stargard Szczeciński	200	87	120	10.0/10.0	711 948	81.3
Uniejów	120	68	7	5.6/3.2	360 977	41.2
Balneological installations						
Ciechocinek	479	27–32	38–58	8.2/8.2	107 770	12.3
Cieplice Zdrój	56.54	22–60	0.6	1.38/1.38	54 167	6.2
Duszniki Zdrój	107.48	16–21	1.6–2.2	0.99/0.99	321 805	36.7
Grudziądz	20	20	78	0.23/0.23	5827	0.7
Lądek Zdrój	59.85	20–44	0.2	1.03/1.03	324 631	37.1
Iwonicz Zdrój	11.7	24		0.13/0.13	6685	0.8
Rabka Zdrój	6.44	28	22	0.11/0.11	6521	0.7
Ustroń	2.2	28	120–135	0.06/0.06	5269	0.6

28°C). They exhibit an increased content of calcium, magnesium, sulfate, sodium, iodine, bromide, boron, iron, strontium and fluorine.

- Group III: medium mineralized waters with medium contents of dissolved minerals (Podhale – Bańska Niżna, Uniejów, Duszniki Zdrój) containing $1.6–9\,g\,L^{-1}$ of dissolved salts, mostly sulfate, chloride, sodium, calcium, magnesium and iron. These geothermal waters are sourced from the Lower Cretaceous sandstones of the Polish Lowlands (Uniejów), limestone, dolomite and sandstone rocks of the Podhale geothermal system (Inner Carpathians, Podhale-Bańska Niżna) and Precambrian shales of the Sudetes (Duszniki Zdrój).
- Group IV: low mineralized waters, freshwaters ($<41\,g\,L^{-1}$), Mszczonów, Podhale – Zakopane-Antałówka, Zakopane-Szymoszkowa, Bukowina Tatrzańska, Cieplice Zdrój and Lądek Zdrój), in which the content of solutes meets the guidelines for drinking water or their treatment does not require the use of advanced technologies.

7.4.2   Choice of desalination technologies

Water desalination costs have decreased in recent years owing to technological advances and the increased scope for using renewable energy end energy efficient techniques.

Different desalination systems may be considered depending on the characteristics of the water. The feasibility of using geothermal waters (directly or indirectly as energy source) to produce drinking water is determined by the physical properties and chemical composition of the waters in question. The most advantageous desalination technologies have been indicated (it has a strictly theoretical nature) with respect to the four groups of waters listed (section 7.4.1).

7.4.2.1   Desalinating very high and high mineralized waters ($22–135\,g\,L^{-1}$)

From an economic perspective, the main objective of desalinating the geothermal waters belonging to group I ($70–135\,g\,L^{-1}$) and group II ($22–58\,g\,L^{-1}$) is the elimination of salts, preferably in the form of commercial products (salt, balneological salt, medical, therapeutics and cosmetics products). This can only be achieved through their concentration with the simultaneous recovery of desalinated water and the subsequent crystallization of the concentrate. This is a highly relevant consideration when examining the process models associated with the treatment of geothermal waters. In order to fully utilize geothermal brines, evaporation methods or hybrid processes should be used; the latter involve membrane-based and evaporation methods, usually including either reverse osmosis and evaporation or nanofiltration, reverse osmosis and evaporation. The most common methods for these water types are multi-stage flash evaporation (MSF) and multi-effect distillation (MED).

In view of the high content of divalent ions (calcium, magnesium and sulfate) in the water, the thermal process should be preceded by water pretreatment, e.g. nanofiltration, chemical (lime-soda) softening or softening using ion exchangers. Depending on the turbidity and organic matter content, water pretreatment should also include other processes such as filtration and even coagulation, sedimentation and filtration.

A general framework for the desalination of highly saline waters is shown in Figure 7.9.

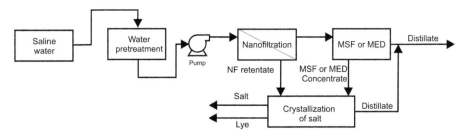

Figure 7.9.   Framework for desalinating highly mineralized waters (groups I and II).

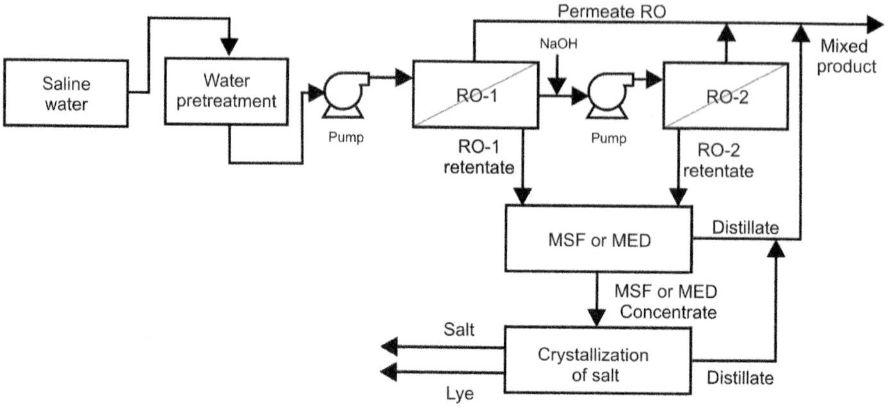

Figure 7.10.   Framework for desalinating medium mineralized waters (group III).

7.4.2.2   *Desalinating medium mineralized waters (1–10 g L⁻¹)*

The chemical composition of medium salinity waters belonging to water group III ($1-10\,\mathrm{g\,L^{-1}}$), suggests a double hybrid system combining reverse osmosis and MSF distillation (multi-stage flash evaporation) with the crystallization of salt or a triple hybrid system combining nanofiltration, reverse osmosis and distillation with the crystallization of salt. The water desalination plant should include the following components (Fig. 7.10):

- a water pretreatment facility; the scope of pretreatment depends on water turbidity and organic solute content;
- a two-stage reverse osmosis setup with NaOH dosing before stage two;
- thermal concentration;
- salt crystallization.

NaOH dosing in the second RO stage is needed due to the high boron content in raw water ($9-13\,\mathrm{mg\,L^{-1}}$). The boron should be removed in the second RO stage.

In addition to the salt produced, permeate from the membrane-based process and distillate from the evaporation process are obtained as products of the technical solution presented. These may be used as drinking or household water or mixed with non-desalinated geothermal water in order to reduce the salinity of water injected into the formation using injection wells. The retentate from this process can be concentrated and crystallized. Obtaining saline of retentate concentrations higher than $70\,\mathrm{g\,L^{-1}}$ (up to $150-160\,\mathrm{g\,L^{-1}}$) without incurring high energy costs is possible thanks to the nanofiltration that precedes the reverse osmosis process (Bodzek and Konieczny, 2005; Magdziorz and Seweryński, 2002; Magdziorz and Motyka, 1994).

Drinking water can be obtained from brackish geothermal waters using a single-stage reverse osmosis setup according to the framework presented in Figure 7.11.

However, this process may prove insufficient depending on the boron content of the water. Where the pH is neutral, the boron is usually present as undissociated boric acid and its retention coefficient usually does not exceed 60%. The pH values of the geothermal waters in Poland is <8. Also, high divalent ion content makes it impossible to adjust the water reaction towards alkaline before the first RO stage, since this would cause the precipitation of sediments and the fouling of the osmosis membranes. For geothermal water with a high boron content, e.g. in Podhale–Bańska Niżna ($10\,\mathrm{mg\,L^{-1}}$), the RO process must be a two-stage one – the first stage takes place at a neutral pH and in the second stage, the pH is raised to around 9–10. At this pH, boron is present as hydroxyborate anions and can be easily removed by the membrane. This solution is presented in Figure 7.12.

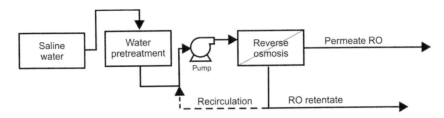

Figure 7.11.   Framework for treating medium and low mineral content waters using reverse osmosis (groups III and IV).

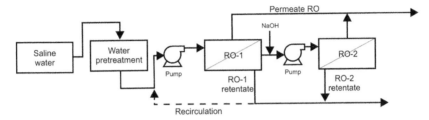

Figure 7.12.   Framework for treating medium and low mineral content waters with a high boron content using reverse osmosis.

7.4.2.3   *Treating freshwaters*

Waters with a mineralization below 1000 mg L^{-1} normally meet the requirements for drinking water. Depending on the turbidity and organic matter content, the water may be subjected to multi-layer bed filtration after optional coagulation or microfiltration/ultrafiltration and disinfection before being introduced into the water supply network. On the other hand, in geothermal waters still boron, fluoride, arsenic, etc. may be too high and requires removal of microelements using a single or a double-stage reverse osmosis (Figs. 7.11 and 7.12).

7.4.3   *Pilot desalination facility*

It was decided that pilot geothermal water desalination tests would be conducted on a semi-industrial scale at two locations. The first desalination project concerned water from the Bańska IG-1 well (the aquifer from which these geothermal waters are extracted lies at a depth of 2565 m b.g.s.), located at the geothermal laboratory of the Mineral and Energy Economy Research Institute of the Polish Academy of Sciences in Kraków (PAS MEERI). Subsequent tests will be conducted at a selected geothermal facility in the Polish Lowlands.

The afore mentioned location was chosen for pilot studies owing to the relatively low mineral-ization of the geothermal waters from the Bańska IG-1 well (2.9 g L^{-1}), high admissible volume of extracted groundwater (120 m^3 h^{-1}) and the deficit of freshwater in the well area (the villages of Bańska Niżna and Szaflary). The capacity of local freshwater intakes only covers 20% of local demand (around 500 houses). The Bańska IG-1 well forms part of the geothermal direct heating circuit operated by the major operator of the Podhale geothermal system – the PEC *Geotermia Podhalańska* S.A. company. The geothermal heating system supplies energy for district heating and heating household water to single- and multi-family houses as well as public buildings. Small volumes of the water are also used at the Termy Podhalańskie geothermal spa which was launched in 2008.

Water with the highest temperature (84–65°C) is used there (Bujakowski, 2010). After being cooled using heat exchangers, usually to around 50°C, some geothermal water (approximately 50%) is injected back into the formation *via* the injection well, and some is discharged into the nearby river (it is allowed by legislation). Efficient use of the thermal energy accumulated in

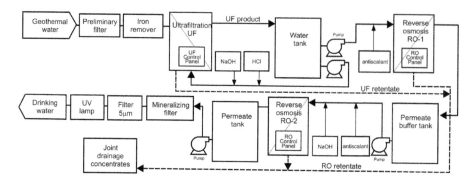

Figure 7.13.   Technological diagram of the pilot geothermal water desalination facility.

the water extracted and the comprehensive utilization of cooled water, for drinking water among other purposes, would make it possible to optimize geothermal water management and improve the balance of freshwater in the area examined.

In order for the pilot research to yield representative results which could provide guidelines for industrial facilities, the desalination facility was fitted with typical components from industrial plant. At the same time the minimum installation capacity, which was set at 1 m^3 h^{-1} of desalinated water, will enable the extrapolation of results from a semi-industrial scale to an industrial one. According to the plan, research work would be conducted at two locations, for 6 to 12 months in each case.

First, part of the water extracted from the Bańska IG-1 geothermal well was subjected to desalination. The pilot facility was constructed within a container and installed at the geother-mal laboratory of the PAS MEERI. Taking into account water salinity (2.9 g L^{-1}) and the increased content of silica (62.5 mg L^{-1}), sulfide (0.085 mg L^{-1}), boron (9.95 mg L^{-1}), barium (0.142 mg L^{-1}), strontium (7.19 mg L^{-1}), ammonium (1.3 mg L^{-1}), fluoride (1.3 mg L^{-1}), bro-mides (1.75 mg L^{-1}) and sulfate (872 mg L^{-1}), a double hybrid setup was selected that combined ultrafiltration and reverse osmosis. The facility is presented on the diagram in Figure 7.13.

The temperature of the water fed into the installation may not exceed 35°C since the iron removal stage, ultrafiltration membranes and reverse osmosis membranes are particularly sensi-tive to high temperature. Exceeding the threshold temperature may cause irreversible damage to hydraulic parts of the equipment included in the water treatment system and thus prevent the further operation of the entire pilot plant. Therefore this factor conditions the operation of the entire installation and requires the efficient cooling of geothermal water.

The simplest solution that was considered for the purposes of the pilot study was to use a fan cooler. This would, however, involve additional purchase cost and an increase in the operating costs of the pilot installation owing to additional electricity consumption.

The alternative closed cooling system used is presented in Figure 7.14. Thermal water with a volume flow of 5 m^3 h^{-1} and a temperature of 80°C was effectively cooled through the use of a plate heat exchanger and a cooling coil laid at the bottom of a 25 × 50 m pool. Water from the geothermal well is sent to the plate heat exchanger which is fed with process water drawn from the pool. The plate heat exchanger has a cooling capacity of 285 kW. After the heat exchanger, the water, now cooled to 31°C, is additionally directed to a coil laid at the bottom of the pool. The coil has a cooling capacity of 47 kW. The closed water circulation system also makes it possible to utilize thermal water pressure to power the desalination plant.

In order to prevent the inflow of water with a temperature exceeding the threshold value into the plant, the feeding pipe was fitted with a temperature sensor coupled with a three-way valve with electric actuator. If the critical value of 37°C is exceeded, the bypass opens automatically while the pipe feeding the plant is cut off and the water is directed to a drain outside the container,

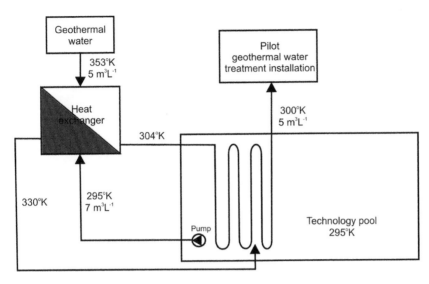

Figure 7.14. Diagram of the system that cools geothermal water for the purposes of the pilot geothermal water desalination facility.

bypassing all the technological stages of the plant. For safety reasons, the entire installation from the connection up to and including the pre-filter was made from materials resistant to temperatures of up to 70°C.

In pilot water desalination tests the renewable energy (heat) is used to drinking water production as:

- the use of geothermal heat for optimal operation of RO at 30°C; geothermal water secures this level of temperature;
- geothermal and solar energy (heat) planned to be used to concentrate retentate obtained from RO, which is a second level recovery of distillate; this does not exclude to gain in this way concentrate for commercial purposes such as medical, therapeutics and cosmetics.

Geothermal energy will be therefore used for production of drinking water in two stages: (i) the use of geothermal waters in a membrane system (optimum temperature of geothermal water for desalination processes) and (ii) concentrate of retentate (using water temperature of 80°C). In the second case geothermal energy (heat) will be supported by solar energy. From the point of view of the function of its individual parts, the water treatment plant can be divided into three main parts: (i) water pretreatment stage, (ii) two-stage reverse osmosis and (iii) final treatment to achieve drinking water parameters (mineralization, sterilization).

The first group of devices includes a mechanical filter, iron removal stage and ultrafiltration module (Fig. 7.15a). These elements are designed to pretreat cooled (raw) geothermal water so as to make it suitable for feeding osmosis membranes, which are very sensitive to solid and sediment-forming contaminants. Following cooling to a temperature lower than 35°C and pre-filtering, geothermal water is directed to the iron removal filter. The concurrent flow of water under pressure through a catalytic bed takes place in this filter and precipitated iron hydroxides are retained there. In order to remove the pollutants accumulated during operation, the filter is regularly rinsed in two stages: backwashing (counter-current rinsing) and concurrent rinsing. The rinse process is initiated and conducted fully automatically.

After filtering major contaminants and removing iron, the water is fed to the ultrafiltration module. Ultrafiltration membranes are used to remove microsuspensions that limit the possibility of feeding the water to the reverse osmosis stage. DOW Chemical OUT–IN flow UF membranes

Figure 7.15.   PAS MEERI geothermal water desalination plant: (a) water pretreatment (mechanical filter, iron removal stage and ultrafiltration module), (b) intermediate tank before the RO module and final treatment (mineralization).

with a diameter of 8″ and a pore size of 0.03 μm were used. Following the UF stage, an intermediate tank with a volume of $2\,m^3$ was fitted; part of the water from this tank is used to rinse UF membranes (Fig. 7.15b). Apart from water, hydrochloric acid and sodium hydroxide are also used during the backwashing of ultrafiltration membranes; these are supplied by independent dosing pumps. All processes related to the operation of the UF module are automated.

Before the water is fed to RO membranes, antiscalant is added. Owing to the relatively high boron content (ca. $10\,mg\,L^{-1}$) in the water supplied, the plant was fitted with a two-stage desalination system based on osmosis membranes connected in series and a system that adjusts the pH of the first permeate. DOW FILMTEC BW30HR–440i osmosis membranes, designed for brackish water with increased silica, boron, ammonium and nitrate content, were used. Two separate assemblies were connected in series. Following antiscalant addition and an increase in pH, the permeate output from the first RO stage is fed to the second RO stage.

After subjecting the water to subsequent processing (mineralization by filtering it through a dolomite bed in order to increase the overall hardness of the water output by the osmosis module and UV sterilization to kill bacteria), the technological cycle of geothermal water treatment is complete.

The plant was fitted with a system controlling and automating its component processes so that the operation of individual components is synchronized. Additionally, in view of the research purposes that the plant is meant to serve, sensors were mounted at certain points that enable the recording of basic hydraulic and water quality parameters, i.e. pressure, temperature and specific electrical conductivity of water. Following each stage, the plant was fitted with special instrument nozzles used to collect water and concentrate samples for laboratory physicochemical measurements.

7.4.4   *Preliminary research results*

During the initial period of plant operation, several operational problems had to be solved that significantly affected the research results.

The basic problem was the optimization of the manner in which the plant was to be fed with water with a temperature not exceeding 35°C at a steady flow rate of $4–5\,m^3\,h^{-1}$ and feed water pressure of 0.3 MPa. The desalination plant is fed with unmodified geothermal waters with a natural content of dissolved gases (mostly carbon dioxide, hydrogen sulfide, nitrogen and oxygen). Owing to its gas content, the water fed into the plant pulsated and problems were encountered at the iron removal stage. Therefore it was necessary to fit an air bleeder system at the feeding stage and remove gas from the catalytic bed iron removal tank.

Antiscalant (a substance complexing divalent ions) is added before the water is fed to RO membranes in order to prevent their fouling and scaling. Theoretical antiscalant dosage was

Table 7.6. Chemical composition of geothermal water and water at individual desalination stages (maximum component content) (water corresponding to drinking water of Figure 7.13).

Parameter	Geothermal water	After UF	After RO–1	After RO–2	Desalin. water	Maximum allowable concentration
Arsenic [mg L^{-1}]	0.00075	0.0028	0.00025	0.00012	0.00010	0.010
Nitrate [mg L^{-1}]	<0.50	<0.50	<0.50	<0.50	<0.50	50
Boron [mg L^{-1}]	9.46	9.45	6.07	0.16	0.01	1.0
Chromium [mg L^{-1}]	0.0019	0.0019	0.0016	0.0016	0.0016	0.050
Cadmium [mg L^{-1}]	0.00018	0.00003	0.00001	0.00001	0.00001	0.005
Copper [mg L^{-1}]	0.0032	0.0016	0.00067	0.00055	0.00088	2.0
Nickel [mg L^{-1}]	0.0029	0.0011	0.00018	0.00002	0.00002	0.020
Lead [mg L^{-1}]	0.001	0.001	0.00096	0.00052	0.00046	0.025
Mercury [mg L^{-1}]	0.00014	0.00014	0.00013	0.0001	0.0001	0.001
Selenium [mg L^{-1}]	0.001	0.001	0.001	0.001	0.001	0.010
Chloride [mg L^{-1}]	527.0	520.0	13.4	13.4	13.4	250
Aluminum [mg L^{-1}]	<0.01	<0.01	<0.01	<0.01	<0.01	0.200
Manganese [mg L^{-1}]	0.018	0.002	0.002	<0.001	<0.001	0.050
pH [–]	7.2	6.87	5.38	9.49	9.49	6.5–9.5
Conductivity [µS cm^{-1}]	3550	3530	195	104	195	2500
Sulfate [mg L^{-1}]	877.1	873.5	3.20	0.30	<0.20	250
Sodium [mg L^{-1}]	545.1	543.9	21.0	19	19	200
Iron [mg L^{-1}]	4.0	0.013	0.009	0.008	0.004	0.200
Magnesium [mg L^{-1}]	42.71	41.7	0.24	<0.10	<0.10	30–125
Silver [mg L^{-1}]	0.0066	0.001	0.00058	0.00013	0.00013	0.010
Hardness mg [CaCO$_3$ L^{-1}]	679.3	674.3	3.8	0.7	53	60–500

calculated on the basis of the chemical composition of the water, but after just a few hours of continuous operation, RO performance dropped and membranes had to be regenerated. A positive aspect was the ease with which membranes could be restored to their initial state by chemical etching (chemical bathing). This indicates that calcium carbonates and not sulfates are deposited on osmosis membranes. Antiscalant dosage was modified and additional doses of diluted hydrochloric acid were used.

The operating parameters of the Bańska IG-1 well geothermal water desalination plant during the current preliminary research stage are as follows: plant capacity 1 m^3 h^{-1} of permeate, recovery ratio 56%, feeding water temperature max. 35°C, RO stage operating pressure 1.2–1.5 MPa. The operating parameters are analyzed and modified on an ongoing basis in order to improve plant capacity and efficiency.

The low permeate recovery ratio achieved is the direct result of the increased boron content in the Bańska IG-1 well geothermal waters. This necessitated the use of a two-stage RO system, which at the same time increased the investment expenditure related to the pilot plant and the operating costs resulting from the need to adjust the pH following the first RO stage.

The results of the measurements of physical properties and the chemical composition of geothermal water and the water collected at several desalination stages (labels as in Fig. 7.13) are presented in Table 7.6.

The results from the preliminary research indicate that, as a result of the desalination of the geothermal water, the permeate obtained following the first RO stage meets the general requirements for drinking water – except as regards boron content. According to its technical specifications, DOW FILMTEC BW30HR–440i osmosis membranes exhibit a boron retention coefficient (minimum salt rejection) of 83%. This means that the maximum boron content in raw water for a single-stage reverse osmosis system may not exceed 1.7–1.8 mg L^{-1}. Therefore, since the boron concentration in water had to be reduced from almost 10 mg L^{-1} to a maximum of 1 mg L^{-1}, a two-stage desalination system was required, which significantly reduced the drinking water recovery ratio. In order to increase plant operation efficiency, part of the concentrate is recirculated after the first and second RO stages.

Preliminary research results are promising – they indicate that the production of drinking water from geothermal waters is feasible, but requires a case-by-case approach. Research will continue on an ongoing basis for at least half a year. During this period, the test results will be analyzed regularly and measures will be undertaken aimed at optimizing plant operation. The research results will enable the economic and environmental aspects of the use of desalinated geothermal waters to be assessed in order to decentralize drinking water production.

The concentrate recovered from the water desalination process is being examined from the point of view of recovering minerals of balneological and economic significance.

7.5   SUMMARY

Membrane-based water desalination technologies and hybrid technologies that combine membrane and thermal processes are widely used to produce drinking water in many regions of the world. They are also considered a technologically and economically viable alternative for desalinating water (mainly seawater), often with the use of renewable (solar, wind, geothermal) energy.

In principle, many solutions exist, but the choice of the specific desalination technology to be implemented is always determined by cost. The cost depends on several factors such as raw water quality, plant technology and scale and energy costs. Therefore there is increasing interest in measures aimed at recovering waste energy and using renewable energy resources.

Each water intake has its specific features and solutions for desalinating water and wastewater disposal must be developed on a case-by-case basis. When engaging in theoretical considerations, one must therefore take into account the physical properties and chemical composition of the water, environmental conditions, the manner in which the concentrate will be utilized and the availability of suitable technologies.

The preliminary results from the geothermal water desalination study conducted at the PAS MEERI are optimistic. In order to validate the assumptions made in the context of industrial applications, further studies have to be conducted in accordance with the project plans. The results of these studies will be the subject of subsequent publications.

It is certain, however, that the utilization of geothermal water for drinking purposes on an industrial scale (greater installation capacity) will require, first of all, better and more efficient water cooling. Optimum utilization of the thermal waters obtained is an important issue for most geothermal plants both in Poland and abroad. Efficient geothermal water management has one main goal: optimizing system operation and improving the economics of geothermal plant operation. Additionally, the use of cooled water for drinking purposes, particularly for open drain installations (without injecting cooled water into the formation) will contribute to the comprehensive utilization of geothermal water and the decentralization of drinking water production. Therefore the present study concerns two significant research areas: geothermics (cooled water utilization methods) and hydrogeology (water management).

Renewable energy resources are the best energy supply option for desalination systems. In pilot studies geothermal energy is used for production of drinking water in two stages: (i) the use of geothermal waters in a membrane system (optimum temperature of geothermal water for desalination processes) and (ii) concentrate of retentate (using water temperature of 80°C). In the second case geothermal energy (heat) will be supported by solar energy.

At this concept stage, it is not possible to determine the amount of renewable energy needed to produce a unit of drinking water, and thus to determine the parameter quantification. The presented geothermal water treatment project is strictly of research nature.

REFERENCES

Aim, R.B. & Vladan, M.: The role of membrane techniques in cleaner production. *Industry and Environment* 12 (1989), pp. 15–18.

Alishiri, M.: The economics of desalination. *Desalination* 223 (2008), pp. 474–482.

Al-Karaghouli, A., Renne, D. & Kazmerski, L.L: Technical and economic assessment of photovoltaic-driven desalination systems. *Renew. Energy* 35 (2010), pp. 323–328.

Alley, W.M.: Desalination of ground water: earth science perspectives. U.S. Geological Survey, October 2003.

Al-Mutaz, I.S.: Water desalination in the Arabian Gulf region. In: F.A. Goosen & W.H. Shayya (eds): *Water management purification and conservation in arid climates*, Vol. 2: *Water purification*. Technomic Publishing Co., Lancaster-Basel, 1999, pp. 245–265.

Awerbuch, L.: Hybrid plants: integration of resources and technology. *Desal. Water Reuse* 15:1 (2005), pp. 18–28.

Baker, R.: Membrane technology in the chemical industry. Future directions. In: S.P. Nunes & K.-V. Peinemann (eds): *Membrane technology in the chemical industry*. Wiley-Vch, Weinheim, Germany, 2001, pp. 268–295.

Banat, F.A. & Simandl, J.: Desalination by membrane distillation: a parametric study. *Separ. Sci. Technol.* 33 (1998), pp. 201–226.

Blank, J.E., Tusel, G.F. & Nisan, S.: The real cost of desalted water and how to reduce it further. *Desalination* 205 (2007), pp. 298–311.

Bodzek, M.: Membrane techniques in water treatment and renovation. In: F.A. Goosen & W.H. Shayya (eds): *Water management purification and conservation in arid climates*, Vol. 2: *Water purification*. Technomic Publishing Co., Lancaster-Basel, 1999, pp. 45–100.

Bodzek, M. & Konieczny, K.: *Wykorzystanie procesów membranowych w uzdatnianiu wody*. Oficyna Wydawnicza Projprzem-EKO, Bydgoszcz, Poland, 2005.

Bodzek, M. & Konieczny, K.: Skojarzone systemy membranowe w uzdatnianiu wody — stan wiedzy. *Materiały VII Międzynarodowej Konferencji: "Zaopatrzenie w wodę, jakość i ochrona wód"*, Zakopane, czerwiec 2006a, tom I, pp. 43–61.

Bodzek, M. & Konieczny, K.: Membrane processes in water treatment — state of art. *Inżynieria i Ochrona Środowiska* 9 (2006b), pp. 129–159.

Bodzek, M., Bohdziewicz, J. & Konieczny, K.: Techniki membranowe w ochronie środowiska. *Wydawnictwo Politechniki Śląskiej*, Gliwice, Poland, 1997.

Bodzek, M., Konieczny, K. & Dudziak, M.: Możliwości wykorzystania technik membranowych w procesach uzdatniania wody do picia. In: M. Szwast (ed.): *Membrany i techniki membranowe — Od pomysłu do przemysłu*. Polymem Ltd. sp. z o.o., Warszawa, Poland, 2009, pp. 5–49.

Brehand, A., Bonnelye, V. & Perez, M.: Assessment of ultrafiltration as a pretreatment of reverse osmosis membranes for surface water desalination. *Proceedings of "Membranes in Drinking and Industrial Water Production MDIW 2002"*, Mülheim an der Ruhr, Germany, B.37a, 2002a, pp. 775–784.

Brehand, A., Bonnelye, V. & Perez, M.: Comparison of MF/UF pretreatment with conventional filtration prior to RO membranes for surface seawater desalination. *Desalination* 144 (2002b), pp. 353–360.

Bujakowski, W.: A review of Polish experiences in the use of geothermal water. *Proceedings World Geothermal Congress* 2005, Antalya, Turkey, 2005.

Bujakowski, W.: The use of geothermal waters in Poland (state in 2009). *Przeglad Geologiczny* 58 (2010), pp. 580–588.

Bujakowski, W. & Barbacki, A.: Potential for geothermal development in Southern Poland. *Geothermics* 33 (2004), pp. 383–395.

Bujakowski, W. & Tomaszewska, B.: Program prac zmierzających do oceny możliwości uzdatniania wód termalnych. *Technika Poszukiwań Geologicznych Geotermia, Zrównoważony Rozwój, Kraków, Poland.* 1/2007 (2007), pp. 3–8.

Bujakowski, W. & Tomaszewska, B.: A conception of geothermal water desalination to improve water balance. *Biuletyn Państwowego Instytutu Geologicznego, Hydrogeologia.* IX/1 436 (2009), pp. 17–21.

Bujakowski, W., Barbacki, A. & Pająk, L.: *Atlas of geothermal water reservoirs in Malopolska*. PAS MEERI Publishers, Krakow, Poland, 2006.

Bujakowski, W., Tomaszewska, B., Kępińska, B. & Balcer, M.: Geothermal water desalination—Preliminary studies. *Proceedings World Geothermal Congress*, Bali, Indonesia, 2010.

Chowaniec, J.: Wody podziemne niecki podhalańskiej. *Współczesne Problemy Hydrogeologii Gdańsk, Poland.* XI/1 (2003), pp. 45–53.

El-Dessouky, H. & Ettouney, H.: MSF developments may reduce desalination costs. *Water Wastewater Int.* 15:3 (2000), pp. 20–21.

El-Bourawi, M.S., Ding, Z., Ma, R. & Khayet, M.: A framework for better understanding membrane distillation separation process. *J. Membr. Sci.* 285 (2006), pp. 4–29.

El-Sayed, E., Ebrahim, S., Al-Saffar, A. & Abdel-Jawad M.: Pilot study of MSF-RO hybrid system. *Desalination* 120 (1998), pp. 121–128.

Eltawil, M., Zhengming, Z. & Yuan, L.: A review of renewable energy technologies integrated with desalination system. *Renew. Sustain. Energy Rev.* 13 (2009), pp. 2245–2262.

Fazle, M., Hawlader, M.N.A. & Mujumda, A.S.: Combined water and power plant (CWPP) – a novel desalination technology. *Desalin. Water Treat.* 5 (2009), pp. 172–177.

Fritzmann, C., Löwenberg, J., Wintgens, T. & Melin, T.: State-of-the-art of reverse osmosis desalination. *Desalination* 216 (2007), pp. 1–76.

Gawroński, R.: Membranowe procesy rozdzielania mieszanin. *Instal.* Volume 7/8 (2004), pp. 12–17.

Gerstandt, K., Peinemann, K.-V., Skilhagen, S.E., Thorsen, T. & Holt, T.: Membrane processes in energy supply for an osmotic power plant. *Desalination* 224 (2008), pp. 64–70.

Godino, M.P., Pena, L., Rincon, C. & Mengual, J.I.: Water production from brines by membrane distillation. *Desalination* 108 (1996), pp. 91–97.

Górecki, W. (ed): *Atlas of geothermal resources of Mesozoic formations in the Polish Lowlands.* GOLDRUK, Poland, 2006.

Górecki, W.: Geothermal waters in the Polish Lowlands. *Przeglad Geologiczny* 58 (2010), pp. 574–579.

Górecki, W., Hajto, M., Strzetelski, W. & Szczepański, A: Lower Cretaceous and Lower Jurassic aquifers in the Polish Lowlands. *Przeglad Geologiczny* 58 (2010), pp. 589–593.

Grater, J.: The early history of reverse osmosis membrane development. *Desalination* 117 (1998), pp. 297–309.

Greenlee, L.F., Lawler, D.F., Freeman, B.D., Marrot, B. & Moulin, P.: Reverse osmosis desalination: water sources, technology and today's challenges. *Water Research* 43:9 (2009), pp. 2317–2348.

Gryta, M.: *Rozdzielanie składników roztworów technika destylacji membranowej.* Prace Naukowe Politechniki Szczecińskiej, nr 577. Wydawnictwo Uczelniane Politechniki Szczecińskiej, Szczecin, Poland, 2003.

Hamada, M.: Brackish water desalination by electrodialysis. *Desalin. Water Reuse* 2:4 (1995), pp. 8–15.

Hamed, O.A.: Overview of hybrid desalination systems – current status and future prospects. *Desalination* 186 (2005), pp. 207–214.

Helal, A.M.: Hybridization – a new trend in desalination. *Desalin. Water Treatm.* 3 (2009), pp. 120–135.

Heshka, D.: Electrodialysis – a viable option for a small Canadian city. *Desalin. Water Reuse* 7:1 (1977), pp. 22–26.

H₂O Applying: Thought to water in the Middle East. Internet site: http://www.h2ome.net/en/2011/09/global-desalination-market-grows-105166/ (accessed 20 October 2011).

IDE Technologies Ltd. Internet site: http://www.ide-tech.com/news/largest-swro-desalination-plant-world-inaugurated-hadera (accessed 20 September 2011).

Karagiannis, C.I. & Soldatos, P.G.: Water desalination cost literature: review and assessment. *Desalination* 223 (2008), pp. 458–466.

Kępińska, B.: Geothermal energy country update report from Poland, 2000–2004. *Proceedings World Geothermal Congress* Antalya, Turkey, 2005.

Kępińska, B.: *Thermal and hydrothermal conditions of the Podhale geothermal system (Poland).* PAS MEERI Publishers, Krakow, Poland, 2006.

Kołtuniewicz, A.: *Wydajność ciśnieniowych procesów membranowych w świetle teorii odnawiania powierzchni.* Oficyna Wydawnicza Politechniki Wrocławskiej, Wrocław, Poland, 1996.

Landais, E.: Desalination treat to Gulf. *Gulfnews*, June 14, 2009, http://gulfnews.com/news/gulf/uae/environment/desalination-threat-to-gulf-1.72060 (accessed 20 September 2011).

Leitner, G. & Leitner-Murnay, F.: U.S. desalination and membrane softening potable water costs. *Desalin. Water Reuse* 7:2 (1977), pp. 44–50.

Lund, J.W.: Development of direct-use projects. *GHC Bulletin.* 29: 2 (2010), pp. 1–7.

Lund, J.W., Freeston, D.H & Boyd T.L.: World-wide direct uses of geothermal energy 2005. *Proceedings World Geothermal Congress* Antalya, Turkey, 2005.

Lund, J.W., Freeston, D.H & Boyd, T.L.: Direct utilization of geothermal energy 2010 worldwide review. *Proceedings World Geothermal Congress* Bali, Indonesia, 2010.

Magdziorz, A.: Utilization of membrane technologies in treatment and desalination of mine water. In: A. Noworyta & A. Trusek-Hołownia (eds): *Membrane Separations*. Wydawnictwo Argi, Wrocław, Poland, 2001, pp. 257–275.

Magdziorz, A. & Motyka, I.: Bezodpadowe technologie odsalania wód kopalnianych z odzyskiem surowców chemicznych: doświadczenie i praktyka Głównego Instytutu Górnictwa. *Materiały I Kongresu Technologii Chemicznej*, Wydawnictwo Uczelniane Politechniki Szczecińskiej, 1994, pp. 728–732.

Magdziorz, A. & Seweryński, J.: *Wykorzystanie technologii membranowych w uzdatnianiu i odsalaniu wód kopalnianych*. Zeszyty Naukowe Politechniki Śląskiej, Seria: Inżynieria Środowiska 46, 2002, pp. 261–273.

Małecka, D.: *Hydrogeologia Podhala*. Prace Hydrogeologiczne. Instytut Geologiczny. Wyd. Geologiczne 14, Warszawa, Poland, 1981.

Malek, A., Hawlader, M.N.A. & Ho, J.C.: Design and economics of RO seawater desalination. *Desalination* 105 (1996) 245–261

Mallevaterialle, J., Odentaal, P.E. & Wiesner, M.R.: The emergence of membranes in water and wastewater treatment. In: *Water Treatment Membrane Processes*. McGraw-Hill, New York-San Francisco-Washington, 1996, pp. 1.1–1.10.

Mathioulakis, E., Belessiotis, V. & Delyannis, E.: Desalination by using alternative energy: review and state-of-the-art. *Desalination* 203 (2007), pp. 346–365.

Matsuura, T.: Progress in membrane science and technology for seawater desalination – a review. *Desalination* 134 (2001), pp. 47–54.

Meller, F.H.: Electrodialysis (ED) & electrodialysis reversal (EDR) technology. *Ionics Incorporated*, 1984.

Mezher, T., Fath, H., Abbas, Z. & Khaled, A.: Techno-economic assessment and environmental impacts of desalination technologies. *Desalination* 266 (2011), pp. 263–723.

Morin, O.J.: Desalination – State of the art, Part 2. *Int. Desalin. Water Reuse Quart.* 4/4 (1995), pp. 32–36.

Mulder, M.: *Basic principles of membrane technology*. Kluwer Academic Publishers, Dordrecht-Boston-London, 1991.

Murrer, J. & Rosberg, R.: Desalting of seawater using UF and RO – results of a pilot study. *Desalination* 118 (1998), pp. 1–4.

Narębska, A. (ed.): *Membrany i membranowe techniki rozdziału*. Wydawnictwo Uniwersytetu Mikołaja Kopernika. Toruń, Poland, 1997.

Ney, R.: Zasoby energii geotermalnej w Polsce I mozliwe kierunki jej wykorzystania. Seminarium Naukowe Problemy Wykorzystania energii geotermalnej i wiatrowej w Polsce, Kraków-Zakopane, Poland, 1997.

Reddy, K.V. & Ghaffour, N.: Overview of the cost of desalinated water and costing methodologies. *Desalination* 205 (2007), pp. 340–353.

Ryabtsev, A.D., Kotsupalo, N.P., Titarenko, V.I., Igumenov, I.K., Gelfond, N.M., Fedotova, N.E., Morozowa, N.B., Shipatchev, V.A. & Tibilov, A.S.: Developmnet of two-stage electrodialysis set-up for economical desalination of sea-type artesian and surface waters. *Desalination* 137 (2001), pp. 207–214.

Sadhukhan, H.K., Misra, B.M. & Tewari, P.K.: Desalination and wastewater treatment to augment water resources. In: F.A. Goosen & W.H. Shayya (eds): *Water management purification and conservation in arid climates* Volume 2: *Water purification*. Technomic Publishing Co., Lancaster-Basel, 1999, pp. 1–29.

Sauvet-Goichon, B.: Ashkelon desalination plant – A successful challenge. *Desalination* 203 (2007), pp. 75–81.

Scott, K.: Handbook of industrial membrane. *Elsevier,* Kidlington, 1997.

Sokołowski, S.: Geologia paleogenu i mezozoicznego podłoża południowego skrzydła niecki podhalańskiej w profilu głębokiego wiercenia w Zakopanem. Biuletyn IG, Warszawa, Poland 265 (1973).

Song, S., Yifang, C.H. & Congije, G.: Electrodialysis in China. *Desalin. Water Reuse* 2:4 (1995), pp. 23–25.

Strathmann, H.: Ion-exchange membrane separation processes. *Membrane science and technology series* 9. Elsevier, Amsterdam, The Netherlands, 2004.

Strathmann, H.: Electrodialysis, a mature technology with a multitude of new applications. *Desalination* 264 (2010), pp. 268–288.

Strathmann, H., Giorno, L. & Drioli, E.: *An introduction to membrane science and technology*. Consiglio Nazionale delle Riservata, Roma, Italy, 2006.

Strzetelski, W.: Geologiczna charakterystyka zbiorników wód geotermalnych na Niżu Polskim. In: W. Górecki (ed.): *Atlas wód geotermalnych Niżu Polskiego*. Archiwum Katedry Surowców Energetycznych AGH, Kraków, Poland, 1990, pp. 49–55.

Surana, T., Atmojo, J.P. & Subandriya, A.: Development of geothermal energy direct use in Indonesia. *GHC Bulletin* 29:2 (2010), pp. 11–15.

Susanto, H.: Towards practical implementations of membrane distillation. *Chem. Eng. Process. Process Intensif.* 50 (2011), pp. 139–150.

Szczepański, A.: Warunki hydrotermalne dolnojurajskiego i dolnokredowego zbiornika geotermalnego–zbiornik dolnokredowy. In: W. Górecki (ed.): *Atlas wód geotermalnych Niżu Polskiego.* Archiwum Katedry Surowców Energetycznych AGH, Kraków, Poland, 1990, pp. 316–322.

Szczepański, A.: Zbiornik dolnokredowy–warunki hydrotermalne. In: W. Górecki (ed.): *Atlas zasobów energii geotermalnej na Niżu Polskim.* Wyd. AGH, Kraków, Poland, 1995, pp. 13.

Thompson, C., Reynolds, T. & Boegli, W.: Compare performance and cost for brackish water treatment. *Desalination & Water Reuse* 7:2 (1977), 34–42.

Tomaszewska, B.: Treatment of geothermal water from Banska IG-1 well to produce drinking water as one of directions of its wide use. *Technika Poszukiwań Geologicznych Geotermia, Zrównoważony Rozwój, Kraków* 2/2009 (2009), pp. 21–28.

Tomaszewska, B. & Pająk, L.: Analysis of treatment possibles of high-mineralized geothermal water (central Poland – Gostynin region). *Proceedings World Geothermal Congress* Bali, Indonesia, 2010.

Tomaszewska, M.: *Destylacja membranowa.* Prace Naukowe Politechniki Szczecińskiej 531, Wydawnictwo Uczelniane Politechniki Szczecińskiej, Szczecin, Poland, 1996.

Tomaszewska, M.: Membrane distillation – direction and development. In: A. Noworyta & I. A. Trusek-Hołownia: *Membrane separations.* Argi, Wrocław, 2001, pp. 181–198.

Tomaszewska, M., Gryta, M. & Morawski, W.A.: Odsalanie wody metodą destylacji membranowej. *Materiały I Ogólnopolskiej Konferencji Naukowej "Membrany i procesy membranowe w ochronie środowiska",* Politechnika Śląska, Wisła, 1995, pp. 63–76.

Truby, R.: Desalination's global growth driven by multiple membrane systems. *Water Wastewater Int.* 15:3 (2000), pp. 26–28.

Tsiourtis, N.X.: Desalination and the environment. *Desalination* 141 (2001), pp. 223–236.

Van der Bruggen, B. & Vandecasteele, C.: Distillation *vs.* membrane filtration: overview of process evolutions in seawater desalination. *Desalination* 143 (2002), pp. 207–218.

Van Hoof, S., Minnery, J.G. & Mack, B.: Dead-end ultrafiltration as pretreatment to seawater reverse osmosis. *Desalin. Water Reuse* 11:3 (2001), pp. 44–48.

Wilf, M.: *The guidebook to membrane desalination technology.* Balaban Desalination Publications, L'Aquila, 2007.

Wirth, D. & Cabassud, C.: Vacuum membrane distillation for seawater desalination: could it compete with reverse osmosis. *Proceedings of Membranes in Drinking and Industrial Water Production MDIW 2002,* Mülheim an der Ruhr, (Germany), B.37a, 2002, pp. 219–226.

Wiśniewski, J.: Elektromembrane processes. In: A. Noworyta & A. Trusek-Hołownia (eds): *Membrane separations.* Argi, Wrocław, Poland, 2001, pp. 147–179.

Xiujuan, C.H., Peigui, C.H. & Yongwen, N.T.: Electrodialysis for the desalination of seawater and high strength brackish water. *Desalin. Water Reuse* 4:4 (1995), pp. 16–22.

Younos, T. & Tulou, K.E.: Overview of desalination techniques. *J. Contemp. Water Res. Educat.* 132 (2005), pp. 3–10.

Zakrzewska-Trznadel, G., Harasimowicz, M. & Chmielewski, A.G.: Uzdatnianie wody do celów kotłowych metodą destylacji membranowej. *Przem. Chem.* 78 (1999), pp. 181–184.

CHAPTER 8

Solar disinfection as low-cost technologies for clean water production

Jorge Martín Meichtry & Marta Irene Litter

> *"I mean, if a thing works, if a thing is right, respect that, acknowledge it, respect it and hold to it."*
>
> Harold Pinter, Party Time

8.1 INTRODUCTION TO LOW-COST TECHNOLOGIES FOR DISINFECTION AND DECONTAMINATION OF DRINKING WATER FOR HUMAN CONSUMPTION

8.1.1 *The problem of water*

Safe drinking water is a human right, as established by the United Nations in 2002 (United Nations, 2002). However, nowadays, more than one of the seven billion people living in the earth – an estimated 1.1 billion people worldwide (U.S. Census Bureau, 2011; World Health Organization, 2011b) – lack access to safe drinking water. Hundreds of million people drink contaminated water because of unsafe water treatments and distribution systems, unsafe water storage and inadequate handling practices.

Although water covers 70% of the surface of the earth, most is saltwater; freshwater covers only 3%, much of it lying frozen in the Antarctic and Greenland polar ice. Freshwater available for human consumption, which comes from rivers, lakes and underground sources and aquifers, accounts for just 1% of all water on earth. The dramatic and immediate consequences of the water scarcity are waterborne diseases, which kill on average more than 6 million children each year (about 20,000 children per day). This constitutes 75% of all diseases in developing countries: contaminated water kills more people than cancer, AIDS, wars or accidents. Besides, although global population is increasing and eight billion people are expected by 2025, the amount of water will remain the same, and the amount of available freshwater per person per year will drop 40% – from more than 8000 to about 5000 m^3 (Third World Academy of Sciences, 2002). Examination of these figures, however, makes perceptible that the problem is not the lack of freshwater but the unequal access to it, taking into account that the daily drinking water requirement per person is 2–4 L. This is a dramatic problem in developing countries, caused by depletion of groundwater aquifers, inefficient irrigation practices, and rapid increase of population, among other reasons.

The World Health Organization (WHO) Drinking-Water Guidelines (WHO, 2006) estimated pathogens in lakes, rivers, streams and groundwater sources, indicating that microbial contamination in different regions is coming from different anthropic and non-anthropic origins. For example, fecal wastewater from human settlements that is not treated before discharge into a river can cause contamination downstream; cattle husbandry close to rivers can cause microbial contamination of a river used later for irrigation, washing and sometimes even drinking. This leads to concentrations of pathogens highly significant to provoke damage in health. Therefore, water disinfection is essential to warrant the safety of water to be consumed.

Water is one of the main transporting vehicles of microorganisms, causing diseases in the digestive system of humans and animals. Pathogens are related with hydric diseases. Typhus, bacterial dysentery and cholera are caused by bacteria, amebiasis by protozoa, schistosomiasis by worms (helminthes) and larvae and some viruses originate infectious hepatitis and poliomyelitis.

Fecal coliforms are a big group of microorganisms, usual inhabitants of the intestines of superior animals. These microorganisms are easy to identify compared to pathogens, which are normally present in a lower number and whose identification is difficult. The presence of coliforms in a water sample not always indicates contamination with pathogens, but their concentration is a parameter to alert about the presence of fecal contamination and pathogens.

Another critical issue is disinfection of water for agriculture, which consumes 70% of fresh water used worldwide, and in developing countries this is more than 95% of the available fresh water. The amount of water necessary per person for food purposes is around 2000–5000 L (FAO, 2008). However, this issue will not be treated in this chapter, which will be restricted only to drinking water.

8.1.2   *Alternative water treatment technologies*

A variety of technologies to bring safe water to the population at municipal or household level have been described, and many of them are widely used in different parts of the world (WHO, 2011a). The principal aim is to improve the microbial quality of water to reduce waterborne diseases. Reliable technologies to make drinking water safe should be accessible, affordable, environmentally friendly, and tailored to the cultural norms of a nation.

Technological options include a number of physical and chemical treatment methods that fall into two broad categories: those used by municipal authorities at centralized points from where water is then distributed, and those that can be practiced in individual homes. Some technologies rely on high technology, and are more suitable for use at city/central point treatment facilities. Others, based on more modest technologies, can be used at the settlement cluster level – in schools, community centers, apartment buildings and villages. Conventional or modified technological options can be used in individual homes or during emergencies. The aim of these technologies, whatever its features, costs, yields, etc., should be excellent, at the most pure level, free from pathogens and toxins.

Among high-technology/high volume methods, those involving sedimentation and filtration followed by the killing of pathogens through chlorination or ozone bubbling are the most used. Such processes remain logistically feasible and even acceptable for relatively low-volume demand (e.g. schools, hospitals and villages), but they can be improved by different strategies. As examples, ultraviolet (UV)-protected granulated activated charcoal bed for adsorption, titanium dioxide plus UV light (TiO_2-heterogeneous photocatalysis, HP), electrochemical activation, can be used. Small portable reverse osmosis water purifiers or adsorbent filters can be also suitable for domestic communities and small industries. Some of these water treatment and storage systems use chemicals and other media and materials that cannot be easily obtained in some places and settings at reasonable cost, and require relatively complex and expensive systems and procedures to treat the water. Such systems may be too inaccessible, complex and expensive.

Application of UV-C lamps (254 nm) is very effective against a wide range of microorganisms. The main advantage of UV-C disinfection is that the storage and transport of reagents is unnecessary. However, its use may be limited depending on water turbidity.

Other water treatment methods that employ simple, low-cost appropriate technologies are available. These methods include filtration, aeration, storage and settlement, disinfection by boiling, chemicals, solar radiation, coagulation and flocculation, and desalination, among others (CARE/CDC Health Initiative, 2010; Skinner and Shaw, 1998, 1999). They will be briefly reviewed in what follows.

Straining: It consists in pouring turbid water through a piece of fine, clean cotton cloth. Suspended solids are removed.

Aeration: It can be accomplished by vigorous shaking in a vessel, or allowing water to trickle down through one or more perforated trays containing small stones. Aeration increases the air content of the water, removes volatile substances such as hydrogen sulfide, which affects odor and taste, and oxidizes iron or manganese so that they form precipitates that can be removed by settlement or filtration.

Storage and settlement: Storage of water for one day results in the die-off of more than 50% of bacteria; longer periods lead to further reduction. The method is especially useful for preventing transmission of diseases like schistosomiasis.

Filtration: It includes mechanical straining, absorption and adsorption, and, particularly with slow sand filters, biochemical processes. Depending on the size, type and depth of filter media, and the flow rate and physical characteristics of the raw water, filters can remove suspended solids, pathogens, and certain chemicals, tastes and odors. Ceramic filters are the most common. Good quality ceramic filter has a pore size of 0.2 μm.

Candles: They can be impregnated with silver to kill pathogens, but in this form, they are relatively expensive. In some systems, a candle filter is preceded by a polypropylene rope filter to remove suspended particles, or packed with activated carbon to remove organic chemicals, tastes, suspended solids and pathogens. In theory, viruses can pass through 0.2 μm, but as they are normally attached to other materials, they are prevented from passing.

Rapid sand filters: They use coarser sand and higher flow rate than slow sand filters to remove impurities by sedimentation, adsorption, straining, chemical and microbiological processes. Suspended solids are removed especially after coagulation and flocculation. They are not effective for removing pathogens.

Slow sand filters: They use relatively fine sand and a low filtration rate to remove impurities by sedimentation, adsorption, straining, chemical and microbiological processes. Removal of pathogens is substantial. However, raw water must have a turbidity of less than 20 NTU.

Sorption or catalytic filters: They use a finely ground filter medium (pore size about 2 μm) composed of zeolite or similar to what impurities chemically bind. These filters remove taste, odor, chlorine, suspended solids, pathogens, volatile organic compounds and heavy metals. They are very simple to use because they are very small filters that are attached to the cap of a water bottle. Sorption filters are relatively expensive.

Mud-pot filtering systems: Those as shown in Figure 8.1 are used in some isolated villages. The top pot serves to feed raw water and contains pre-washed gravel and sand. The water exits through a hole into the second pot, whose mouth is covered with a cloth filter while a crushed coal bed lies on a pad below. Clean water is collected in the third pot. This kind of devices removes toxins and germs. The collection rate is slow, but it can be done at night to allow clean, cool drinking water to be available throughout the day.

Desalination: Desalination by distillation or inexpensive reverse osmosis (with low-cost membranes) produces water without chemical salts. Various methods can be used at household level, for example to treat seawater. Desalination is also effective in removing fluoride, arsenic and iron from water.

Figure 8.1.   Scheme of a mud-pot filtering system.

Coagulation-flocculation: This technique removes fine suspended solids. In coagulation, a substance is added to the water to cause particles to be attracted towards each other, or towards the added material. The flocculation process follows coagulation, and usually consists of slow gentle stirring; as the particles come into contact, they form larger particles that can be removed by settlement or filtration. Alum (aluminum sulfate) is a usual coagulant both at the household level and in water treatment plants. There are also natural coagulants such as powdered seeds of *Moringa olifeira* tree and clays such as bentonite.

Disinfection: This assures that drinking water is free from pathogens. Boiling, chemical and solar disinfection are cheap, simple methods to be applied in isolated areas. The effectiveness of these methods is reduced by the presence of organic matter and suspended solids (see section 8.2).

Disinfection by boiling: A typical recommendation for disinfecting water by boiling is to bring the water to a progressive boil for 10–12 min. One minute at 100°C will kill most pathogens including those causing cholera; many are killed already at 70°C. The main disadvantages of boiling water are that it uses up fuel, affects the taste of water and causes threat of fires.

Chemical disinfection: Chlorination is the most common, effective and widely used method for disinfecting drinking water. Iodine and ozone are other excellent chemical disinfectants.

Other low-cost methods: Harvesting of rainwater on rooftops, soak pits and village ponds, palm leaves, trunk of trees and rocks is very often used in tropical humid areas where precipitation is abundant.

In isolated rural communities of developing countries and during emergencies, there should be access to rapid but reliable methods of purification that supply small volumes of water (10 to 1000 L). The systems should rely as much as possible on local labor and material. Use of alum, permanganate and chlorine tablets for quick purification is easy and practicable at the domestic level as well as solar disinfection in bottles (SODIS, see section 8.3.1).

Technologies based on the use of solar light: They can be employed where sunlight is an abundant costless source: solar disinfection (SODIS), HP, and those based on iron (an abundant species in many waters), such as photolysis of iron complexes, photo-Fenton, etc. are the main solar technologies. These will be described in what follows, comprising their basic concepts, research at the laboratories, development of technologies, and when appropriate, real examples of application.

8.2   DRINKING WATER DISINFECTION

Drinking water disinfection is the destruction of microorganisms causing hydric diseases, from which cholera and typhoid fever are examples. Most resistant infectious agents include prions, coccidian (*Cryptosporidium*) and bacterial spores (*Bacillus*), mycobacteria (*M. tuberculosis*), viruses (poliovirus), fungi (*Aspergillus*), Gram-negative (*Pseudomonas*) and Gram-positive bacteria (*Enterococcus*). Some are highly infectious and can multiply in water supply systems; for example, *Pseudomonas aeruginosa* is responsible for infections in eyes and ears; *Enterobacter cloacae* is a common soil and aquatic microorganism that causes infections in the urine and respiratory systems. *Shigella* species are related to diarrhea and dysentery, and *Salmonella enteritidis* causes gastroenteritis. Viruses (adenoviruses, enteroviruses, hepatitis A and hepatitis E virus, noroviruses and saproviruses, rotaviruses) can easily cause infections highly detrimental to health. Protozoa (*Acanthamoeba* spp. – common in soil and aquatic environments – *Cryptosporidium parvum, Cyclospora cayetanensis, Entamoeba histolytica, Giardia intestinalis, Naegleria fowleri, Toxoplasma gondii*) are very hazardous and infectious at very low concentrations and moderately to highly persistent in water systems. Prions are proteinaceous infectious agents causing fatal neurodegenerative diseases known as transmissible spongiform encephalopathies (TSEs) and responsible for different hospital illnesses.

It is important to distinguish between sterilization and disinfection. Sterilization of a medium implicates a process of removal of all forms of life in this medium, i.e., make it completely free of alive germs at the end of the process. On the other hand, disinfection only destroys pathogenic microorganisms (infectious), but not necessarily all the microorganisms. For example, spores can survive and reactivate.

Pretreatment processes such as coagulation, flocculation, and sedimentation remove around 90% of bacteria, 70% of viruses and 90% of protozoa. Filtration by granular, slow sand, precoat and membranes, with proper design and adequate operation, can act as consistent and effective barriers for microbial pathogens, leading to around 99% bacteria removal.

The commonly used drinking water disinfection techniques, e.g., chlorination (chlorine and derivates), UV-C, and ozonation are the safest methods against most of the pathogens, although the last two are expensive. On the other hand, chlorination and ozonation can produce harmful disinfection byproducts (DBPs).

Chlorine is a very effective disinfectant for most microorganisms. The source of chlorine can be sodium hypochlorite (such as household bleach or chlorine electrolytically generated from a solution of salt and water), chlorinated lime, or high test hypochlorite or trichloroisocyanuric acid (chlorine tablets). Chlorine doses depend on the microorganisms and the necessary amount varies in the order: bacteria < viruses < protozoa. *Cryptosporidium* cannot be safely inactivated with chlorine at all (WHO, 2006). Even under poor sanitary and hygienic conditions, in which people collect water from tanks, wells, pumps and taps for use in their homes, chlorination of water derives in a dramatic decline in the incidence of water-borne diseases. It is relatively easy to distribute and use, particularly in emergencies, and it has residual effect for days. However, an alarming disadvantage of chlorine is the appearance of DBPs such as organohalides, especially trihalomethanes (THMs), chlorophenols and aldehydes. Organohalides were first found in the 1970s, and later fully characterized. These findings have led to severe criticism for use of chlorination in drinking water and even in irrigation water (Malato *et al.*, 2009; WHO, 2011b).

Iodine is another excellent chemical disinfectant but it should not be used longer than a few weeks.

Both chlorine and iodine must be added in sufficient quantities to destroy all pathogens but not so much that taste is adversely affected.

Ozone bubbling is also an effective treatment but costs are unaffordable for low-income populations due to expensive installation, electricity and maintenance operation.

8.3   USE OF SOLAR ENERGY FOR DISINFECTION

8.3.1   *The solar disinfection method (SODIS)*

Solar water disinfection, SODIS, is a method of disinfecting water in small (0.5–2 L) plastic or glass bottles using only sunlight as disinfection reagent. It is an easy, small-scale and cost effective technique to provide safe water at homes or small communities, and uses solar radiation to inactivate pathogens present in water.

The treatment consists of filling throw-away, colorless and transparent containers (generally PET – polyethyleneterephthalate – plastic bottles) with water and exposing them to full sunlight for about five hours (or two consecutive days under 100% cloudy sky). Disinfection (i.e., inactivation of pathogenic microorganisms) occurs by a combination of UV radiation and thermal treatment (infrared radiation, IR); the temperature of the water does not need to rise much above 50°C. Before filling the bottles, water is filtered to remove possible solids and particulate matter.

Although the method does not sterilize water, pathogens responsible for serious illnesses (cholera, diarrhea, intestinal infections, etc.) are inactivated. PET bottles present several advantages with respect to open containers: there is no danger of pollution by airborne microbes, loss of CO_2 is minimal and it is possible to reuse them for several months without contamination coming from the plastic material (Litter, 2002, 2006; Litter and Jiménez González, 2004; Meierhofer *et al.*, 2002; Wegelin *et al.*, 1994, 1998, 2000).

The first study on this method was described by Acra *et al.* (1980), who reported that enteric bacteria were inactivated after exposure to 6 h of sunlight. The main results reported were (Acra *et al.*, 1990; Martín-Domínguez *et al.*, 2005):

- UV radiation (320–400 nm) is the main component of solar light involved in bacterial destruction.
- *Escherichia coli* strains are more resistant to lethal effects of sunlight than other bacteria, like *P. aeruginosa*, *S. flexneri* and *S. typhi*, therefore serving as indicators of the disinfection degree reached on enteric bacteria.

The resistance of microorganisms to solar radiation was found to vary among them (Gill and McLoughlin, 2007), and this subject will be treated in section 8.3.2.

Solar radiation is believed to contribute to the inactivation of pathogenic organisms by the action of (EAWAG/SANDEC, 2008):

- UV-A radiation (wavelength range: 320–400 nm), which damages pathogens through two different mechanisms: i) direct interference on the metabolism, destroying cell structures of bacteria, or ii) production of highly reactive forms of oxygen, called ROS, reactive oxygen species (e.g., oxygen free radicals and hydrogen peroxide, H_2O_2) coming from oxygen present in the water. However, see section 8.3.2.
- Cumulative solar energy (including the IR component), which heats the water. If the temperature of the water rises above 50°C, the disinfection process is three times faster than at 30°C.

For SODIS to be efficient (at a water temperature of about 30°C), a threshold solar radiation intensity of at least 500 W m^{-2} (all spectral light) and about 5 h are required. This dose contains energy of 2000 kJ m^{-2} in the UV-A and violet light range (350–450 nm), corresponding to about 6 h of mid-latitude (European) midday summer sunshine (Wegelin *et al.*, 1994). Berney *et al.* (2006a) reported that it is needed a UV-A dose of 1500 kJ m^{-2} to achieve one log reduction on *E. coli* at 37°C. At water temperatures higher than 45°C, synergistic effects of UV radiation and temperature further enhance the disinfection efficiency (Wegelin *et al.*, 1994); it has been reported that at 45°C the bacteria protection toward H_2O_2 drops-off dramatically (Fisher *et al.*, 2008), an effect that was related to the inactivation of catalase and peroxidase, the enzymes responsible for H_2O_2 scavenging. There are many reports indicating that the effect of temperature on solar disinfection can only be appreciated at more than 45°C (Fisher *et al.*, 2008; McGuigan *et al.*, 1998).

In Figure 8.2, a description of the process is indicated (Meierhofer, 2006; Meierhofer *et al.*, 2002). To maximize the effectiveness, the bottle should lie on the ground, and its surface should be blackened to better absorb the light and generate heat. The treated water can be stored in a cool mud pot for drinking. This method has been successfully tested in e.g. Bolivia, Burkina Faso, China, Colombia, Indonesia, Thailand and Togo (CARE/CDC Health Initiative, 2010), and it is presently applied by more than 2 million people in 31 countries (Schmidt *et al.*, 2008). SODIS was found an appropriate household water treatment technology also for use as an emergency intervention in natural or anthropic disasters, useful not only to fight against bacterial but also against protozoan pathogens (McGuigan *et al.*, 2006).

The results of SODIS guarantee the safety of drinking water not only in remote poor areas without access to clean drinking water but also in peri-urban slums less than 1 km from the centers of modern cities. Therefore, SODIS is now used by millions of people throughout the world. SODIS is especial for countries receiving sufficient sunlight at latitudes between 30°N and 30°S. The effectiveness of the process depends on the quality of the water source, temperature, turbidity, and resistance of the specific microorganisms, irradiance and dissolved oxygen. Like other disinfection methods such as boiling, chlorination or filtration, SODIS is not suitable for the treatment of chemically polluted water. In comparison with other methods for point-of-use water treatment, the advantages of SODIS lie in low maintenance costs and its independence

Exposing Procedure

1 Wash the bottle well the first time you use it

2 Fill the bottle 3/4 full with water

3 Shake the bottle for 20 secondes

4 Now fill up the bottle fully and close the lid

5 Place the bottles on a corrugated iron sheet

6 or put them on the roof...

7 Expose the bottle to the sun from morning until evening for at least six hours

8 The water is now ready for consumption

2.7. Application Procedure

Preparation

1. Check if the climate and weather conditions are suitable for SODIS.

2. Collect plastic PET-bottles of 1-2 litre volume. At least 2 bottles for each member of the family should be exposed to the sun while the other 2 bottles are ready for consumption. Each family member therefore requires 4 plastic bottles for SODIS.

3. Check the water tightness of the bottles, including the condition of the screw cap.

4. Choose a suitable underground for exposing the bottle, for example a CGI (corrugated iron) sheet.

5. Check if the water is clear enough for SODIS (turbidity < 30 NTU). Water with a higher turbidity needs to be pre-treated before SODIS can be applied.

6. At least two members of the family should be trained in the SODIS application.

7. A specific person should be responsible for exposing the SODIS bottles to the sun.

8. Replace old and scratched bottles.

Figure 8.2.   SODIS method. With permission of EAWAG, Switzerland (Meierhofer *et al.*, 2002).

from materials other than the PET bottles, the acceptance due to natural odor and fresh taste of the water after treatment (often not the case for chlorine), its sustainability because no chemicals are consumed and no need of post-treatment. Due to these advantages, on World Water Day on 22 March 2001, SODIS has been recommended by the WHO as a viable method for household water treatment and safe storage for the reduction of health hazards related to drinking water (Meierhofer, 2006). Disadvantages include the relatively long treatment time (6 h to 2 days), the dependence on a minimum dose of sunlight (SODIS is not recommended under conditions of continuous rainfall), the relatively small volumes of treated water (not higher than 3 L per bottle) and the decreased disinfection efficiency in the case of highly turbid water (see section 8.3.3). Water should be consumed before 48 h following the treatment to avoid bacteria regrowth. Several pathogens (especially resistant spores and viruses) have been not tested yet. Recent research on SODIS tried to decrease the length of time required for inactivation: in cloudy days, 2 consecutive days of exposure are recommended (Malato *et al.*, 2007, 2009).

SODIS is not adequate for disinfection of large volumes of water; for this purpose, photoreactor with compound parabolic collectors (CPCs) are under study. For example, McLoughlin *et al.* (2004a) tested water disinfection using low cost CPCs to assess the effectiveness of using near UV light to disinfect water supplies for potential applications in developing countries. A pilot scale photoreactor comprised of non-tracking compound parabolic collectors was installed at *Plataforma Solar de Almería* (PSA, Spain) and a comparison of disinfection efficiency using *E. coli* K-12 was carried out with a reactor configuration of 3 and 1 m^2 illuminated area. Removal from initial bacterial concentrations to below the limit of detection was achieved in 30 min. In a following paper, McLoughlin *et al.* (2004b) compared three different collector shapes for the disinfection of water heavily contaminated with *E. coli* K-12 under real sunlight using laboratory scale reactors to determine the performance of different reflector profiles. The reactors were constructed using Pyrex tubing and aluminum reflectors of compound parabolic, parabolic and V-groove profiles and results showed that the CPC promoted a more successful inactivation than the other two profiles. Solar disinfection can be achieved even for real water sources in such reactors, but the effect has been noticeably improved by the presence of photocatalysts (see section 8.4.5).

Navntoft *et al.* (2008) studied the inactivation kinetics of suspensions of *E. coli* in natural well-water and natural sunlight (cloudy and cloudless) conditions using CPC mirrors to enhance the efficiency of SODIS. On clear days, a faster inactivation rate is obtained in the system with CPC compared to that without reflectors. On cloudy days, inactivation to below the detection limit is only achieved using systems fitted with CPCs. It was observed that the reflectivity of CPC systems that had been in use outdoors for at least 3 years deteriorated in a non-homogeneous fashion. Reflectivity values for these older systems were found to vary between 27 and 72% compared to uniform values of 87% for new CPC systems. This means that even with degradation, a CPC provides a major advantage and therefore, the use of CPC has been proven to be a good technological enhancement in clear and cloudy days for the treatment of higher volumes of water by SODIS on the same day.

Caslake *et al.* (2004) designed and tested a portable, low-cost and low-maintenance solar unit to disinfect non-potable water. The solar disinfection unit was composed of two parts, a rectangular base, made in dark gray PVC plate machined with snake-shaped grooves running across the plate from one end to the other, and a cover, made of UV transparent acrylic, to keep the heat inside and to prevent air to enter the system when the unit is in operation. The volume of the disinfection unit was approximately 1 L. The disinfection unit was tested with river waters and water from two wastewater treatment plants in the USA. The solar disinfection unit has been tested also by *the Centro Panamericano de Ingeniería Sanitaria y Ciencias del Ambiente* (CEPIS) in Lima, Peru. At moderate light intensity, the solar disinfection unit was capable of reducing the bacterial load in a controlled contaminated water sample more than 99.99% and disinfected approximately 1 L of water in 30 min.

8.3.2 *Fundamentals of SODIS and mechanisms of disinfection*

Deleterious effects of UV radiation on bacterial cells have been long recognized and its applications on antimicrobial process have received great attention. The most energetic fraction of the UV spectrum, corresponding to the UV-C range (200–290 nm), is commonly used as an antibacterial agent in water and air treatments, allowing effective disinfection rates by the employment of germicidal lamps (emission wavelength: 254 nm). Irradiation of microorganisms with UV light of high energy causes photochemical reactions in the fundamental components of the cell, disrupting their normal functioning, e.g., interrupting the cell division mechanism or provoking cell death (Bolton, 1999; White, 1999). The description of these mechanisms is beyond the scope of this chapter.

On the other hand, photo-induced bacterial inactivation caused by UV-A (320–400 nm) is well known, and its lethal and sublethal effects have been studied by several workers (Favre *et al.*, 1985; Jagger, 1981; Pizarro, 1995; Pizarro and Orce, 1988; Webb, 1977). Direct UV-A irradiation produces deleterious effects on bacteria cells, with different sensitivity to the radiation depending on the type of bacteria and light doses. In particular, natural sunlight as a killing agent for microorganisms was first described in 1877 (Downes and Blunt, 1877). The exposure of bacteria to UV-A radiation can cause severe alterations to the membrane structure including changes in membrane-bound enzyme activities, metabolic pathways, transport systems and permeability alterations, leading to bacterial cell death (Fernández and Pizarro, 1996; Pizarro, 1995; Pizarro and Orce, 1998). It has been found, for example, that UV-A irradiation causes cell death in *Pseudomonas aeruginosa* at doses at which cell viability of *E. coli* or *Enterobacter cloacae* is not affected (Fernández and Pizarro, 1996). Oppezzo and Pizarro (2001) reported sublethal effects of UV-A on *E. cloacae* in comparison with those produced in *E. coli*.

Sunlight used during the SODIS process consists mainly of UV-A, and hence the main inactivation mechanism is a photooxidative process as well as the generation of ROS, such as the superoxide anion radical ($O_2^{\bullet-}$) or non-radicals such as H_2O_2. When ROS interact with DNA, single strand breaks occur as well as nucleic base modifications, which may be lethal and mutagenic. Oxidation of proteins and membrane damage is also induced (Ubomba-Jaswa *et al.*, 2009). However, many aspects of UV-A inactivation are still unclear. The most destructive wavelengths for the forms of microbial life are those of the near UV-A spectrum (320–400 nm), whereas the spectral band from 400 to 490 nm is the least harmful. Whereas differences in the speed of bacteria inactivation at temperatures between 12 and 40°C are negligible, when the temperature rises to 50°C, the bactericidal action is enhanced by a factor of 2, due, to the synergistic effect between radiation and temperature as discussed earlier.

E. coli is often chosen as the reference enteric coliform bacterium in many SODIS studies, because it is a very well-known bacterium from many points of view (structure and composition, morphology, types, metabolism, behavior under different nutrient media, pathogenicity, etc.) and it is relatively simple to inactivate by solar radiation (Cho *et al.*, 2002; Gill *et al.*, 2007; Lonnen *et al.*, 2005; Martín-Domínguez *et al.*, 2005; McGuigan *et al.*, 1999). The Gram-negative bacteria are of similar sensitivity as *E. coli*, with the exception of *Vibrio cholerae*, which is much more resistant.

The species of *Shigella* and *Salmonella enteritidis* are relatively easy to inactivate under solar light in comparison to *E. coli*, although *S. typhimurium* and *Sh. sonnei* are slightly more resistant. Other organisms have been tested, including *Sh. dysenteria* (Berney *et al.*, 2006b; Kehoe *et al.*, 2004; Smith *et al.*, 2000). Gram-positive bacteria, represented by *Enterococci* sp. (*E. faecalis*) and *Bacillus subtilis*, are more difficult to be treated because they cannot be inactivated under solar radiation. Several phytopathogenic fungi of the *Fusarium* genera (Sichel *et al.*, 2007b) have also been studied.

Rincón and Pulgarín (2004a) found that the inactivation rates under simulated sunlight of several bacteria followed the sequence: *E. coli* > Gram-negative > *Enterococcus* species, a similar result found by Sinton *et al.* (1994). Boyle *et al.* (2008) reported the order *Campylobacter jejuni* < *Staphylococcus epidermidis* < *E. coli* < *Yersinia enterocolitica* for the exposure time required for complete inactivation by SODIS. The solar disinfection technique is not effective for *B. subtilis* spores. *Acanthamoeba polyphaga* is quite resistant, even compared with other protozoa. *Bacteriophage* poliovirus and MS-2 behave like virus and presents a higher resistance than *E. coli* to solar radiation. For this reason, recent reports indicate that *E. coli* and total coliforms may not be adequate indicators of the efficacy of SODIS on enteric bacteria disinfection, as pathogenic bacteria like *Salmonella* sp. and others are more resistant (Berney *et al.*, 2006b; Sciacca *et al.*, 2010).

Other types of organisms like bacterial endospores cannot be inactivated only by solar radiation, as in the case of *B. cereus*, used as a surrogate for the *B. anthracis* (the bacterium that causes anthrax), which has been proven resistant to simulated solar-UV radiation (Lee *et al.*, 2005). On the other hand, Lonnen *et al.* (2005) demonstrated that SODIS is unfeasible for the inactivation of cysts of the protozoan pathogen *Acanthamoeba polyphaga* and for *B. subtilis* spores. Nevertheless, the study also showed that fungal pathogens such as *Candida albicans* and *Fusarium solani* are readily inactivated using the solar disinfection technology. *C. albicans* and *F. solani* are more resistant than *E. coli* although their main route of infection is not *via* water, but *via* other media such as soil. Heaselgrave *et al.* (2006) found that the fungus *F. solani* and poliovirus are more resistant than *E. coli*. Interestingly, Kehoe *et al.* (2004) proved that Type I *Sh. dysenteriae* is very sensitive to SODIS and is easily inactivated even under cloudy conditions; this process was found appropriate for use in developing countries during *S. dysenteriae* Type I epidemics. Méndez-Hermida *et al.* (2005) showed that batch solar disinfection was effective for *Cryptosporidium parvum* oocysts in water.

Despite the effect of UV-A radiation on bacteria is not completely clear, Berney *et al.* (2006a), in a study with flow-cytometry on *E. coli*, reported a series of effects that are dose-dependent, such as the loss of membrane potential, glucose uptake activity and loss of culturability, which take place at a dose of 1500 kJ m^{-2} (more than 80% of the cells affected); loss of membrane potential was found to be the most critical parameter. Cells exposed to doses higher than 1500 kJ m^{-2} were not able to recover. In a recent study on *E. coli*, Bosshard *et al.* (2010a) reported that even under limited exposition, exposed cells were affected in their energy metabolism, in particular the respiratory chain, influencing also their potential to generate ATP and essential enzymes for carbon metabolism and defense against oxidative stress, causing the membrane dysfunction responsible for the loss of membrane potential previously reported. They also reported that the UV-A light main target are structural and enzymatic proteins (Bosshard *et al.*, 2010b), being the damage pattern very similar to those caused by reactive oxygen stress; the authors proposed that light absorption by proteins, especially those containing active iron-sulfur clusters, was the origin of these oxygen species. Proteins are oxidized with the generation of carbonyl groups, causing covalent protein cross-linking that produces aggregation and subsequent cell death (Maisonneuve *et al.*, 2008). At the beginning of irradiation, ATP depletion takes place, being this the reason for the deactivation of ATP-depending enzymes, responsible for avoiding protein misfolding and aggregation; however, Berney *et al.* (2007) reported that, under low irradiation intensities (50 W m^{-2}), *E. coli* showed an increase in ATP concentration, but the authors could not explain if this effect was due to an increase in the rate of ATP synthesis or to a decrease in ATP consumption.

It was proved that the bacterial decontamination rate by solar radiation is proportional to the intensity of radiation and the temperature, and inversely proportional to the depth of the water, due to the dispersion of the light. The amount of radiation attenuated by this effect depends on the range of wavelengths; for example, at 375 nm, the reduction does not attain 5% per meter of depth (Acra *et al.*, 1990), and at longer wavelengths, it can reach up to 40% per meter (Caslake *et al.*, 2004).

The effect of the mode of exposure on the damage induced on cells by UV-A radiation has been studied by Merwald *et al.* (2005); they reported that, for the same dose, the probability of cell death is increased if the exposure is fractioned in intervals of 10 to 120 min, while longer intervals produce a decrease in cell death probability. This work makes clear that not only the dose is an important parameter for the determination of the exposure time, but that environmental conditions should be also determined at the moment of setting a safety dose for disinfection, like the minimum amount of continuous irradiation.

It is important to emphasize that many investigations are concentrated in the reduction of a single microorganism, but they do not study the real water disinfection in the presence of turbidity, chemical compounds and other microbial atmospheres.

Once removed from sunlight, remaining bacteria may again reproduce in the dark (process called regrowth, Sciacca *et al.*, 2010). It has been proposed that, after SODIS treatment, some cells can enter into a state called viable but nonculturable (Colwell, 2000). In this situation, bacteria may not grow in a normal medium, but with the addition of pyruvate and in anaerobic conditions, higher counts can be obtained (Berney *et al.*, 2006a); then, it is possible that normal culturable media may produce an oxidative burst in the cells during incubation, resulting in counts lower than the real number of viable cells.

The possibility of bacterial adaptation to UV light has been studied by some authors (Berney *et al.*, 2007; Caldeira de Araujo *et al.*, 1985; Hoerter *et al.*, 2005). With cells grown under batch conditions, Hoerter *et al.* (2005) reported that cells increase the activity of oxidative stress related proteins, resulting in an increase in UV-A resistance, and that at a total dose of 135 kJ m^{-2} at a fluence of 50 W m^{-2} would result in cell death. Berney *et al.* (2007) reported that, under continuous culture and 50 W m^{-2} UV-A irradiation intensities, *E. coli* expressed an adaptive response, reflected on larger cells, lower growth yields and also lower efflux pump activity (not lethal for *E. coli* under stationary phase, Berney *et al.*, 2006a); only 1% of the initial cells were able to adapt, with growth rates between 0.7 and 0.3 h^{-1}, the adaptation being maintained even after 21 generations. However, it must be taken into account that this adaptation experiments were carried out under irradiation intensities that are at least half the minimum recommended for the application of SODIS; besides, Berney *et al.* (2007) indicated that the growth rates under which adaptation takes place are very unlikely, being generally much lower under environmental conditions, decreasing the probability of UV-A adaptation.

Acra *et al.* (1990) studied the effect of the irradiation wavelength on bacteria inactivation, and observed that blue-violet light, up to 490 nm, can also cause inactivation. They described the wavelength relative efficiency for coliforms according to equation (8.1), referenced to the inactivation rate at 260 nm:

$$\% \text{ Inactivation} = 398e^{-0.0051\lambda} \tag{8.1}$$

According to equation (8.1), UV-A efficiency is 1.5 times higher than that of violet-blue radiation (400–490 nm), calculated for the same doses. The presence of natural organic matter (NOM) has almost no effect at concentrations lower than 3.2 mg L^{-1} of dissolved organic carbon, but at higher values, a decrease on the lethal efficiency of the light was observed. This decrease was explained as an inner filter effect of NOM (Wegelin *et al.*, 1994), whereas no photosensitization was observed for NOM.

The inactivation of bacteria as a function of light intensity (I)[1] is usually assumed to comply with a first-order kinetics (Acra *et al.*, 1990; McGuigan *et al.*, 1998), which can be approximated by the following equation:

$$N/N_0 = e^{-KIt} \tag{8.2}$$

where N is the bacterial concentration after exposure (colony-forming units (CFU) per mL of aqueous treated solution), N_0 is the initial bacterial concentration (CFU per mL), K is the intensity-dependent rate constant $(m^2 \, W^{-1} \, h^{-1})$, I is the intensity of received radiation $(W \, m^{-2})$ and t the exposure time. The value of K is a direct measure of the bacterial sensitivity to the used radiation. It should be taken into account that this exponential relationship is not always observed over the entire period (Acra *et al.*, 1990). Equation (8.2) can be applied to pure bacteria population; if there are mixed population, due to the presence of different species of different growing periods, multiple first-order kinetics should be used, with particular kinetic rate constants for each population (Wegelin *et al.*, 1994).

Fisher *et al.* (2008) listed several first-order kinetic constants for bacteria inactivation (*E. coli* and coliforms) under both artificial and solar irradiation.

Fluence (F) or radiation dose is the product of the radiation intensity I (see footnote) and the exposure time t; when solar irradiation is used, I is not a constant parameter; then, in these cases, survival curves are plotted as percentage of bacterial survival as a function of irradiation dose, in place of irradiation time.

Simple exponential kinetics can be associated with a single-hit single-target mechanism; it implies that one single "event" accounts for the cell death. This simple model is the most frequently used in studying SODIS disinfection kinetics; however, experimental results can often be better fitted to a multi-target model (Wegelin *et al.*, 1994), which implies that more than one hit is needed to cause cell death (accumulative damage). The equation that expresses this kinetics is equation (8.3):

$$N/N_0 = 1 - (1 - e^{-KF})^m \tag{8.3}$$

where m accounts for the number of hits needed to cause cell death. This type of kinetics is particularly useful because it considers the observed lag period before cells concentration starts

---

[1]In recent times, the IUPAC indicated that intensity is a traditional term indiscriminately used for photon flux, fluence rate, irradiance, or radiant power and that concerning an object exposed to radiation. The term should now be used only for qualitative descriptions (see S. Braslavsky *et al.*, *Pure Appl. Chem.*, 79 (2007) 293–465). However, we will keep this notation in equations (8.1) and (8.2) in accordance with that used by the authors.

to decrease. Finally, a more complex model that involves both types of mechanism can be used, as expressed by equation (8.4):

$$N/N_0 = e^{-K1F} \times \lfloor 1 - (1 - e^{-K2F})^m \rfloor \tag{8.4}$$

where $K1$ and $K2$ are the inverse of the dose required for 37% of survival from the single-hit single-target mechanism and the inverse of the dose required for 37% of survival from accumulative damage, respectively. Equation (8.4) has been used for modeling the deactivation kinetics of *Salmonella enterica* by solar irradiation at a constant temperature of 26°C (Oppezzo *et al.*, 2011).

8.3.3   *Experimental conditions for SODIS*

Solar disinfection has been studied using different experimental setups. Acra *et al.* (1990) proposed the one-step continuous reactor in bottles.

Bottle material: some glass or PVC materials may prevent UV light from reaching the water. Commercially available bottles made of PET are recommended, with a much more convenient handling. Polycarbonate blocks all UV-A and UV-B rays, and therefore should not be used.

Aging of plastic bottles: SODIS efficiency depends on the physical condition of the plastic bottles, with scratches and other signs of wear reducing the efficiency. Heavily scratched or old, blind bottles should be replaced (Meierhofer *et al.*, 2002). PET soft drink bottles are often easily available and thus most practical for SODIS application.

Turbidity: SODIS requires relatively clear water. EAWAG/SANDEC (2008) recommends a turbidity value not above 30 NTU for assuring SODIS effectiveness. With turbidities higher than 100 NTU, exposure times are long and disinfection is not always guaranteed for all pathogens. The intensity of the UV radiation decreases rapidly with increasing water depth. At a water depth of 10 cm (4 inches) and moderate turbidity of 26 NTU, UV-A radiation is reduced to 50%. Wegelin *et al.* (1994) reported that a turbidity of 25 NTU has little influence on the disinfection efficiency, despite the authors' recommendation of reducing turbidity as much as possible before the solar treatment. On studies carried out with the pathogenic protozoan parasite *Cryptosporidium*, Gómez-Couso *et al.* (2009) reported that turbidity is a statistically significant parameter on survival of this microorganism, but radiation intensity is an even more important factor.

Oxygen: Under sunlight, oxygen produces ROS in the water. As said before, these reactive molecules contribute to the destruction process of the microorganisms. Under normal conditions (rivers, creeks, wells, ponds, tap), water contains sufficient oxygen (more than $3 \, mg \, L^{-1}$) and it does not have to be aerated before SODIS application.

Irradiation conditions: The irradiance (i.e., the radiant energy per unit of time and of cross surface) is very important. Disinfection results of a bacterial suspension exposed to UV-A radiation of $25 \, W \, m^{-2}$ (average irradiance) during 2 h are different from those found when the same water is exposed to $50 \, W \, m^{-2}$ of UV-A during 1 h.

Continuous light exposure (without interruptions) makes the bactericidal effect faster and more efficient than when the light is applied intermittently (Rincón and Pulgarin, 2003). It was suggested that this effect is due to bacterial dark-repair mechanisms that allow bacteria to reactivate after the treatment. This fact can also be attributed to the partial damage produced by the radiation, which leads to partial inactivation of the bacteria colonies. It is also known that some cells damaged by UV radiation (but incompletely died or destroyed) can recover viability by means of two DNA repair mechanisms. One of them is the so-called "photo-reactivation" or "photo-repairing", and it takes place when cells are exposed to 300–500 nm radiation. The other one is a repair mechanism based on excision (resynthesis and postreplication of cells). As we will see, this is different in the case of photocatalytic disinfection (see section 8.4.2).

Ubomba-Jaswa *et al.* (2009) studied the effect on SODIS of solar UV-A irradiance and solar UV-A dose for the inactivation of *E. coli K*-12 seeded in natural well-water in borosilicate glass tubes, in PET bottles and in a continuous flow system ($10 \, L \, min^{-1}$), to determine the effect of an interrupted and uninterrupted solar dose.

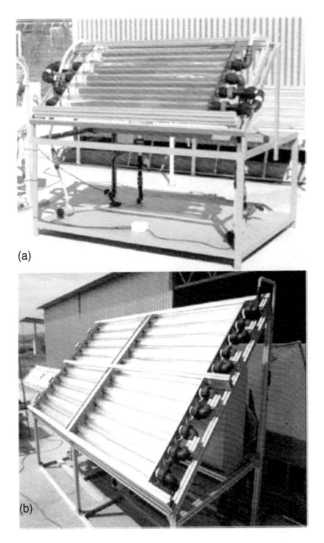

Figure 8.3. Picture of the 14 L solar CPC reactor (a) and 70 L solar CPC reactor (b) at PSA facilities (Almería, Spain) (Ubomba-Jaswa *et al.*, 2009); reproduced by permission of The Royal Society of Chemistry (RSC) for the European Society for Photobiology, the European Photochemistry Association, and the RSC.

One of the main goals of this work was to investigate also whether larger volumes of water can be successfully treated by SODIS. For this purpose, 14 L (Fig. 8.3a) and 70 L (Fig. 8.3b) solar CPC reactors were designed and built, consisting of borosilicate glass tubes, with an optical cut-off at 280 nm and higher transmittance than that of PET bottles. The borosilicate glass tubes were placed in the focus of CPC reflectors and held by aluminum frames mounted on platforms tilted at 37° local latitude at PSA facilities. The 70 L reactor consisted of four panels of solar CPC collectors, each having 5 CPC reflectors and 5 glass tubes of a similar design to those used in the 14 L reactor. The solar collectors were connected in series to a tank and a pump in a whole system that follows a similar flow diagram.

Results showed that inactivation from approximately 106 CFU mL^{-1} to below the detection level (4 CFU mL^{-1}) for *E. coli* K-12 is a function of the total uninterrupted dose delivered to the bacteria and that the minimum dose should be higher than 108 kJ m^{-2} for the conditions described (spectral range of 295–385 nm). For complete inactivation, this dose needs to be received regardless of the incident solar UV intensity and needs to be delivered in a continuous and uninterrupted manner. This is illustrated by a continuous flow system in which bacteria were not fully inactivated (residual viable concentration $\sim$102 CFU mL^{-1}) even after 5 h of exposure to strong sunlight and a cumulative dose lower than 108 kJ m^{-2}.

Ubomba-Jaswa et al. (2010a) studied the effect of irradiation time and weather conditions on SODIS applicability. They designed a simple SODIS reactor (25 L), constructed from a methacry-late tube placed along the linear focus of a CPC and mounted at 37° inclination. Experiments were carried out for 7 months by seeding a 10^6 CFU mL^{-1} concentration of E. coli K-12 in well water or turbid water to mimic field conditions and determine the microbial effectiveness of the reactor. During periods of strong sunlight, complete inactivation of bacteria occurred in 6 h, even with water temperatures lower than 40°C. Under cloudy and low solar intensity conditions, prolonged exposure was needed. Turbid water (100 NTU) was disinfected in 7 h with water temperatures higher than 50°C. No regrowth of bacteria occurred within 24 h and 48 h following SODIS.

Concerning the small SODIS reactor bottles, as they are only illuminated on the upper side of the bottle, a large fraction of the available radiation cannot reach the water. To improve the efficiency and concentrate solar radiation, reflecting surfaces have been tested, such as aluminum foil attached to the back of the bottles (Kehoe et al., 2004), wall reflectors (Rijal and Fujioka, 2003) or reflective solar boxes (Martín Domínguez et al., 2005).

Leaching of bottle material: There has been some concern on whether plastic drinking contain-ers can release chemicals or toxic components into water, a process possibly accelerated by heat. The Swiss Federal Laboratories for Materials Testing and Research (EMPA) have examined the diffusion of plasticizers di(2-ethylhexyl)adipate (DEHA) and di(2-ethylhexyl)phthalate (DEHP) from new and reused PET-bottles in the water during solar exposure. The concentration for both DEHA and DEHP in the treated water after 17 h of exposure to the sun at 60°C was three and one order of magnitude, respectively, below the WHO guidelines for drinking water (WHO, 2008) and in the same magnitude as the concentrations of these plasticizers in commercial bottled water, as reported by Kohler and Wolfensberger (2003) and Schmidt et al. (2008). Schmidt et al. (2008) also reported that the most decisive factor in the leaching of plasticizers was the country of origin of the bottles; to a lesser extent, temperature values around 60°C or higher can be accounted for an increase in the concentration of plasticizers. A study of Nathan and Philip (2009) also reported that the amount of DEHA and DEHP leached by PET bottles is independent whether the bottles are heated up to 60°C and/or exposed under solar light, being the concentration of these plasticizers well below the WHO guidelines. Though microbially safe, concerns have been raised about the genotoxic/mutagenic quality of solar-disinfected drinking water, which might be compromised as a result of photodegradation of PET bottles used as SODIS reactors. Studies using the Ames fluctuation test gave negative genotoxicity results for water samples that had been in PET bottles and exposed to normal SODIS conditions (6 h of strong natural sunlight, followed by overnight room temperature storage) over 6 months. They were then emptied and refilled the following day and exposed to sunlight again. Genotoxicity was detected after 2 months in water stored in PET bottles and exposed continuously (without refilling) to sunlight for a period ranging from 1 to 6 months. However, similar genotoxicity results were also observed for the dark control (without refill) samples at the same time-point and in no other samples after that time; therefore, it is unlikely that this genotoxicity event is related to solar exposure (Ubomba-Jaswa et al., 2010b).

Concerns about the general use of PET bottles were also expressed after a report published by Shotyk et al. (2006) on antimony being released from the PET bottles for soft drinks and mineral water stored over several months in supermarkets. However, the antimony concentrations found in the bottles are orders of magnitude below the WHO (2008) guidelines for concentrations in drinking water.

Studies on leaching of PET demonstrated that time is a dominant factor governing the release of organic substances like aldehydes, independently from light exposure (Nawrocki et al., 2002; Wegelin et al., 2001). However, as water treated by SODIS is not expected to be stored over

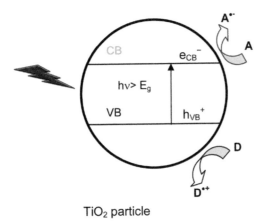

Figure 8.4.   Simplified diagram of the HP processes occurring at an illuminated TiO_2 particle.

extended periods in the bottles, high concentration of organic compounds should not be a risk associated with this technology.

Additives: SODIS can be combined with other disinfection agents such as H_2O_2. Hartman and Eisenstark (1980) reported the lethal effect of H_2O_2 and near-UV light for *E. coli* K12, while Anathaswamy and Eisenstark (1977) reported the same effect on phage T7. The addition of $10\,mg\,L^{-1}$ H_2O_2 was effective in preventing the regrowth of wild *Salmonella* (Sciacca *et al.*, 2010). Sichel *et al.* (2009) reported the lethal synergy of solar UV-radiation and H_2O_2 on wild *F. solani* spores in distilled and natural well water, and H_2O_2 could be reduced from 500 to $5\,mg\,L^{-1}$ with better disinfection results than disinfection tests in the dark.

SODIS can be improved by the addition of small quantities of other additives like citrus fruit juices (Fisher *et al.*, 2008); however, this procedure implies the use of chemicals, and is detrimental to the simplicity of this method. The use of additives can be helpful when low irradiation intensity can be expected.

8.4   HETEROGENEOUS PHOTOCATALYSIS

8.4.1   *Fundamentals of HP*

Heterogeneous photocatalysis with TiO_2 is a very convenient option for purification and remediation of water and air. There are excellent revisions about this technology (see e.g., Blesa and Sánchez Cabrero, 2004; Fujishima and Zhang, 2006; Hoffmann *et al.*, 1995; Linsebigler *et al.*, 1995; Mills and Le Hunte, 1997). HP uses the excitation of a wide band semiconductor, suspended in water, by light of energy equal or higher than its bandgap[2] (E_g), to generate electrons in the conduction band (e_{CB}^-) and holes in the valence band (h_{VB}^+). TiO_2 is the most used semiconductor material for photocatalytic purposes, due to its excellent optical and electronic properties, chemical stability, lack of toxicity and low cost. The bandgaps of the photocatalytic forms of TiO_2, anatase and rutile, are 3.23 eV (corresponding to 384 nm) and 3.02 eV (corresponding to 411 nm), respectively, and these values correspond to the minimal energy that has to be given to the material to excite it and to allow a photocatalytic reaction. The German company Evonik (previously Degussa) produces, under the name of P25, the most popular commercial photocatalytic form of TiO_2. Photogenerated holes and electrons can recombine or migrate to the surface where they can react with donor (D) or acceptor (A) species (Fig. 8.4). The energy level at the bottom

---

[2]In semiconductors and insulators, the bandgap is the difference of energy between the bottom of the conduction band and the top of the valence band.

of the CB is actually the reduction potential of photoelectrons and the energy level at the top of the VB determines the oxidizing ability of photoholes, each value reflecting the ability of the system to promote reductions and oxidations. For P25, the values of the edges of the conduction band (CB) and valence band (VB) at pH 0 have been calculated as -0.3 and $+2.9$ V *vs.* NHE, respectively (Martin *et al.*, 1994). In this case, VB holes are strong oxidants that may attack directly oxidizable species D or form $HO^\bullet$ from water or surface hydroxide ions, while e_{CB}^- are mild reducing acceptors. From a thermodynamic point of view, couples can be photocatalytically reduced by TiO_2 e_{CB}^- if they have redox potentials more positive than the edge of the CB, and can be oxidized by TiO_2 h_{VB}^+ if they have redox potentials more negative than edge of the VB.

The above-described scheme is completed by the following basic equations:

$$TiO_2 + h\nu \rightarrow e_{CB}^- + h_{VB}^+ \tag{8.5}$$

$$e_{CB}^- + A \rightarrow A^{\bullet-} \tag{8.6}$$

$$h_{VB}^+ + H_2O \rightarrow HO^\bullet + H^+ \tag{8.7}$$

$$h_{VB}^+ + D \rightarrow D^{\bullet+} \tag{8.8}$$

In particular, O_2 adsorbed to TiO_2 can be reduced by e_{CB}^-, generating $O_2^{\bullet-}$ in a thermodynamically feasible but rather slow electron transfer reaction (Hoffmann *et al.*, 1995). This cathodic pathway is an additional source of $HO^\bullet$ through several steps:

$$O_{2(ads)} + e_{CB}^-(+H^+) \rightarrow O_2^{\bullet-}(HO_2^\bullet) \rightarrow H_2O_2 \rightarrow HO^\bullet \tag{8.9}$$

If in solution there is a metal or metalloid ion of convenient redox potential, e_{CB}^- can reduce the species to a lower oxidation state:

$$M^{n+} + e_{CB}^- \rightarrow M^{(n-1)+} \tag{8.10}$$

Alternatively, the metal or metalloid can be oxidized by h_{VB}^+ or $HO^\bullet$:

$$M^{n+} + h_{VB}^+/HO^\bullet \rightarrow M^{(n+1)+} \tag{8.11}$$

The potential applications of the HP technology include organic matter degradation, abatement of metal toxic ions and water disinfection (Blesa and Sánchez Cabrero, 2004; Fujishima and Zhang, 2006; Hoffmann *et al.*, 1995; Ibáñez *et al.*, 2003; Linsebigler *et al.*, 1995; Litter, 1999, 2009; Mills and Le Hunte, 1997).

8.4.2 *Use of heterogeneous photocatalysis in water disinfection*

Heterogeneous photocatalysis with TiO_2 has been proposed as one of the best disinfection technologies because no dangerous (carcinogenic or mutagenic) or malodorous halogenated compounds are formed, in contrast with other disinfection techniques, e.g. those that use halogenated reagents (see section 8.2). Reviews on the subject have been published by Blake *et al.* (1999), Blanco-Gálvez *et al.* (2007), Block and Goswami (1995), Guimarães *et al.* (2004), Malato *et al.* (2007), Malato *et al.* (2009) and Mills and Le Hunte (1997).

The first contribution to photocatalytic disinfection was performed by Matsunaga *et al.* in 1985, and since then much research has been made. Matsunaga *et al.* (1985) showed inactivation of the Gram-positive *Lactobacillus acidophilus* bacteria, the Gram-negative *E. coli* bacteria, the *Saccharomyces cerevisiae* yeast and the *Chlorella vulgaris* algae, all at 10^3 cells mL^{-1} concentration after 120 min of incubation with irradiated TiO_2/Pt powders. Inactivation to the detection limit was reported, except for the algae, which after 120 min still showed 55% survival. Table 8.1 shows the differences.

Table 8.1.   Inactivation efficiency of TiO_2/Pt–UV for several microorganisms.

Microorganism	Exposure time (min)	Survival rate (%)
S. cerevisiae	60	54
	120	0
E. coli	60	20
	120	0
L. acidophilus	60	0
C. vulgaris	60	85
	120	55

The antimicrobial activity of UV/TiO_2 has been essayed in other bacteria and viruses including *E. coli* (Armon *et al.*, 1998; Block and Goswami, 1995; Cooper *et al.*, 1998; Ibáñez *et al.*, 2003; Ireland *et al.*, 1993; Matsunaga *et al.*, 1985; Matsunaga *et al.*, 1988; Salih, 2002; Wei *et al.*, 1994), *Lactobacillus acidophilus* (Matsunaga *et al.*, 1985), *Serratia marcescens* (Block and Goswami, 1995; Armon *et al.*, 1998), *Pseudomonas aeruginosa* (Armon *et al.*, 1998; Ibáñez *et al.*, 2003), *Pseudomonas stutzeri* (Biguzzi and Sham,1994), *Bacillus pumilus* (Pham *et al.*, 1995), *Streptococus mutans*, *Streptococus rattus* and *Streptococus cricetus* (references in Blake *et al.*, 1999), *Streptococus sobrinus AHT* (Saito *et al.*, 1992), *Deinococcus radiophilus* (Matsunaga *et al.*, 1985; Laot *et al.*, 1999), *Serratia marcescens*, *Streptococcus aureus* (Bekbölet, 1997; Block *et al.*, 1997), *Streptococcus mutans* (Morioka *et al.*, 1988), *S. typhimurium*, *P. aeruginosa*, *E. cloacae* (Ibáñez *et al.*, 2003).

Yeasts as *Saccharomyces cerevisiae* (Matsunaga *et al.*, 1988; Seven *et al.*, 2004), *Cryptosporidium parvum* oocysts (Méndez-Hermida *et al.*, 2007), algae as *Chlorella vulgaris* (Matsunaga *et al.*, 1988) and viruses such as phage MS2 (Matsunaga *et al.*, 1988; Laot *et al.*, 1999; Sjogren and Sierka, 1994), *B. fragilis* bacteriophage (Matsunaga *et al.*, 1988) and poliovirus type 1 (Watts *et al.*, 1995) were tested. The inactivation of three different *Fusarium* sp. spores (microconidia, macroconidia and chlamydospores, Polo-López *et al.*, 2010) and of five wild strains of the *Fusarium* genus (*F. equiseti*, *F. oxysporum*, *F. anthophilum*, *F. verticillioides*, and *F. solani*, Fernández-Ibáñez *et al.*, 2009; Sichel *et al.*, 2007b), in distilled and well water by solar photocatalysis in a solar bottle reactor was reported. More resistant microorganisms were also successfully inactivated such as *Staphylococcus aureus* and *C. albicans*, but the fungus *Aspergillus niger* was resistant to the treatment (Seven *et al.*, 2004). Lonnen *et al.* (2005) confirmed inactivation of *C. albicans* and found decrease of *F. solani*.

Inactivation of prions was also reported (Paspaltsis *et al.*, 2006). In this work, initial *in vitro* experiments were followed by a bioassay with the scrapie strain 263 K in Syrian hamsters. The results obtained from this study indicate that the HP treatment reduces infectivity titers significantly. The method could also be adapted for decontamination of surgical instruments.

Nagame *et al.* (1989) inactivated microorganisms of the oral cavity (*cricetus HS-6* and *Actinomyces viscosus*, ATCC 19246, ca. 10^4 cells mL^{-1}). The species were treated under a fluorescent lamp ($\lambda = 578$ nm) and rutile TiO_2 (0.1% (m/v)). Notably, *Streptococcus rattus* FA-1 cultures were also totally inactivated either in the dark or under irradiation, but no effect was observed on the viability of *Streptococcus rattus* BHT and *C. albicans* IFO 1060.

Sjorgren and Sierka (1994) described the photocatalytic inactivation (1 g L^{-1} P25, pH 7.2) of a bacteriophage (MS2 phage = ATCC15597B1) using *E. coli* (ATCC15597) as host and a UV lamp (15 W, $\lambda = 365$ nm, 2 mW cm^{-2}, close to that of solar light in clear days).

Comparison of experiments in different seasons, early and later in the day, and under cloudy and sunny conditions, led to the conclusion that solar photocatalytic disinfection of *E. coli*, *F. solani*, and *F. antophilum* does not depend proportionally on solar UV irradiance as long as enough photons have been received for disinfection. The minimum UV energy necessary to

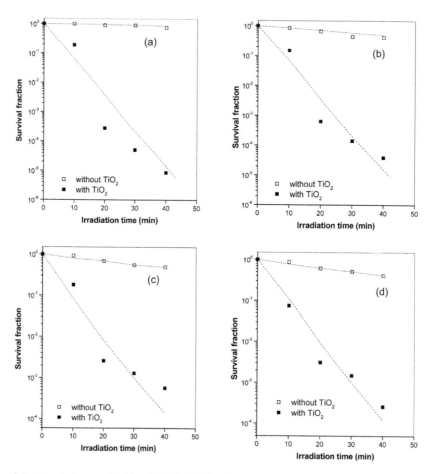

Figure 8.5.   Survival curves for *E. coli* K-12 (a), *S. typhimurium* (b), *P. aeruginosa* (c) and *E. cloacae* (d) exposed to UV-A irradiation (365 nm) with and without TiO_2. Conditions: $[TiO_2] = 0.1\,gL^{-1}$, irradiance $= 5.5\,mW\,cm^{-2}$ in (a), (b) and (d), and $1.4\,mW\,cm^{-2}$ in (c). Dashed lines are first-order fittings (Ibáñez *et al.*, 2003). With permission of Elsevier.

reach complete inactivation depends more on the microorganism and the reactor configuration (Sichel *et al.*, 2007a).

Use of optical fibers (Matsunaga and Okochi, 1995) or intermittent and variable irradiation (Laot *et al.*, 1999) has been also recommended to improve the application.

Ibáñez *et al.* (2003) studied the HP degradation of several bacteria. In particular, *E. coli* showed a very important loss of cell viability (5 orders of magnitude in 40 min) when exposed to a $5.5\,mW\,cm^{-2}$ UV-A photon flux in the presence of TiO_2 (Fig. 8.5a). Similar results (more than 4 orders of magnitude in 40 min) were obtained also when *S. typhimurium* (Fig. 8.5b) was exposed to the same conditions. The great sensitivity (Fig. 8.5c) of *P. aeruginosa* to UV-A (note that a four-fold lower irradiance is used) has been reported previously (Fernández and Pizarro, 1996). For this reason, suspensions of these bacteria were exposed to a lower UV-A intensity than the other bacteria, i.e., $1.4\,mW\,cm^{-2}$, assuring in this way the absence of cellular death due only to irradiation effects. The results showed a rapid decrease in the colony forming ability with time (more than 3 orders of magnitude after 40 min) in the presence of TiO_2. The control without TiO_2 confirmed the absence of lethality (Fig. 8.5c). *E. cloacae*, found previously very resistant to UV-A exposure in the absence of TiO_2 (Oppezzo and Pizarro, 2001), when exposed to a $5.5\,mW\,cm^{-2}$

of UV-A photon flux in the presence of TiO_2, showed a very important bacterial cell lethality, reaching a reduction of almost 4 orders of magnitude after 40 min exposure (Fig. 8.5d).

Concerning real applications, Herrera Melián *et al.* (2000) performed disinfection experiments on urban wastes, monitoring total coliforms and *Streptococcus faecalis*. UV irradiation (800 W, P25) produced a two order reduction in 1 h. Solar light was also effective but slower. However, no significant differences were detected with and without photocatalyst. No further bacterial growth was observed.

However, Wist *et al.* (2002) reported strong *E. coli* regrowth in treated water from the Cauca River (Cali, Colombia) after 24 h. They found that the bacteria showed good inactivation but, after the treatment, cells increased rather rapidly compared with *E. coli* in distilled water. The authors attributed this effect to the presence of different compounds in the real water samples. On the other hand, Rincón and Pulgarin (2004b) investigated the behavior of *E. coli* K12 and bacterial consortia present in real wastewater from a biological wastewater treatment plant in Lausanne (Switzerland). They observed that in the presence of TiO_2, the decrease of bacteria was faster than in its absence, that the inactivation continued in the dark, and that no regrowth was observed within the following 60 h. They concluded that *Enterococcus* species are less sensitive to the photocatalytic treatment than coliforms and other Gram-negative bacteria, a similar result obtained when using SODIS (see section 8.3.2). In addition, they found differences in real waters depending on the date of collection of the samples. Similar results were obtained later (Rincón and Pulgarín, 2005) with other types of microorganisms. Due to differences among the different microbial species or characteristics of the samples, Rincón and Pulgarin (2004a) proposed the "effective disinfection time" parameter (EDT), as a method of assessing the bactericidal inactivation rate in solar photocatalytic processes for drinking water, defined as the time necessary to avoid bacterial regrowth after 24 or 48 h in the dark after the phototreatment. The EDT depends on the residence time of water in the illuminated part of the system, the light intensity, and the specific time of the day selected for the treatment when solar radiation is used.

Watts *et al.* (1995) studied the potential of HP on inactivation of bacteria and viruses present in an effluent coming from a secondary treatment of wastewaters using solar light and artificial black light. Results showed that poliovirus type 1 was inactivated more rapidly than coliform bacteria: 150 min exposure under the artificial lamp was needed to reach a two-order reduction for bacteria, while the same reduction for poliovirus type 1 was attained after 30 min.

The photocatalytic water disinfection of biologically pretreated municipal wastewater in Hannover (Germany) was studied (Dillert *et al.*, 1998, 1999), and it was demonstrated that the TiO_2 photocatalytic treatment of *E. coli* suspensions enables the reuse of municipal wastewater.

The technology could even be applied to destroy bioaerosols in air (Goswami *et al.*, 1997, 1999).

As TiO_2 photocatalysis can make use of the UV part of the solar spectrum, it becomes promising to potabilize waters in developing tropical countries with scarce hydric resources and high availability of solar irradiation. In this sense, Gelover *et al.* (2006) exposed to sunlight spring water of Mexico naturally polluted with coliform bacteria in plastic bottles with TiO_2 using simple homemade solar collectors consisting of five wooden squares covered by aluminum foil. Inactivation of total as well as fecal coliforms took place in periods of 15 to 30 min on a sunny day (more than $1000 \, W \, m^{-2}$ irradiance).

Combined technologies were also evaluated. The efficacy to inactivate *E. coli* in sterile water of HP compared with sonophotocatalysis induced by UV-A irradiation and low frequency (24–80 kHz) ultrasound irradiation in the presence of TiO_2 and peracetic acid (PAA) as an additional disinfectant was studied (Drosou *et al.*, 2009). PAA-assisted UV/TiO_2 photocatalysis led to nearly complete *E. coli* inactivation in 10–20 min of contact time, with the extent of inactivation depending on the type and load of photocatalyst and PAA concentration. Sonophotocatalysis in the presence of PAA resulted in increased *E. coli* inactivation compared with the efficiencies of the individual processes, what was attributed to the increased generation of ROS serving as disinfectants, as well as the ultrasound driven enhanced catalytic activity of titania.

8.4.3   *Mechanisms of photocatalytic disinfection processes and disinfection kinetics*

The photochemical mechanism of the TiO_2 biocidal action seems to proceed through the involvement of ROS such as $HO^\bullet$ and $O_2^{\bullet-}$, the main generated species in HP in the presence of oxygen (eq. 8.7 and 8.9). Cho *et al.* (2002) proposed that $HO^\bullet$ acts independently or in collaboration with other ROS, at least for *E. coli* inactivation, with a linear correlation between their steady-state concentrations and the rates of *E. coli* degradation. Salih (2002) also proposed that $HO^\bullet$ were the main species responsible for the degradation of the bacteria.

Other ROS, such as H_2O_2, hydroperoxyl radical ($HO_2^\bullet$) and singlet oxygen ($^1O_2$) have been also proposed to be involved. Kikuchi *et al.* (1997) proposed that long-range interactions between the active species and the cells, due to the large bacteria size, are necessary for the occurrence of photocatalytic processes in bacterial systems. A large mineralization extent was measured (evidenced by CO_2 production, SEM and ^{14}C radioisotope labeling experiments) during the photocatalytic process, which suggests a main role of $HO^\bullet$ on the microorganism (Jacoby *et al.*, 1998).

It is proposed that the first target of the oxidative radicals is the surface of the external membrane of the cell wall, where the TiO_2 photocatalytic surface first makes contact with the cells. Matsunaga *et al.* (1985) attributed disinfection to the photochemical oxidation of the intracellular coenzyme A, which caused inhibition of respiration and a decrease of the metabolic activity, leading to cell death. Saito *et al.* (1992) demonstrated that TiO_2 photocatalysis provoked rupture of the cell membrane in *Streptococcus sobrinus*, as shown by electron microscopy and demonstrated by intracellular leakage of K^+ ions and slow release of protein and DNA, leading ultimately to cell wall breakdown and complete cell death. Chromosomal aberration by DNA lesion caused by the photoexcited TiO_2 was also reported (Nakagawa, 1997; Huang *et al.*, 2000). Further evidence of this mechanism was found by Sunada *et al.* (1998), who proved destruction of endotoxin, an integral component of the outer membrane of *E. coli*, mediated by TiO_2. Later, Sunada *et al.* (2003) suggested that the photokilling reaction was initiated by partial decomposition of the outer membrane, followed by cytoplasmic membrane disorganization, resulting in cell death. Maness *et al.* (1999) showed that lipid peroxidation of the cell membrane is the underlying mechanism of inactivation of *E. coli* K-12 cells in the presence of irradiated TiO_2, leading to a loss of essential cell functions, e.g., respiratory activity, followed by cell death. The photocatalytic action progressively increased the permeability of cells and, thereafter, the open flow of intracellular components, allowing TiO_2 particles easier access and causing photooxidation of intracellular elements, with acceleration of cell death. Figure 8.6 shows a schematic view.

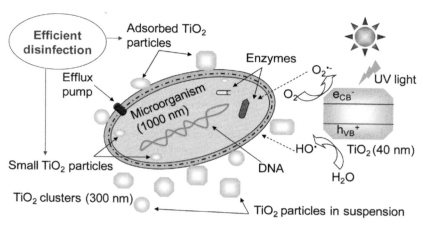

Figure 8.6.   Schematic illustration of the solar photocatalytic process for bacteria inactivation in the presence of an aqueous TiO_2 suspension.

Nadtochenko *et al.* (2004) studied *E. coli* photokilling in HP conditions using attenuated total reflection Fourier transform infrared spectroscopy (ATR-FTIR) and atomic force microscopy (AFM), relating the results to the bacteria viability. Peroxidation products and changes on the membranes of the bacteria walls were reported as the precursor events leading to bacterial lysis. The appearance and growth kinetics of these products were followed through formation of conjugated double bonds, malondialdehyde, peroxides, and CO_2 evolution. It was affirmed that the significant part of the photocatalytic peroxidation takes place at the TiO_2 surface (Kiwi and Nadtochenko, 2004, 2005).

The photocatalytic inactivation rate as a function of the initial bacterial concentration generally obeys to a first order kinetics, similar to that found for inactivation under light in the absence of TiO_2 (section 8.3.2, eq. 8.2). This has been already proven with total coliforms, spores, etc., as for example found by Wei *et al.* (1994) and by Ibáñez *et al.* (2003) for the four types of bacteria studied: *E. coli, S. typhimurium, P. aeruginosa* and *E. cloacae*. The range of concentration of microorganisms for which the first order kinetics is valid depends, however, on the nature of the microorganism (Malato *et al.*, 2009).

8.4.4   *Effect of photocatalyst: immobilization*

In most research papers, TiO_2 is used as a slurry in aqueous suspensions. However, if the system is designed for drinking water purification at household level (point-of-use water treatment), this configuration is not feasible because the particles of the photocatalyst have to be removed before consumption. TiO_2 powder cannot be left in drinking water because it changes the organoleptic properties and because of possible health risks still not assessed. Therefore, immobilization of the photocatalyst is mandatory. It is known that in the case of removal of organic contaminants, immobilization of the photocatalyst causes a decrease in efficiency compared to the use of suspensions, mainly due to a less available active surface. However, this is not always true in the case of bacteria and viruses because the size of the cells is orders of magnitude larger than the size of the photocatalyst particles, leading to a totally different adsorption kinetics (Malato *et al.*, 2009).

Several supports were tested for the immobilization of the photocatalyst using *E. coli* as the model microorganisms for disinfection. Among other examples, Kikuchi *et al.* (1997), and Sunada *et al.* (1998) used transparent TiO_2 films prepared on soda-lime glass plates previously coated with silica thin film and dipped in titanium isopropoxide solution, with further calcination at 500°C; Matsunaga *et al.* (1988) impregnated acetylcellulose membranes; Laot *et al.* (1999) entrapped TiO_2 in silicon tetramethoxide by a sol-gel technique; Gumy *et al.* (2006) coated TiO_2 on nets of natural and synthetic fibers and Alrousan *et al.* (2009) dip-coated borosilicate glass sheets with P25.

Curtis *et al.* (2002) treated water containing the pathogen *C. parvum* using two forms of immobilized TiO_2 (sol-gel and thermal-film) in simple Petri dishes. Better results for *C. parvum* were achieved by Méndez-Hermida *et al.* (2007) with TiO_2 fixed on flexible plastic inserts inside of 1.5 mL bottles under natural sunlight.

TiO_2 was immobilized also on several resistant flexible materials, like cylinders, pills, balls, mesh, etc. One example is TiO_2 deposited on fiberglass (Ahlstrom, 1999), which was then inserted in a tubular photoreactor with a CPC.

In a thorough work, Meichtry *et al.* (2007) immobilized P25 to different cheap materials – glass rings, glass rods and porcelain beads – by dip coating, or directly to the plastic wall of the bottles, with the aim of using common plastic PET bottles under solar irradiation for disinfection. The adherence and stability of TiO_2 on the supports and the photocatalytic activity in bottles under solar irradiation were evaluated using model organic compounds (4-chlorophenol and 2,4-dichlorophenoxyacetic acid). Rings were found to be the best glass supports, but PET bottles were superior for this specific application, as no fragile fillings are used, and the materials can be easily fabricated on site.

To improve the efficiency of the immobilized photocatalysts, some work has been done with photoelectrocatalysis. Butterfield *et al.* (1997) prepared an adherent defect TiO_2 film in which the performance to disinfect water containing *Clostridium perfringens* spores and *E. coli* was greatly improved by the application of a small positive bias (electric field enhancement). When TiO_2 powder was immobilized electrophoretically on electrodes and electric fields were applied, *E. coli* K12 disinfection rates were reported to increase by 40% using P25 and by 80% using Aldrich TiO_2 (Dunlop *et al.*, 2002). A similar publication by Christensen *et al.* (2003), reported the photoelectrocatalytic and photocatalytic disinfection of *E. coli* suspensions by TiO_2 in a sparked photoelectrochemical reactor with "thermal" electrodes (oxidized metallic titanium mesh) and "sol-gel" electrodes (layers of deposited and heated titania gel on titanium mesh). The authors reported that the photoelectrochemical system with "thermal" electrodes was more efficient than the photocatalytic TiO_2 slurry.

Another way to increase the efficiency of TiO_2 coatings is to modify their chemical composition by doping with other elements, e.g., iron, silver, copper. For example, Yu *et al.* (2003), after preparing TiO_2 dip-coatings on stainless steel, noticed a diffusion of iron (Fe^{3+} and Fe^{2+}) ions from the stainless steel substrate into the TiO_2 films during high-temperature calcinations. For this reason, the iron behaved as a dopant and conferred a significantly higher activity to the materials compared with those prepared on glass. In addition, this TiO_2 film showed photoinduced hydrophilicity and could be used for the sterilization of *Bacillus pumilus*. The authors proved that the survival ratio for *B. pumilus* on the TiO_2 film with 3 h calcination under UV illumination decreased to 50% within 2 h.

Yu *et al.* (2005) reported positive results also for doping with non-metals, finding that sulfur-doped TiO_2 nanoparticles showed strong visible-light induced activity to disinfect water containing *Micrococcus lylae* (Gram-positive bacteria).

8.4.5   *Photoreactors for heterogeneous photocatalytic disinfection*

In 1993, Ireland *et al.* built a flow-through photoreactor formed by several photocatalytic modules, each one composed of a UV lamp (300–400 nm) coaxially wrapped with TiO_2 (anatase) coated fiberglass mesh, connected in series. The authors tested recycled dechlorinated tap water to which bacteria cultures were added, achieving very good results for the abatement of *E. coli* and total coliforms in less than 10 min irradiation. In highly colored surface waters of a pond close by to their laboratory, the authors obtained a reduction of viable cells of up to 7 orders of magnitude after 6 min exposure. In another experiment, 25 ($\sim$95 L) of an *E. coli* HB101 suspension was irradiated with a set of a fluorescent lamps and another tungsten halide lamp in a thermostatted glass reactor in the presence of suspended P25. After 30 min irradiation ($\sim$1660 μeinstein s^{-1} m^{-2}) the total inactivation of the cells took place.

Matsunaga *et al.* (1988) constructed a continuous sterilization reactor with a mercury lamp (1800 μeinstein s^{-1} m^{-2}) and TiO_2 powder immobilized on acetylcellulose membranes (see section 8.4.4). With this reactor, an *E. coli* culture was destroyed in 16 min until less of 1% survival.

Vidal *et al.* (1999, 2000) reported the evaluation of a low-cost CPC prototype using TiO_2 photocatalysis under solar light for water disinfection. The pilot plant was composed of a combined flow-through and batch reactor system low-cost CPC prototype. This solar photo-reactor could be tilted at local latitude to maximize the available solar radiation. Results with *E. coli* and *E. faecalis* and TiO_2 suspensions (average solar UV value ca. 25 W m^{-2}) showed a 5-log decrease after 30 min irradiation. The estimated total costs for photocatalytic degradation was estimated to be ca. 0.7 US$ m^{-3} of treated water.

As said before, McLoughlin *et al.* (2004a,b) tested water disinfection using CPCs and other collectors to assess the effectiveness of disinfection under UV light. In all cases, the photoreactors were tested without and with TiO_2 suspensions finding an enhancement only with doses of TiO_2 higher than 3 mg L^{-1}. TiO_2 coated Pyrex rods fixed within the reactors were also used.

A further study proved that the irradiated area in the CPC collector plays a key role in the bacteria inactivation by solar irradiation when a CPC solar photoreactor was used for *E. coli* inactivation under solar photocatalysis with both TiO_2 slurries and TiO_2 deposited on fiberglass (Ahlstrom, 1999) and inserted in a tubular photoreactor (Fernández-Ibáñez *et al.*, 2005). A similar

CPC solar photoreactor was used later to study disinfection of water contaminated by *F. solani* spores (Polo-López *et al.*, 2010; Fernández-Ibáñez *et al.*, 2009).

Sichel *et al.* (2007a) studied the quantification of operating parameters, such as flow rate, water quality and initial bacterial concentration, and osmolarity on bacterial viability in solar photocatalysis for disinfection of *E. coli* K12 in a CPC solar reactor, using a TiO_2 coated paper matrix fixed on a tubular support in the focus of the CPC. Between 10 and 2 L min^{-1}, photocatalytic disinfection was found to be more efficient at lower flow rates, indicating that low-power pumps, of low energy consumption, are convenient and of special interest for rural water disinfection systems. Mechanical inactivation increased with rising flow rates and was notably reduced in saline solution (water with 9 wt.% NaCl). It was found that photocatalytic disinfection of distilled water at bacterial concentrations below 10^5 CFU mL^{-1} is caused entirely by mechano-osmotic effects, which, if not taken into account, might be attributed to solar photocatalysis.

Two projects for developing a cost effective technology based on solar photocatalysis for water decontamination and disinfection in rural areas of developing countries, SOLWATER and AQUACAT, were financed in last years by the European Union International Cooperation, INCO (Malato *et al.*, 2009). Field tests with the final prototypes were carried out to validate operation under real conditions (Herrmann *et al.*, 2002), and similar tests have been performed in photoreactors installed in Argentina (Navntoft *et al.*, 2007), Egypt, France, Greece, Mexico, Morocco, Peru, Spain, Switzerland and Tunisia. The prototypes use a series of CPCs, and the water from a feed tank is pumped through connected illuminated tubes. Electricity is provided by a solar panel and water was recirculated using a centrifugal pump powered by a battery connected to the solar panel.

8.5 FENTON AND PHOTO-FENTON PROCESSES IN WATER DISINFECTION

In Fenton or Fenton-like processes, Fe salts (generally Fe^{2+} and sometimes Fe^{3+}) or other metals are used in conjunction with H_2O_2 to generate $HO^\bullet$:

$$Fe^{2+} + H_2O_2 \rightarrow Fe^{3+} + HO^{\bullet+} + HO^- \tag{8.12}$$

Combination of copper and iron combined with H_2O_2 in the dark have been found to be rather good alternative disinfectants. Although metals alone can inactivate some microorganisms, addition of H_2O_2 enhances the process (Sagripanti, 1992; Sagripanti and Bonifacino, 1997; Cross *et al.*, 2003). Cross *et al.* (2003) reported results for a modified Fenton reaction to deactivate spores of *Bacillus globigii*. This modified Fenton reaction does not use H_2O_2 but involves the conversion of aqueous dissolved oxygen to H_2O_2 by the action of cupric chloride (Cu^{2+}) on ascorbic acid and subsequent generation of $HO^\bullet$.

Fenton processes have to be driven under acid pH (around 3) to maintain iron in solution and to avoid formation of iron flocs. However, at this pH 3, most microorganisms are no longer viable and the use of these processes for disinfection is questionable. On the other hand, in Fenton reaction, H_2O_2 and iron released inside the cells can cause damage to cell functions *in vivo* and *in vitro* by catalyzing the production of ROS and initiation of lipid peroxidation (Malato *et al.*, 2009). Therefore, for less acidic solutions (i.e., natural waters), the use of photo-Fenton reactions might be an excellent treatment for disinfection.

Fenton processes are enhanced under UV irradiation, what can be explained by complex mechanisms (Litter, 2005). Organic ligands forming strong complexes with Fe^{3+} are usually added in Fenton processes to maintain the iron in solution or are present in natural or residual waters; under irradiation, photoreduction of these complexes (eq. 8.13) can take place at longer wavelengths than those of aquo- and hydroxyl ligands and the complexes are more active for the photoreduction:

$$Fe^{3+} - L + h\nu \rightarrow Fe^{2+} + L^\bullet \tag{8.13}$$

These processes help photo-Fenton reactions and improve the use of wavelengths present in the solar spectrum.

As examples of use of Fenton reactions for disinfection, Sjorgren and Sierka (1994), in their studies of photocatalytic inactivation of a MS2 phage at pH 7.2 using P25 (section 8.4.2), added small amounts of $FeSO_4 \cdot H_2O$ ($2\,\mu mol\,L^{-1}$) in one of the experiments, obtaining a significant increase in the disinfection efficiency, attributed to the contribution of Fenton reactions.

Sciacca *et al.* (2010) studied disinfection of surface water of the urban dams in Ouagadougou, Burkina Faso, *via* SODIS treatment in PET bottles, where the waters contained dissolved iron ($0.3\,mg\,L^{-1}$) with pH around 7.5. In these conditions, the use of SODIS alone (6 h of sunlight irradiation) could not avoid regrowth of *Salmonella* sp. after 72 h of dark post-treatment, meanwhile the addition of $10\,mg\,L^{-1}$ H_2O_2 showed a strong enhancement of the inactivation rate without any regrowth of the bacteria; this was attributed to the operation of a homogeneous photo-Fenton reaction, probably enhanced by the presence of NOM in the waters.

Paspaltsis *et al.* (2006) demonstrated the capability of the photo-Fenton reaction to degrade prion proteins (PrP). Photo-Fenton-mediated homogeneous photocatalytic oxidation of 1% w/v scrapie-infected brain tissue with $500\,mg\,mL^{-1}$ Fe^{3+} and $5000\,mg\,mL^{-1}$ H_2O_2 at pH 3 eliminated PrP immunoreactivity in less than 2 min, reducing the initial PrP content at least 50-fold. For the 10% w/v homogenate, PrP was efficiently degraded under the same conditions within 5 h of treatment, indicating the convenience of the treatment even for more concentrated microorganisms.

In laboratory experiments with a UV-Vis solar simulator, Rincón *et al.* (2006) compared the efficiency of Fenton and photo-Fenton processes in comparison with TiO_2 photocatalysis and direct UV, and their combinations, on *E. coli* K12 disinfection. The addition of Fe^{3+} or H_2O_2 to the homogeneous UV-Vis irradiation system increased the disinfection rate. The Fe^{3+} effect was explained by the photoactivity of the aquo-complexes that generate active radical species (mainly $HO^\bullet$) and Fe^{2+}, and furthermore to an increase in Fe^{3+} concentration in the bacteria, enhancing intracellular Fenton and photo-Fenton like reactions. H_2O_2 addition increased the disinfection rate to a lower extent than Fe^{3+}, being its effect associated with an increase in the sensitivity of bacteria to UV-Vis light; the photo-Fenton process ($UV/Fe^{3+}/H_2O_2$) was much more effective, reducing to 50% the time needed for total disinfection with Fe^{3+} alone. The comparison between HP and photo-Fenton showed that HP was more efficient; the combination TiO_2/H_2O_2 improved significantly the efficiency of HP, while no significant improvement was observed from the combination HP/Fe^{3+}. In the dark controls, H_2O_2 showed no effect on *E. coli*, while the combination H_2O_2/Fe^{3+} rendered some degree of disinfection, but in a far lower extent than under irradiation. It was observed that both the photolytic and photocatalytic disinfection was accelerated in natural waters (Leman Lake in Switzerland) in comparison to Milli-Q water, confirming that solar disinfection in the presence and absence of catalyst depends on the chemical matrix of water. These results are promising for practical application with real waters of tropical countries where the surface water contains often iron levels near to $0.3\,mg\,L^{-1}$. Concerning the post-irradiation growth of bacteria stored 24 and 48 h in the dark and after complete culturability loss was achieved, direct UV-Vis could not avoid bacteria growth as soon as 24 h after the treatment, but bacteria growth was not observed 48 h in the dark, after photo-Fenton and HP treatments.

Later, Rincón and Pulgarin (2007a,b) tested field disinfection of natural water of the Leman Lake spiked with *E. coli* K12 in a large solar CPC photoreactor (35–70 L at 30–35°C in Lausanne, Switzerland) by direct sunlight alone, sunlight/TiO_2, sunlight/TiO_2/Fe^{3+}, and sunlight/Fe^{3+}/H_2O_2. Under 5 h solar irradiation, total disinfection could not be reached, together with bacterial recovery during the subsequent 24 h in the dark. The addition of TiO_2, TiO_2/Fe^{3+} or Fe^{3+}/H_2O_2 accelerated the bactericidal action of sunlight, leading to total disinfection. The addition of Fe^{3+} to TiO_2 suspensions enhanced the photocatalytic inactivation. However, the Fe^{3+}/H_2O_2/sunlight process showed a more effective disinfecting action compared to the TiO_2/sunlight and TiO_2/Fe^{3+}/sunlight systems. The fact that a low Fe^{3+} concentration ($0.3\,mg\,L^{-1}$) at natural (not acidic) pH enhanced the solar photocatalytic disinfection is promising for the disinfection of water sources containing Fe^{3+}, as it is often the case in tropical countries. No bacterial regrowth was observed after Fe^{3+} and TiO_2 solar photo-assisted disinfection after 24 h in the dark and, in some samples, the decrease of bacteria continued even in the dark, indicating a residual disinfection effect.

8.6 CONCLUSIONS

Solar disinfection technologies are environmentally friendly technologies useful for improving drinking water quality, especially at the household level. Use of bottles is convenient for treatment in individual houses of small communities, and they are costless materials. CPC photoreactors can be constructed for the disinfection of larger volumes.

The SODIS technology is used widely in many places of the world and in emergencies when pure water is urgently needed. In the technology, the combination of UV and IR radiation leads to the inactivation of several microbiological species, responsible for different serious illnesses. The use of SODIS was found to improve the microbiological quality of household water, with the consequence of an important reduction of cases of diarrhea and other hydric diseases, especially in developing countries.

HP can improve SODIS technology, TiO_2 being a cheap, non-toxic material. TiO_2 disinfection is possible even with microorganisms very resistant to UV-A irradiation, and simple methods for immobilization of the photocatalyst have been tested.

However, for the successful application of disinfection using solar technologies, some considerations should be kept in mind. Natural (wild) microorganisms cannot be considered and treated as molecules because of their larger size and inherent traits of living organisms. Moreover, the influence of environmental conditions in their sensitivity or resistance to stressing agents such as UV radiation should be taken into account. Besides, during light processes, decomposition compounds initially not present in the waters can originate and can be toxic and/or inhibitory to the microorganisms. The presence of inorganic-radical scavengers or organic matter competing with cells by the oxidative photogenerated species can inhibit cell lethality, provoking light-filter effects or even acting as nutrients. Regrowth of bacteria has been observed usually in most of the processes after some days; therefore, prolonged storage of the treated water is not advisable and water should be consumed in the next following days or retreated. Even residual effects of photocatalytic processes (HP or photo-Fenton) are scarce.

ACKNOWLEDGEMENTS

This work was performed as part of Agencia Nacional de Promoción Científica y Tecnológica PICT-512. J.M.M. thanks Consejo Nacional de Investigaciones Científicas y Técnicas (CONICET) for a postdoctoral fellowship.

REFERENCES

Acra, A., Karahagopian, Y., Raffoul, Z. & Dajani, R.: Disinfection of oral rehydration solutions by sunlight. *The Lancet* 316 (1980), pp. 1257–1258.

Acra, A., Jurdi, M., Mu'allem, H., Karahagopian, Y. & Raffoul, Z.: *Water disinfection by solar radiation-assessment and application.* IDRC, Ottawa, Canada, 1990.

Ahlstrom Research and Competence Center: *Photocatalytic composition*, European Patent EP1069950B1. Pont Eveque, France, granted 1999. Available at: http://worldwide.espacenet.com/numberSearch?locale=en_EP (accessed October 2011).

Alrousan, D.M.A., Dunlop, P.S.M., McMurray, T.A. & Byrne, J.A.: Photocatalytic inactivation of *E. coli* in surface water using immobilised nanoparticle TiO_2 films. *Water Res.* 43 (2009), pp. 47–54.

Anathaswamy, H.N. & Eisenstark, A.: Near-UV-induced breaks in phage DNA: sensitization by hydrogen peroxide (a tryptophan photoproduct). *Photochem. Photobiol.* 24 (1977), pp. 439–442.

Armon, R., Laot, N., Narkis, N. & Neeman, I.: Photocatalytic inactivation of different bacteria and bacteriophages in drinking water at different TiO_2 concentration with or without exposure to O_2. *J. Adv. Oxid. Technol.* 3 (1998), pp. 145–150.

Bekbölet, M.: Photocatalytic bactericidal activity of TiO_2 in aqueous suspensions of *E-coli*. *Water Sci. Technol.* 35 (1997), pp. 95–100.

Berney, M., Weilenmann, H.U. & Egli, T.: Flow-cytometric study of vital cellular functions in *Escherichia coli* during solar disinfection (SODIS). *Microbiology* 152 (2006a), pp. 1719–1729.

Berney, M., Weilenmann, H.U., Simonetti, A. & Egli, T.: Efficacy of solar disinfection of *Escherichia coli, Shigella flexneri, Salmonella typhimurium* and *Vibrio cholerae*. *J. Appl. Microbiol.* 101:4 (2006b), pp. 828–836.

Berney, M., Weilenmann, H.U. & Egli, T.: Adaptation to UVA radiation of *E. coli* growing in continuous culture. *J. Photochem. Photobiol.* B 86 (2007), pp. 149–159.

Biguzzi, M. & Shama, G.: Effect of titanium dioxide concentration on the survival of *Pseudomonas stutzeri* during irradiation with near ultraviolet light. *Lett. Appl. Microbiol.* 19 (1994), pp. 458–460.

Blake, D.M., Maness, P.C., Huang, Z., Wolfrum, E.J. & Huang, J.: Application of the photocatalytic chemistry of titanium dioxide to disinfection and the killing of cancer cells. *Sep. Purif. Methods* 28 (1999), pp. 1–50.

Blanco-Gálvez, J. Fernández-Ibáñez, P. & Malato-Rodríguez, S.: Solar photocatalytic detoxification and disinfection of water: recent overview. *J. Solar Energy Eng.* 129 (2007), pp. 4–15.

Blesa, M.A. & Sánchez Cabrero, B. (eds): *Eliminación de contaminantes por fotocatálisis heterogénea. Texto colectivo elaborado por la Red CYTED VIII-G.* CIEMAT, Madrid, Spain, 2004.

Block, S.S. & Goswami, D.W.: Chemically enhanced sunlight for killing bacteria. *Sol. Eng.* 1 (1995), pp. 431–437.

Block, S.S., Seng, V.P. & Goswami, D.W.: Chemically enhanced sunlight for killing bacteria. *J. Sol. Energy Eng.* 119 (1997), pp. 85–91.

Bolton, J.R.: Light compendium-ultraviolet: principles and applications. USEPA – Newsletter 66 (1999), pp. 9–36.

Bosshard, F., Bucheli, M., Meur, Y. & Egli, T.: The respiratory chain is the cell's Achilles' heel during UVA inactivation in *Escherichia coli*. *Microbiology* 156 (2010a), pp. 2006–2015.

Bosshard, F., Riedel, K., Schneider, T., Geiser, C., Bucheli, M. & Egli, T.: Protein oxidation and aggregation in UVA-irradiated *Escherichia coli* cells as signs of accelerated cellular senescence. *Environ. Microbiol.* 12 (2010b), pp. 2931–2945.

Boyle, M., Sichel, C., Fernández-Ibáñez, P., Arias-Quiroz, G.B., Iriarte-Puña, M., Mercado, A., Ubomba-Jaswa, E. & McGuigan, K.G.: Bactericidal effect of solar water disinfection under real sunlight conditions. *Appl. Environ. Microbiol.* 74 (2008), pp. 2997–3001.

Butterfield, I.M., Christensen, P.A, Curtis, T.P. & Gunlazuardi, J.: Water disinfection using an immobilised titanium dioxide film in a photochemical reactor with electric field enhancement. *Water Res.* 31 (1997), pp. 675–677.

Caldeira de Araujo, A. & Favre, A.: Induction of size reduction in *Escherichia coli* by near-ultraviolet light. *Eur. J. Biochem.* 146 (1985), pp. 605–610.

CARE/CDC Health Initiative, the Estes Park Rotary Club and the Gangarosa International Health Foundation Safe Water Systems for the Developing World: *A handbook for implementing household-based water treatment and safe storage projects.* Department of Health & Human Services Centers for Disease Control and Prevention.Centers for Disease Control and Prevention, Atlanta, USA, 2010; www.cdc.gov/safewater/manual/sws_manual.pdf (accessed April 2011).

Caslake, L.F., Connolly, D.J., Menon, V., Duncanson, C.M., Rojas, R. & Tavakoli, A.: Disinfection of contaminated water by using solar irradiation. *Appl. Environ. Microbiol.* 70 (2004), pp. 1145–1151.

Cho, I.-H., Moon, I.-Y., Chung, M.-H., Lee, H.-K. & Zoh, K.-D.: Disinfection effects on *E. coli* using TiO_2/UV and solar light system. *Water Sci. Technol.: Water Supply* 2 (2002), pp. 181–190.

Christensen, P.A., Curtis, T.P., Egerton, T.A., Kosa, S.A.M. & Tinlin, J.R.: Photoelectrocatalytic and photocatalytic disinfection of *E. coli* suspensions by titanium dioxide. *Appl. Catal.* B: *Environ.* 41 (2003), pp. 371–386.

Colwell, R.R.: Viable but nonculturable bacteria: a survival strategy. *J. Infect. Chemother.* 6 (2000), pp. 121–125.

Cooper, A.T., Goswami, D.Y. & Block, S.S.: Solar photochemical detoxification and disinfection for water treatment in tropical developing countries. *J. Adv. Oxid. Technol.* 3 (1998), pp. 151–154.

Cross, J.B., Currier, R.P., Torraco, D.J., Vanderberg, L.A., Wagner, G.L. & Gladen, P.D.: Killing of *Bacillus* spores by aqueous dissolved oxygen, ascorbic acid, and copper ions. *Appl. Environ. Microbiol.* 69 (2003), pp. 2245–2252.

Curtis, T.P., Alker, G.W., Dowling, B.M. & Christensen, P.A.: Fate of *Cryptosporidium* oocysts in an immobilised titanium dioxide reactor with electric field enhancement. *Water Res.* 36 (2002), pp. 2410–2413.

Dillert, R., Siemon, U. & Bahnemann, D.: Photocatalytic disinfection of municipal wastewater. *Chem. Eng. Technol.* 21 (1998), pp. 356–358.

Dillert, R., Siemon U. & Bahnemann, D.: Photocatalytic disinfection of municipal wastewater. *J. Adv. Oxid. Technol.* 4 (1999), pp. 55–59.

Downes, A. & Blunt, T.P.: Researches on the effect of light upon bacteria and other organisms. *Proc. R. Soc. London* 26 (1877), pp. 488–500.

Drosou, C., Coz, A., Xekoukoulotakis, N.P., Moya, A., Vergara, Y. & Mantzavinos, D.: Peracetic acid-enhanced photocatalytic and sonophotocatalytic inactivation of *E. coli* in aqueous suspensions. *J. Chem. Technol. Biotechnol.* 85 (2009), pp. 1049–1053.

Dunlop, P.S.M., Byrne, J.A., Manga, N. & Eggins, B.R.: The photocatalytic removal of bacterial pollutants from drinking water. *J. Photochem. Photobiol. A: Chem.* 148 (2002), pp. 355–363.

EAWAG (Swiss Federal Institute for Environmental Science and Technology)/SANDEC (Water & Sanitation in Developing Countries): SODIS solar water disinfection; http://www.sandec.ch (accessed October 2008).

FAO (Food and Agriculture Organization): AQUASTAT: FAO's Information System on Water and Agriculture, NRL, Rome, Italy, 2008; http://www.fao.org/nr/water/aquastat/main/index.stm (accessed October 2011).

Favre, A., Hajnsdorf, E., Thiam, K. & Caldeira de Araujo, A.: Mutagenesis and growth delay induced in *Escherichia coli* by near-ultraviolet radiations. *Biochimie* 67 (1985), pp. 335–342.

Fernández, R.O. & Pizarro, R.A.: Lethal effect induced in *Pseudomonas aeruginosa* exposed to ultraviolet-A radiation. *Photochem. Photobiol.* 64 (1996), pp. 334–339.

Fernández-Ibáñez, P., Blanco, J., Sichel, C. & Malato, S.: Water disinfection by solar photocatalysis using compound parabolic collectors. *Catal. Today* 101 (2005), pp. 345–352.

Fernández-Ibáñez, P., Sichel, C., Polo-López, M.I., de Cara-García, M. & Tello, J.C.: Photocatalytic disinfection of natural well water contaminated with *Fusarium solani* using TiO_2 slurry in solar CPC photo-reactors. *Catal. Today* 144 (2009), pp. 62–68.

Fisher, M.B., Keenan, C.R., Nelson, K.L. & Voelker, B.M.: Speeding up solar disinfection (SODIS): effects of hydrogen peroxide, temperature, pH, and copper plus ascorbate on the photoinactivation of *E. coli*. *J. Water Health* 6 (2008), pp. 35–51.

Fujishima, A. & Zhang, X.: Titanium dioxide photocatalysis: present situation and future approaches. *C.R. Chimie* 9 (2006), pp. 750–760.

Gelover, S., Gómez, L.A., Reyes, K. & Leal, M.T.: A practical demonstration of water disinfection using TiO_2 films and sunlight. *Water Res.* 40 (2006), pp. 3274–3280.

Gill, L.W. & McLoughlin, O.A.: Solar disinfection kinetic design parameters for the continuous flow reactors. *J. Sol. Energy Eng.* 129 (2007), pp. 111–118.

Gómez-Couso, H., Fontán-Sainza, M., McGuigan, K.G. & Ares-Mazás, E.: Effect of the radiation intensity, water turbidity and exposure time on the survival of *Cryptosporidium* during simulated solar disinfection of drinking water. *Acta Tropica* 112 (2009), pp. 43–48.

Goswami, D.Y., Trivedi, D.M. & Block, S.S.: Photocatalytic disinfection of indoor air. *J. Sol. Energy Eng.* 119 (1997), pp. 92–96.

Goswami, T.K., Hingorani, S.K. Greist, H., Goswami, D.Y. & Block, S.S.: Photocatalytic system to destroy bioaerosols in air. *J. Adv. Oxid. Technol.* 4 (1999), pp. 185–188.

Guimarães, J.R., Ibáñez, J., Litter, M.I. & Pizarro, R.: Desinfección de agua. In M.A. Blesa & B. Sánchez Cabrero (eds): *Eliminación de contaminantes por fotocatálisis heterogénea. Texto colectivo elaborado por la Red CYTED VIII-G.* CIEMAT, Madrid, Spain, 2004, pp. 305–316.

Gumy, D., Rincón, A.G., Hajdu, R. & Pulgarin, C.: Solar photocatalysis for detoxification and disinfection of water: Different types of suspended and fixed TiO_2 catalysts study. *Solar Energy* 80 (2006), pp. 1376–1381.

Hartman, P.S. & Eisenstark, A.: Killing of *Escherichia coli* K-12 by near-ultraviolet radiation plus hydrogen peroxide: role of double-strand DNA breaks in the absence of recombinational repair. *Mutat. Res.* 72 (1980), pp. 31–42.

Heaselgrave, W., Patel, N., Kilvington, S., Kehoe, S.C. & McGuigan, K.G.: Solar disinfection of poliovirus and *Acanthamoeba polyphaga* cysts in water – a laboratory study using simulated sunlight. *Lett. Appl. Microbiol.* 43 (2006), pp. 125–130.

Herrera Melián, J.A., Doña Rodríguez, J.M., Vieira Suárez, A., Tello Rendón, E., Valdés do Campo, C., Arana, J. & Pérez Peña, J.: The photocatalytic disinfection of urban waste waters. *Chemosphere* 41 (2000), pp. 323–327.

Herrmann, J.M., Guillard, Ch., Disdier, J., Lehaut, C., Malato, S. & Blanco, J.: New industrial titania photocatalysts for the solar detoxification of water containing various pollutants. *Appl. Catal.* B: *Environ.* 35 (2002), pp. 281–294.

Hoerter, J.D., Arnold, A.A., Kuczynska, D.A., Shibuya, A., Ward, C.S., Sauer, M.G., Gizachew, A., Hotchkiss, T.M., Fleming, T.J. & Johnson, S.: Effects of sublethal UVA irradiation on activity levels of oxidative defense enzymes and protein oxidation in *Escherichia coli*. *J. Photochem. Photobiol.* B 81 (2005), pp. 171–180.

Hoffmann, M.R., Martin, S.T., Choi, W. & Bahnemann, D.W.: Environmental applications of semiconductor photocatalysis. *Chem. Rev.* 95 (1995), pp. 69–96.

Huang, Z., Maness, P., Blake, D.M., Wolfrum, E.J., Smolinski, S.L. & Jacoby, W.A.: Bactericidal mode of titanium dioxide photocatalysis. *J. Photochem. Photobiol.* A 130 (2000), pp. 163–170.

Ibáñez, J.A., Litter, M.I. & Pizarro, R.A.: Photocatalytic bactericidal effect of TiO_2 on *Enterobacter cloacae*. Comparative study with other Gram (–) bacteria. *J. Photochem. Photobiol.* A 157 (2003), pp. 81–85.

Ireland, J.C., Klostermann, P., Rice, E.W. & Clark, R.M.: Inactivation of *Eschericchia coli* by titanium dioxide photocatalytic oxidation. *Appl. Environ. Microbiol.* 59 (1993), pp. 1668–1670.

Jacoby, W.J., Maness, P.C., Wolfrum, E.J., Blake, D.M. & Fennell, J.A.: Mineralization of bacterial cell mass on a photocatalytic surface in air. *Environ. Sci. Technol.* 32 (1998), pp. 2650–2653.

Jagger, J.: Near-UV radiation effects on microorganisms. *Photochem. Photobiol.* 34 (1981), pp. 761–768.

Kehoe, S.C., Barer, M.R., Devlin, L.O. & McGuigan, K.C.: Batch process solar disinfection is an efficient jeans of disinfecting drinking water contaminated with *Shigella dysenteriae* type I. *Lett. Appl. Microbiol.* 38 (2004), pp. 410–414.

Kikuchi, Y., Sunada, K., Iyoda, T., Hashimoto, K. & Fujishima, A.: Photocatalytic bactericidal effect of TiO_2 thin films: dynamic view of the active oxygen species responsible for the effect. *J. Photochem. Photobiol.* A 106 (1997), pp. 51–56.

Kiwi, J. & Nadtochenko, V.: New evidence for TiO_2 photocatalysis during bilayer lipid peroxidation. *J. Phys. Chem.* B 108 (2004), pp. 17, 675–17, 684.

Kiwi, J. & Nadtochenko, V.: Evidence for the mechanism of photocatalytic degradation of the bacterial wall membrane at the TiO_2 interface by ATR-FTIR and laser kinetic spectroscopy. *Langmuir* 21 (2005), pp. 4631–4641.

Kohler, M. & Wolfensberger, M.: Migration of organic components from polyethylene terephthalate (PET) bottles to water. Swiss Federal Institute for Materials Testing and Research (EMPA), Report 429670 (2003), pp. 1–13; http://www.sodis.ch/methode/forschung/publikationen/papers/kohler_pet_2003.pdf (accessed April 2011).

Laot, N., Narkis, N., Neeman, I., Vilanovic, D. & Armon, R.: TiO_2 photocatalytic inactivation of selected microorganisms under various conditions. Sunlight, intermittent and variable irradiation intensity, CdS augmentation and entrapment of TiO_2 into sol-gel. *J. Adv. Oxid. Technol.* 4 (1999), pp. 97–102.

Lee, S.H., Pumprueg, S., Moudgil, B. & Sigmund, W.: Inactivation of bacterial endospores by photocatalytic nanocomposites. *Colloids Surf.* B: *Biointerfaces* 10 (2005), pp. 93–98.

Linsebigler, A.L., Lu, G. & Yates, J.T. Jr.: Photocatalysis on TiO_2 surfaces: principles, mechanisms, and selected results. *Chem. Rev.* 95 (1995), pp. 735–758.

Litter, M.I.: Heterogeneous photocatalysis: Transition metal ions in photocatalytic systems. *Appl. Catal.* B 23 (1999), pp. 89–114.

Litter, M.I. (ed.): *Prospect of rural Latin American communities for application of low-cost technologies for water potabilization, OAS Project AE141/2001.* Digital Grafic, La Plata, Argentina, 2002; http://www.cnea.gov.ar/xxi/ambiental/agua-pura/default.htm (accessed October 2011).

Litter, M.I.: Introduction to photochemical advanced oxidation processes for water treatment. In: P. Boule, D.W. Bahnemann, P.K.J. Robertson (eds): *The handbook of environmental chemistry*, Vol. 2, Part M, *Environmental photochemistry* Part II. Springer-Verlag, Berlin, Heidelberg, Germany, 2005, pp. 325–366.

Litter, M.I. (ed.): *Final results of the OAS/AE/141 Project: research, development, validation and application of solar technologies for water potabilization in isolated rural zones of Latin America and the Caribbean,* OAS Project AE141.OEA, Buenos Aires, Argentina, 2006; http://www.cnea.gov.ar/xxi/ambiental/agua-pura/default.htm (accessed October 2011).

Litter, M.I.: Treatment of chromium, mercury, lead, uranium and arsenic in water by heterogeneous photocatalysis. In: H. De Lasa & B. Serrano (eds): *Adv. Chem. Eng.* 36 —*Photocatalytic Technologies*, 2009, pp. 37–67.

Litter, M.I. & Jiménez González, A. (eds): *Advances in low-cost technologies for disinfection, decontamination and arsenic removal in waters from rural communities of Latin America (HP and SORAS methods)*, *OAS Project AE141*. Digital Grafic, La Plata, Argentina, 2004; http://www.cnea.gov.ar/xxi/ambiental/agua-pura/default.htm (accessed October 2011).

Lonnen, J., Kilvington, S., Kehoe, S.C., Al-Touati, F. & McGuigan, K.G.: Solar and photocatalytic disinfection of protozoan, fungal and bacterial microbes in drinking water. *Water Res.* 39 (2005), pp. 877–883.

Maisonneuve, E., Ezraty, B. & Dukan, S.: Protein aggregates: an aging factor involved in cell death. *J. Bacteriol.* 190 (2008), pp. 6070–6075.

Malato, S., Blanco, J., Alarcón, D.C., Maldonado, M.I., Fernández-Ibáñez, P. & Gernjak, W.: Photocatalytic decontamination and disinfection of water with solar collectors. *Catal. Today* 122 (2007), pp. 137–149.

Malato, S., Fernández-Ibáñez, P., Maldonado, M.I., Blanco, J. & Gernjak, W.: Decontamination and disinfection of water by solar photocatalysis: Recent overview and trends. *Catal. Today* 147 (2009), pp. 1–59.

Maness, P.C., Smolinski, S., Blake, D.M., Huang, Z., Wolfrum, E.J. & Jacoby, W.A.: Bactericidal activity of photocatalytic TiO_2 reaction: towards an understanding of its killing mechanism. *Appl. Environ. Microbiol.* 65 (1999), pp. 4094–4098.

Martin, S.T., Herrmann, H. & Hoffmann, M.R.: Time-resolved microwave conductivity. Part 2—Quantum-sized TiO_2 and the effect of adsorbates and light intensity on charge-carrier dynamics. *J. Chem. Soc., Faraday Trans.* 90 (1994), pp. 3323–3330.

Martín-Domínguez, A., Alarcón-Herrera, M.T., Martín-Domínguez, I.R. & González-Herrera, A.: Efficiency in the disinfection of water for human consumption in rural communities using solar radiation. *Solar Energy* 78 (2005), pp. 31–40.

Matsunaga, T. & Okochi, M.: TiO_2-mediated photochemical disinfection of *Escherichia coli* using optical fibers. *Environ. Sci. Technol.* 29 (1995), pp. 501–505.

Matsunaga, T., Tomoda, R., Nakajima, T. & Wake, H.: Photoelectrochemical sterilization of microbial cells by semiconductor powders. *FEMS Microbiol. Lett.* 29 (1985), pp. 211–214.

Matsunaga, T., Tomoda, R., Nakajima, T., Nakamura, N. & Komine, T.: Continuous-sterilization system that uses photosemiconductor powders. *Appl. Environ. Microbiol.* 54 (1988), pp. 1330–1333.

McGuigan, K.G., Joyce, T.M., Conroy, R.M., Gillespie, J.B. & Elmore-Meegan, M.: Solar disinfection of drinking water contained in transparent plastic bottles: characterizing the bacterial inactivation process. *J. Appl. Microbiol.* 84 (1998), pp. 1138–1148.

McGuigan, K.G., Joyce, T.M. & Conroy, R.M.: Solar disinfection: use of sunlight to decontaminate drinking water in developing countries. *J. Med. Microbiol.* 48 (1999), pp. 785–787.

McGuigan, K.G., Méndez-Hermida, F.,. Castro-Hermida, J.A , Ares-Mazas, E., Kehoe, S.C., Boyle, M., Sichel, C., Fernandez-Ibanez, P., Meyer, B.P., Ramalingham, S. & Meyer, E.A.: Batch solar disinfection inactivates oocysts of *Cryptosporidium parvum* and cysts of *Giardia muris* in drinking water. *J. Appl. Microbiol.* 101 (2006), pp. 453–463.

McLoughlin, O.A., Fernández-Ibáñez, P., Gernjak, W., Malato, S. & Gill, L.W.: Photocatalytic disinfection of water using low cost compound parabolic collectors. *Solar Energy* 77 (2004a), pp. 625–634.

McLoughlin, O.A., Kehoe, S.C., McGuigan, K.G., Duffy, E.F., Al Touati, F., Gernjak, W. Oller Alberola, I., Malato Rodríguez, S. & Gill, L.W.: Solar disinfection of contaminated water: a comparison of three small-scale reactors. *Solar Energy* 77 (2004b), pp. 657–664.

Meichtry, J.M., Lin, H., de la Fuente, L., Levy, I.K., Gautier, E.A., Blesa, M.A. & Litter, M.I.: Low-cost TiO_2 photocatalytic technology for water potabilization in plastic bottles for isolated regions. Photocatalyst fixation. *J. Solar Energy Eng.* 129 (2007), pp. 119–126.

Meierhofer, R.: Establishing solar water disinfection as a water treatment method at household level. *Madagascar Conservation & Development* 1 (2006), pp. 25–30; http://journalmcd.com/index.php/mcd/article/view/263/217 (accessed September 2011).

Meierhofer, R., Wegelin, M., Torres, X.R., Gremion, B., Mercado, A., Mäusezahl, D., Hobbins, M., Indergand-Echeverria, S., Grima, B. & Aristanti, C.: *Solar water disinfection: a guide for the application of SODIS*. Department of Water and Sanitation in Developing Countries (SANDEC), Dübendorf, Switzerland, 2002, http://www.sodis.ch/methode/anwendung/ausbildungsmaterial/dokumente_material/manual_e.pdf (accessed October 2011).

Méndez Hermida, F., Castro, J.A., Ares, E., Kehoe, S.C. & McGuigan, K.G.: Effect of batch-process solar disinfection on survival of *cryptosporidium parvum* oocysts in drinking water. *Appl. Environ. Microbiol.* 71 (2005), pp. 1653–1654.

Méndez-Hermida, F., Ares-Mazás, E., McGuigan, K.G., Boyle, M., Sichel, C. & Fernández-Ibáñez, P.: Disinfection of drinking water contaminated with *Cryptosporidium parvum* oocysts under natural sunlight and using the photocatalyst TiO_2. *J. Photochem. Photobiol.* B 88 (2007), pp. 105–111.

Merwald, H., Klosner, G., Kokesch, C., Der-Petrossian, M., Hönigsmann, H. & Trautinger, F.: UVA-induced oxidative damage and cytotoxicity depend on the mode of exposure. *J. Photochem. Photobiol.* B 79 (2005), pp. 197–207.

Mills, A. & Le Hunte, S.: An overview of semiconductor photocatalysis. *J. Photochem. Photobiol.* A 108 (1997), pp. 1–35.

Morioka, T., Saito, T., Nara, Y. & Onoda, K.: Antibacterial action of powdered semiconductor on a serotype g *Streptococcus mutans*. *Caries Res.* 22 (1988), pp. 230–231.

Nadtochenko, V.A., Rincón, A.G., Stanca, S.E. & Kiwi, J.: Dynamics of *E. coli* membrane cell peroxidation during TiO_2 photocatalysis studied by ATR-FTIR spectroscopy and AFM microscopy. *J. Photochem. Photobiol.* A 169 (2004), pp. 131–137.

Nagame, S., Oku, T., Kambara, M. & Konishi, K.: Antibacterial effects of the powdered semiconductor TiO_2 on the viability of oral micro-organisms. *J. Dental Res.* 68 (1989), pp. 1696–1697.

Nakagawa, Y., Wakuri, S., Sakamoto, K. & Tanaka, N.: The photogenotoxicity of titanium dioxide particles. *Mutation Res.* 394 (1997), pp. 125–132.

Nathan, J.S. & Philip, L.: Leaching of DEHA and DEHP from PET bottles to water. Indian Institute of Technology, Chennai, India, 2009; http://www.sodis.ch/news_documents/deha_dehp_indien.pdf (accessed October 2011).

Navntoft, C., Araujo, P., Litter, M.I., Apella, M.C., Fernández, D., Puchulu, M.E., Hidalgo, M.V. & Blesa, M.A.: Field tests of the solar water detoxification SOLWATER reactor in Los Pereyra, Tucumán, Argentina. *J. Solar Energy Eng.* 129 (2007), pp. 127–134.

Navntoft, C., Ubomba-Jaswa, E., McGuigan, K.G. & Fernández-Ibáñez, P.: Effectiveness of solar disinfection using batch reactors with non-imaging aluminium reflectors under real conditions: natural well water and solar light. *J. Photochem. Photobiol.* B 93 (2008), pp. 155–161.

Nawrocki, J., Dabrowska, A. & Borcz, A.: Investigation of carbonyl compounds in bottled waters from Poland. *Water Res.* 36 (2002), pp. 4893–4901.

Oppezzo, O.J. & Pizarro, R.A.: Sublethal effects of ultraviolet-A radiation on *Enterobacter cloacae*. *J. Photochem. Photobiol.* B 62 (2001), pp. 158–165.

Oppezzo, O.J., Costa, C.S. & Pizarro, R.A.: Influence of rpoS mutations on the response of *Salmonella enterica* serovar Typhimurium to solar radiation. *J. Photochem. Photobiol.* B 102 (2011), pp. 20–25.

Paspaltsis, I., Kotta, K., Lagoudaki, R., Grigoriadis, N., Poulios, I. & Sklaviadis, T.: Titanium dioxide photocatalytic inactivation of prions. *J. Gen. Virol.* 87 (2006), pp. 3125–3130.

Pham, H.N., Mc Dowell, T. & Wilkins, E.: Photocatalytically-mediated disinfection of water using TiO_2 as a catalyst and spore-forming *Bacillus pumilus* as a model. *J. Environ. Sci. Health* 3 (1995), pp. 627–636.

Pizarro, R.A.: UV-A oxidative damage modified by environmental conditions in *Escherichia coli*. *Int. J. Radiat. Biol.* 68 (1995), pp. 293–299.

Pizarro, R.A. & Orce, L.V.: Membrane damage and recovery associated with growth delay induced by near-UV radiation in *Escherichia coli* K–12. *Photochem. Photobiol.* 47 (1988), pp. 391–397.

Polo-López, M.I., Fernández-Ibáñez, P., García-Fernández, I., Oller, I., Salgado-Tránsito, I. & Sichel, C.: Resistance of *Fusarium* sp. spores to solar TiO_2 photocatalysis: influence of spore type and water (scaling-up results). *J. Chem. Tech. & Biotech.* 85 (2010), pp. 1038–1048.

Rijal, G.K. & Fujioka, R.S.: Use of reflectors to enhance the synergistic effects of solar heating and solar wavelengths to disinfect drinking water sources. *Water Sci. Technol.* 48 (2003), pp. 481–488.

Rincón, A.G. & Pulgarin, C.: Photocatalytical inactivation of *E. coli*: effect of (continuous–intermittent) light intensity and of (suspended–fixed) TiO_2 concentration. *Appl. Catal.* B: *Environ.* 44 (2003), pp. 263–284.

Rincón, A.G. & Pulgarin, C.: Field solar *E. coli* inactivation in the absence and presence of TiO_2: is UV solar dose an appropriate parameter for standardization of water solar disinfection? *Solar Energy* 77 (2004a), pp. 635–648.

Rincón, A.G. & Pulgarin, C.: Bactericidal action of illuminated TiO_2 on pure *E. coli* and natural bacteria consortia: post-irradiation events in the dark assessment of the effective disinfection time. *Appl. Catal.* B: *Environ.* 49 (2004b), pp. 99–112.

Rincón, A.G. & Pulgarin, C.: Use of coaxial photocatalytic reactor (CAPHORE) in the TiO_2 photo-assisted treatment of mixed *E. coli* and *Bacillus* sp. and bacterial community present in wastewater. *Catal. Today* 101 (2005), pp. 331–344.

Rincón, A.G. & Pulgarin, C.: Comparative evaluation of Fe^{3+} and TiO_2 photoassisted processes in solar photocatalytic disinfection of water. *Appl. Catal.* B: *Environ.* 63 (2006), pp. 222–231.

Rincón, A.G. & Pulgarin, C.: Fe^{3+} and TiO_2 solar-light-assisted inactivation of *E. coli* at field scale: implications in solar disinfection at low temperature of large quantities of water. *Catal. Today* 122 (2007a), pp. 128–136.

Rincón, A.G. & Pulgarin, C.: Absence of *E. coli* regrowth after Fe^{3+} and TiO_2 solar photoassisted disinfection of water in CPC solar photoreactor. *Catal. Today* 124 (2007b), pp. 204–214.

Sagripanti, J.L.: Metal-based formulations with high microbicidal activity. *J. Appl. Environ. Microbiol.* 58 (1992), pp. 3157–3162.

Sagripanti, J. & Bonifacino, A.: Effects of salt and serum on the sporicidal activity of liquid disinfectants. *J. AOAC Int.* 80 (1997), pp. 1198–1207.

Saito, T., Iwase, T., Horie, J. & Morioka, T.: Mode of photocatalytic bactericidal action of powdered semiconductor TiO_2 on mutans streptococci. *J. Photochem. Photobiol.* B 14 (1992), pp. 369–379.

Salih, F.M.: Enhancement of solar inactivation of *Escherichia coli* by titanium dioxide photocatalytic oxidation. *J. Appl. Microbiol.* 92 (2002), pp. 920–926.

Schmidt, P., Kohler, M., Meierhofer, R., Luzi, S. & Wegelin, M.: Does the reuse of PET bottles during solar water disinfection pose a health risk due to the migration of plasticisers and other chemicals into the water? *Water Res.* 42 (2008), pp. 5054–5060.

Sciacca, F., Rengifo-Herrera, J.A., Wéthé, J. & Pulgarin, C.: Dramatic enhancement of solar disinfection (SODIS) of wild *Salmonella* sp. in PET bottles by H_2O_2 addition on natural water of Burkina Faso containing dissolved iron. *Chemosphere* 78:9 (2010), pp. 1186–1191.

Seven, O., Dindar, B., Aydemir, S., Metin, D., Ozinel, M.A. & Icli, S.: Solar photocatalytic disinfection of a group of bacteria and fungi aqueous suspensions with TiO_2, ZnO and Sahara desert dust. *J. Photochem. Photobiol.* A 165 (2004), pp. 103–107.

Shotyk, W., Krachler, M. & Chen, B.: Contamination of Canadian and European bottled waters with antimony from PET containers. *J. Environ. Monit.* 8 (2006), pp. 288–292.

Sichel, C., Fernández-Ibáñez, P., Blanco, J. & Malato, S.: Effects of experimental conditions on *E. coli* survival during solar photocatalytic water disinfection. *J. Photochem. Photobiol.* A 189 (2007a), pp. 239–246.

Sichel, C., Tello, J., de Cara, M. & Fernández-Ibáñez, P.: Effect of UV-intensity and dose on the photocatalytic disinfection of bacteria and fungi under natural sunlight. *Catal. Today* 129 (2007b), pp. 152–160.

Sichel, C., Fernández-Ibáñez, P., Tello, J. & de Cara, M.: Lethal synergy of solar UV-radiation and H_2O_2 on wild *Fusarium solani* spores in distilled and natural well water. *Water Res.* 43 (2009), pp. 1841–1850.

Sinton, L.W., Davies-Colley, R.J. & Bell, R.G.: Inactivation of enterococci and fecal coliforms from sewage and meatworks effluents in seawater chambers. *Appl. Environ. Microbiol.* 60 (1994), pp. 2040–2048.

Sjogren, J.C. & Sierka, R.A.: Inactivation of phage MS2 by iron-aided titanium dioxide photocatalysis. *Appl. Environ. Microbiol.* 60 (1994), pp. 344–347.

Skinner, B. & Shaw, R.: Household water treatment 1 & 2, technical briefs #58 & #59. *Waterlines* Loughborough University, Loughborough, England 1998, 1999; http://www.lboro.ac.uk/well/resources/technical-briefs/58-household-water-treatment-1.pdf (accessed October 2011).

Smith, R.J., Kehoe, S.C., McGuigan, K.G. & Barer, M.R.: Effects of simulated solar disinfection on infectivity of *Salmonella typhimurium*. *Lett. Appl. Microbiol.* 31 (2000), pp. 284–288.

Sunada, K., Kikuchi, Y., Hashimoto, K. & Fujishima, A.: Bactericidal and detoxification effects of TiO_2 thin film photocatalyst. *Environ. Sci. Technol.* 32 (1998), pp. 726–728.

Sunada, K., Watanabe, T. & Hashimoto, K.: Studies on photokilling bacteria on TiO_2 thin film. *J. Photochem. Photobiol.* A 156 (2003), pp. 227–233.

Third World Academy of Sciences: *Safe drinking water, the need, the problem, solutions and an action plan.* Third World Academy of Sciences, Trieste, Italy, 2002; http://twas.ictp.it/publications/twas-reports/safedrinkingwater.pdf/at_download/file (accessed October 2011).

Ubomba-Jaswa, E., Navntoft, C., Polo-López, M.I., Fernández-Ibáñez, P. & McGuigan, K.G.: Solar disinfection of drinking water (SODIS): An investigation of the effect of UVA dose on inactivation efficiency. *Photochem. Photobiol. Sci.* 8 (2009), pp. 587–595.

Ubomba-Jaswa, E., Fernández-Ibáñez, P., Navntoft, C., Polo-López, M.I. & McGuigan, K.G.: Investigating the microbial inactivation efficiency of a 25 L batch solar disinfection (SODIS) reactor enhanced with a compound parabolic collector (CPC) for household use. *J. Chem. Technol Biotechnol.* 85 (2010a), pp. 1028–1037.

Ubomba-Jaswa, E., Fernández-Ibáñez, P. & McGuigan, K.G.: A preliminary Ames fluctuation assay assessment of the genotoxicity of drinking water that has been solar disinfected in polyethylene terephthalate (PET) bottles. *J. Water Health* 8 (2010b), pp. 712–719.

United Nations: *Substantive issues arising in the implementation of the General Comment No. 15 (2002), The right to water (arts. 11 and 12 of the International Covenant on Economic, Social and Cultural Rights),*

E/C.12/2002/11; http://www.unhchr.ch/tbs/doc.nsf/0/a5458d1d1bbd713fc1256cc400389e94/$FILE/ G0340229.pdf (accessed October 2011).

U.S. Census Bureau: World POPClock Projection; http://www.census.gov/ipc/www/popclockworld.html (accessed April 2011).

Vidal, A., Díaz, A.I., El Hraiki, A., Romero, M., Muguruza, I., Senhaji, F. & González, J.: Solar photocatalysis for detoxification and disinfection of contaminated water: pilot plant studies. *Catal. Today* 54 (1999), pp. 283–290.

Vidal, A. & Díaz, A.J.: High-performance, low-cost solar collectors for disinfection of contaminated water. *Water Environ. Res.* 72 (2000), pp. 271–276.

Watts, R., Kong, S., Orr, M.P., Miller, G.C. & Henry, B.Y.: Photocatalytic inactivation of coliform bacteria and viruses in secondary waste-water effluent. *Water Res.* 29 (1995), pp. 95–100.

Webb, R.B.: Lethal and mutagenic effects of nearultraviolet radiation. In K.C. Smith (ed.): *Photochem. Photobiol. Rev.* 2, New York, 1977, pp. 169–268.

Wegelin, M., Canonica, S., Mechsner, K., Fleischmann, T., Pesaro, F. & Metzler, A.: Solar water disinfection: scope of the process and analysis of radiation experiments. *J. Water SRT-Aqua* 43:3 (1994), pp. 154–169.

Wegelin, M. & Sommer, B.: Solar water disinfection (SODIS) — Destined for worldwide use? *Waterlines* 16 (1998), pp. 30–32.

Wegelin, M., Gechter, D., Hug, S., Mahmud, A. & Motaleb, A.: SORAS—a simple arsenic removal process. *Water, Sanitation, Hygiene: Challenges of the Millennium. 26th WEDC Conference*, Dhaka, 5–9 November 2000, Dhaka, Bangladesh, 2000, pp. 379–382; http://www.fischer.eawag.ch/ organisation/abteilungen/sandec/publikationen/publications_ws/downloads_ws/SORAS.pdf (accessed October 2011).

Wegelin, M., Canonica, S., Alder, A.C., Marazuela, D., Suter, M.J.-F., Bucheli, T.D., Haefliger, O.P., Zenobi, R., McGuigan, K.G., Kelly, M.T., Ibrahim, P. & Larroque, M.: Does sunlight change the material and content of polyethylene terephthalate (PET) bottles? *J. Water SRT-Aqua* 50:3 (2001), pp. 125–133.

Wei, Ch., Lin, W.-Y., Azinal, Z., Williams, N.E., Zhu, K., Kruzic, A.P., Smith, R.L. & Rajeshwar, K.: Bactericial activity of TiO_2 photocatalyst in aqueous media: toward a solar-assisted water disinfection system. *Environ. Sci. Technol.* 28 (1994), pp. 934–938.

White, G.C. (ed): *Handbook of chlorination and alternative disinfectants*. 4th edn., Wiley-Interscience Publication, John Wiley & Sons, NJ, 1999.

WHO (World Health Organization): *Guidelines for drinking-water quality*. First addendum to third edition. Vol. 1: recommendations. WHO Library Cataloguing-in-Publication Data, Geneva, Switzerland, 2006; http://www.who.int/water_sanitation_health/dwq/gdwq0506. pdf (accessed October 2011).

WHO (World Health Organization): *The world health report 2007: a safer future*. WHO Library Cataloguing-in-Publication Data, Geneva, Switzerland, 2007; http://www.who.int/whr/2007/whr07_ en.pdf (accessed April 2011).

WHO (World Health Organization): *Guidelines for drinking-water quality, third edition incorporating the first and second addenda*. WHO Library Cataloguing-in-Publication Data, Geneva, Switzerland, 2008; http://www.who.int/water_sanitation_health/dwq/fulltext.pdf (accessed October 2011).

WHO (World Health Organization): *Household water treatment and safe storage*. Internet site: http:// www.who.int/household_water/research/technologies_intro/en/index.html (accessed October 2011a).

WHO (World Health Organization): http://www.who.int/water_sanitation_health/hygiene/en/ (accessed August 2011b).

Wist, J., Sanabria, J., Dierolf, C., Torres, W. & Pulgarin, C.: Evaluation of photocatalytic disinfection of crude water for drinking-water production. *J. Photochem. Photobiol. A.* 147 (2002), pp. 241–246.

Yu, J.C., Ho, W., Lin, J., Yip, H. & Wong, P.K.: Photocatalytic activity, antibacterial effect, and photoinduced hydrophilicity of TiO_2 films coated on a stainless steel substrate. *Environ. Sci. Technol.* 37 (2003), pp. 2296–2301.

Yu, J.C., Ho, W., Yu, J., Yip, H., Wong, P.K. & Zhao, J.: Efficient visible-light-induced photocatalytic disinfection on sulfur-doped nanocrystalline titania. *Environ. Sci. Technol.* 39 (2005), pp. 1175–1179.

Subject index

Sustainable Energy Developments

Series Editor: Jochen Bundschuh

Publisher: CRC/Balkema, Taylor & Francis

1. Global Cooling – Strategies for climate protection
   Hans-Josef Fell
   2012
   ISBN: 978-0-415-62077-2

2. Renewable Energy Applications for Freshwater Production
   Editors: Jochen Bundschuh & Jan Hoinkis
   2012
   ISBN: 978-0-415-620895

Printed and bound by CPI Group (UK) Ltd, Croydon, CR0 4YY

18/10/2024

01776253-0003